UIA Berlin 2002
XXI World Congress of Architecture

Resource Architecture

Main Congress
Report and Outlook

Annex: The Future of Architecture

UIA White Book 2002

UIA Berlin 2002
XXI World Congress of Architecture

Resource Architecture

Main Congress
Report and Outlook

Annex: The Future of Architecture

UIA White Book 2002

UIA Berlin 2002 e.V. (Ed.)

Birkhäuser – Publishers for Architecture
Basel · Boston · Berlin

Editor / Herausgeber
UIA Berlin 2002 e.V.
on behalf of / im Auftrag von:
Union Internationale des Architectes (UIA), Paris
Bund Deutscher Architekten BDA, Berlin

Editing / Redaktion
Andreas Denk
Anne Ruf
Carl Steckeweh
assisted by Sophie Pache and Esther Schwöbel

English Translation / Englische Übersetzung
Robert Bryce
Robin Benson
Ruslan Tulburg
Annette Wiethüchter
Deirdre Winter

Graphic Design / Layout
Philipp Dörner
Kerstin Peick
for / für Runze & Casper Werbeagentur GmbH, Berlin

Print / Druck
Schwarzdruck, Berlin

A CIP catalogue record for this book is available from the Library of Congress, Washington D.C., USA

Die Deutsche Bibliothek – CIP-Einheitsaufnahme

Resource architecture / XXI World Congress of Architecture. UIA Berlin 2002 e.V. (ed.).
[Ed.: Carl Steckeweh. Engl. Transl.: Robert Bryce].
- Basel ; Boston ; Berlin : Birkhäuser

Main congress : report and outlook / ed.: Andreas Denk .... - 2002
ISBN 3-7643-6752-0

Printed on acid-free paper produced from chlorine-free pulp. TCF ∞
Printed in Germany
ISBN 3-7643-6752-0

9 8 7 6 5 4 3 2 1

www.birkhauser.ch

Index of Content
Inhaltsverzeichnis

Innovation and Tradition – Dialogue between Cultures
Innovation und Tradition – Dialog der Kulturen

Space and Identity – Architecture as the Focus of Social Value Systems
Raum und Identität – Architektur als Fokus der Gesellschaftlichen Wertorientierung

Forewords
Vorworte

Foreword of the President of the UIA

UIA World Congresses have always been great events that have meant a lot to architects and to the role that we architects have to play in the continually changing context in which we are asked to practice.

It is now becoming more important than ever before to address the critical issues that face our profession. Now that we know that the world is heading towards an unsustainable built environment which is becoming vulnerable in more ways than one.

It is a clear that our profession is at a crossroads and that Berlin, this July, will be the ideal forum for dialogue. It is easy to say that the world, especially in the light of recent events, needs a dialogue of cultures and civilisations. Yet we know that dialogues are most effective when the players meet face to face. It is precisely this kind of opportunity that the UIA Berlin Congress is offering. We cannot afford to let it slip by.

The challenge that we are facing clearly is this: Can we as a profession lead society towards what we believe should be the architectural, urban, social and environmental status quo of the future ?

We believe than we can, and that the UIA Berlin Congress, with the contributions of all, will successfully respond to this challenge.

We know that architecture cannot be practised in a political, economic or social vacuum. In Berlin representatives of all facets of society will come together. This will be one of the major strengths of this Congress.

But it will also give us the unique opportunity to open up towards the public at large, the unique opportunity to make society aware of who we are and what we aspire to. This becomes crucially important in the knowledge that, without a society that is receptive to our perception of what constitutes quality architecture, there can be no hope of ever attaining the built environment that we envisage.

We also know that, while it takes great effort to encourage society to respect the architectural heritage of our respective cultures, it takes greater effort still to lead society towards a contemporary architecture that is respectful of tradition.

Berlin with its many examples of "best practice" on show in the urban context, will provide pointers to a better future for our cities and answers to the greenfield versus brownfield debate. There is another parameter that is crucial. We will need to show that all our aims, missions and concerns, be they aesthetic, social, cultural or environmental, are not independent of one another, that it is

Vorwort des Präsidenten der UIA

UIA-Weltkongresse waren immer großartige Veranstaltungen, die viel bedeutet haben – sowohl für uns Architekten als auch für die Rolle, die wir Architekten zu spielen haben in einem sich stetig ändernden Kontext, in dem wir arbeiten sollen.

Heute wird es mehr denn je wichtig, die kritischen Fragen anzusprechen, denen sich unsere Profession gegenübergestellt sieht. Jetzt, da wir wissen, dass wir auf eine nicht nachhaltig gebaute Umwelt zusteuern, die in mehr als einer Hinsicht verletzlich ist.

Es ist eindeutig, dass unsere Profession an einem Scheideweg steht und dass Berlin in diesem Juli das ideale Forum zur Kommunikation sein wird. Es ist leicht gesagt, gerade angesichts der jüngsten Ereignisse, dass die Welt einen Dialog der Kulturen und der Zivilisationen braucht. Wir wissen jedoch, dass Dialoge am fruchtbarsten sind, wenn die Beteiligten sich von Angesicht zu Angesicht gegenüberstehen. Das genau ist die Gelegenheit, die der Berliner UIA-Kongress bietet. Wir können uns nicht leisten, diese zu verpassen.

Die Herausforderung, der wir uns eindeutig stellen, ist: Können wir als eine Berufsgruppe die Gesellschaft zu etwas führen, was wir als architektonischen, städtebaulichen, sozialen und umweltgerechten Status quo der Zukunft bezeichnen?

Wir glauben, dass wir das können, und sind der festen Überzeugung, dass der Berliner UIA-Kongress mit den Beiträgen aller erfolgreich auf diese Herausforderung antworten wird.

Wir wissen, dass Architektur nicht in einem politischen, ökonomischen oder sozialen Vakuum praktiziert werden kann. Eine der größten Stärken dieses Kongresses wird es sein, dass in Berlin Vertreter aller Facetten der Gesellschaft zusammenkommen werden.

Aber er wird uns auch die einzigartige Chance geben, uns in großem Maße der Allgemeinheit zu öffnen, die einzigartige Chance, der Gesellschaft bewusst zu machen, wer wir sind und was wir vorhaben. Dieses wird umso bedeutender in der Gewissheit, dass ohne eine Gesellschaft, die Verständnis zeigt für unsere Auffassung von dem, was qualitätvolle Architektur ausmacht, es nicht die geringste Hoffnung dafür gibt, jemals die gebaute Umwelt unserer Vorstellung zu erschaffen.

Darüber hinaus wissen wir, dass es nicht nur eine große Herausforderung bedeutet, die Gesellschaft dazu zu bringen, das architektonische Erbe unserer jeweiligen Kulturen zu respektieren, sondern eine noch viel größere Aufgabe ist, die Gesellschaft zu einer zeitgenössischen Architektur zu führen, die Traditionen respektiert und integriert.

possible to address them in a holistic way, and that progress can be compatible with these concerns.

For me, one of the major tasks of this Congress will be to establish the need to develop what I would term a "*realpolitik* for sustainable development", meaning to further explore ways to link sustainability with affordability, ands to make sustainable architecture feasible for the whole world. The needs of our planet know no distinction between developed and less developed countries.

It is with great expectations that we architects await UIA Berlin 2002, but it is also with great expectations that the world awaits a message from us architects, a message of hope that there can be a better and more equitable built environment. This is what makes coming to Berlin so particularly exciting for us architects. It is the feeling that we will not be mere spectators, but active participants in an event that will be generating new ideas, an event where architects will show that they do not follow, but that they lead.

After all, legacies don't just happen. They need to be planned in advance if they are to be meaningful.

The Berlin Congress is a Congress that, I am convinced, we will all immensely enjoy. Colleagues from all over the world, meeting together in what will surely be an architectural happening with a common purpose, and an unforgettable experience for everyone involved. This time in particular, as it will be taking place in the architectural showcase setting of Berlin.

The future is unknown. But to some extent it is predictable. What is certain is that we can help shape it.

In our Berlin Congress we are being asked to define the preferred future – for the built environment and for our profession. The ball is in our hands.

Berlin mit seinen vielen Beispielen von „Best Practice", die in der Stadt zu sehen sind, bietet genug Hinweise auf eine bessere Zukunft für unsere Städte und Antworten auf die „greenfield versus brownfield debat" (Grünfläche versus Industriebrache).

Ein weiterer Parameter ist von einschneidender Bedeutung. Wir werden jedem zeigen müssen, dass all unsere Ziele, Aufgaben und Interessen, ob ästhetisch, sozial, kulturell oder umweltbezogen, nicht unabhängig voneinander sind, sondern dass es möglich ist, sie ganzheitlich anzugehen und dass Fortschritt nicht unvereinbar mit diesen Interessen ist.

Für mich wird eine der Hauptaufgaben dieses Kongresses sein, das Bedürfnis nach einer – wie ich es nennen würde – „Realpolitik für eine nachhaltige Entwicklung" zu etablieren, d. h. Wege der Verbindung von Nachhaltigkeit und Finanzierbarkeit zu ersinnen, in anderen Worten: nachhaltige Architektur für die ganze Welt möglich zu machen. Die Bedürfnisse unseres Planeten machen keinen Unterschied zwischen entwickelten und weniger entwickelten Ländern.

Nicht nur wir Architekten erwarten UIA Berlin 2002 mit großer Spannung, sondern auch die Welt erwartet ein Zeichen von uns Architekten, ein Zeichen der Hoffnung darauf, dass es eine bessere und gerechtere gebaute Umwelt geben kann. Darum ist es für uns Architekten so besonders spannend, nach Berlin zu kommen. Das Gefühl, nicht nur bloße Zuschauer zu sein, sondern aktive Teilnehmer an einem Ereignis, das neue Ideen hervorbringen wird und wo Architekten zeigen, dass sie nicht hinterherlaufen, sondern anführen.

Letztendlich entsteht ein großes Vermächtnis nicht einfach. Man muss es gut vorbereiten, wenn es bedeutsam sein soll.

Der Berliner Kongress ist ein Kongress, den wir alle – da bin ich mir sicher – sehr genießen werden. Kollegen aus der ganzen Welt, alle zusammen bei einer Sache, die sicher ein für jeden Beteiligten nicht so leicht zu vergessendes architektonisches Happening wird. Besonders diesmal, da es in dem architektonischen Schaufenster Berlin stattfinden wird.

Die Zukunft ist ungewiss. Aber in einem gewissen Maße vorhersehbar. Was sicher ist: dass wir mithelfen können, sie zu gestalten.

Mit unserem Berliner Kongress werden wir gefragt sein, die wünschenswerteste Zukunft zu definieren – für die gebaute Umwelt und für unseren Beruf. Wir haben es in den Händen.

Vassilis Sgoutas

Foreword of the President of the BDA

"Resource Architecture" – the guiding theme of the XXI UIA World Congress of Architecture Berlin 2002 offers participants exceptional opportunities. In dramatically changing times, this subject is particularly relevant and important. Be they local, regional, national or global – architects must find answers to numerous questions concerning societal, economic and ecological transformation. This challenge cannot be dealt with as a sideline, it is central to our profession, because architects are committed to the shaping of our environment in a special way. Their main activities focus on creating new material realities while making the most of the existing context. Thanks to their creativity, imagination and urge to design 'livable' environments, architects are able to contribute substantially to using our natural and social resources with care. Responsible architects have never been high-handed individualists, their design propositions have never been exclusive. On the contrary, aware of the complexity of the task, they think in networks, behave openly and are ready to cooperate with others. They face discussion and criticism, and respect the experience of their partners. To them, consideration for all the social groups is a moral imperative, and the commitment to beauty the highest principle. Architects feel co-responsible for the humane and ecological development of the natural and the built environment and invest a lot of energy in discussing its shape. We need this debating about architecture, we need to be aware of our responsibility and the consequences of our actions if we wish to improve the public perception of quality architecture.

If architects wish to demonstrate that they are indispensable to society, that their knowledge and talents enable them to design and develop the quality of our built environment as trustees of all its users – then the subject of "Resource Architecture" must increasingly be brought to public awareness. This is why we architects should 'intervene' in society more often and in new and different ways, penetrating social discourse more deeply with our themes, also with a view to creating the right conditions or, in other words, to changing the framework of our work. Although the framework conditions for our 'will to design' are fixed by society, we architects have to influence society by emphasising the importance and responsibility of our profession towards society so that it creates the right conditions. If we use this opportunity consciously and with all our energy, all of us together will succeed in meeting the challenges of the future, by offering 'resource architecture' as an essential model to society.

Vorwort des Präsidenten des BDA

„Ressource Architektur" – das Leitthema des XXI. Architektur-Weltkongresses UIA Berlin 2002 birgt für alle Beteiligten besondere Chancen. Es gewinnt gerade in den sich dramatisch ändernden Zeiten außergewöhnliche Bedeutung. Ob lokal, regional, national oder global – die Architekten müssen Antworten auf die vielen Fragen des gesellschaftlichen, ökonomischen und ökologischen Wandels suchen. Diese Herausforderungen sind nicht nebenbei zu lösen, sondern sie stehen im Mittelpunkt unseres Berufsbildes. Denn der Architekt ist der Gestaltung der Umwelt ganz besonders verpflichtet. In seinen Kerntätigkeiten steht die Schaffung neuer Wirklichkeiten unter bestmöglicher Berücksichtigung des Vorhandenen im Mittelpunkt des Schaffens. Architekten können dank ihrer Kreativität, ihrer Vorstellungskraft und ihres Willens zur Gestaltung einer lebenswerten Umwelt, wesentlich zur Schonung der natürlichen und gesellschaftlichen Ressourcen beitragen. Verantwortungsbewusste Architekten waren daher niemals selbstherrliche Individualisten, ihr Gestaltungsanspruch hatte niemals ausschließenden Charakter, sondern im Bewusstsein um die Komplexität der Sache denken sie vernetzt, offen und partnerschaftlich. Sie stellen sich der Diskussion und achten die Erfahrungen ihrer Partner. Die Rücksicht gegenüber allen sozialen Gruppen ist für sie moralischer Anspruch, die Verpflichtung zur Schönheit oberstes Gebot. Architektinnen und Architekten fühlen sich mitverantwortlich für die humane und ökologische Entwicklung der gebauten Umwelt und initiieren mit viel Energie die Diskussion über ihre Gestalt. Es bedarf gerade dieser Diskussion über Architektur, im Bewusstsein der Verantwortung und den sich daraus ergebenden Konsequenzen, um eine Verbesserung der gesellschaftlichen Wahrnehmung qualitätvoller Architektur zu erreichen.

Wenn Architekten zeigen wollen, dass sie für die Gesellschaft unabkömmlich sind, dass sie dank ihres Wissens und ihres Talentes in der Lage sind, die Qualität der gebauten Umwelt treuhänderisch für alle Nutzerinnen und Nutzer zu gestalten, dann muss das Thema „Ressource Architektur" verstärkt in das Bewusstsein der Öffentlichkeit gebracht werden. Deshalb müssen wir uns stärker und anders als gewohnt einmischen und mit unseren Themen mehr als bisher in den gesellschaftlichen Diskurs eindringen, auch um die Rahmenbedingungen für unsere Arbeit zu verändern. Die Rahmenbedingungen unseres Gestaltungswillens werden zwar von der Gesellschaft festgelegt, aber wir müssen die Gesellschaft hinsichtlich dieser Rahmenbedingungen beeinflussen, indem wir die Bedeutung und Verantwortung der Architektur und der Architekten gegenüber der Gesellschaft hervorheben. Wenn wir diese Chance bewusst und mit aller Kraft wahrnehmen, können wir – indem wir die „Ressource Architektur" als wesentliches gesellschaftliches Leitbild etablieren – gemeinsam die Herausforderungen der Zukunft meistern.

Kaspar Kraemer

Foreword of the President of the Congress

Every three years, an UIA World Congress takes stock of the international architectural debate. A Scientific Committee was assigned the task of structuring discussions during this year's Congress entitled "Resource Architecture". The congress president wishes to thank the committee members most warmly for their service. Their deliberations have been summarised and are published in this documentation. The Scientific Committee has formulated ten questions (also stating detailed reasons for asking them) for the Congress to answer. Towards the end of the Congress, we shall try to find and formulate answers from participants' contributions. The Scientific Committee assisted in selecting the main speakers for the plenary sessions of the four Congress days. Our aim was to win important, well-known personalities from different countries, not only architects, but also politicians, scientists, managers, artists and other friends and critics of architectural culture. Instead of a professional monologue, the congress is intended as a dialogue between different disciplines, civilisations and cultures. The richness in the diversity of traditions, self-interpretations and regional characteristics around the globe should be the basis of new architectural strategies for building peace with the environment. Over 700 architects responded to the call to participate in the Congress in this spirit and to contribute ideas, projects and designs. It was not easy to select the speakers at forums, workshops and seminars from this large number. So as not to lose any of the other outstanding contributions, the authors were asked to present them in poster format. The poster exhibition in the International Congress Centre will therefore be one of the 'informative highlights' of the World Congress.

This documentation also includes the UIA White Book, 'The Future of Architecture: Propositions for the Built Environment', which contains potential answers from congress participants to the questions asked by the Scientific Committee. It is designed to state the grounds for an architectural policy based on responsibility for architectural culture, society and the environment and reflects the current state of discussions within the UIA.

With this documentation, we hope to offer participants of the XXI UIA World Congress of Architecture in Berlin 2002 stimulating new ideas for their own work towards a humane built environment.

Vorwort des Kongresspräsidenten

Die Weltkongresse der UIA fassen alle drei Jahre den Stand der internationalen Diskussion über Architektur zusammen. Diese Diskussion inhaltlich zu strukturieren und dem Thema des Kongresses „Ressource Architektur" eine Richtung zu geben, war Aufgabe des Wissenschaftlichen Komitees. Den Mitgliedern des Komitees sei an dieser Stelle besonderer Dank für ihre Arbeit ausgesprochen. Die Diskussionen wurden zusammengefasst und sind in dieser Dokumentation als Arbeitspapiere veröffentlicht. Das Wissenschaftliche Komitee stellt als Resümee 10 Fragen – mit ausführlicher Begründung – an den Kongress. Aus den Kongressbeiträgen sollen Antworten auf diese Fragen formuliert werden. Es war Aufgabe des Wissenschaftlichen Komitees, an der Auswahl der Hauptredner der Plenarveranstaltungen der vier Kongresstage mitzuwirken. Absicht war es, bedeutende und bekannte Persönlichkeiten aus aller Welt zu gewinnen: nicht nur Architekten, sondern auch Politiker, Wissenschaftler, Manager, Künstler und andere Freunde und Kritiker der Baukultur. Ziel des Kongresses soll nicht der Monolog einer Profession, sondern der Dialog zwischen den Disziplinen, den Zivilisationen und Kulturen sein. Der Reichtum der Verschiedenheit der Traditionen, der Selbstverständnisse und der Regionen dieser Welt soll Grundlage einer neuen Architekturpolitik für den Frieden mit unserer Umwelt werden. Dem Aufruf, sich in diesem Sinne mit Ideen, Projekten und Entwürfen am Kongress zu beteiligen, folgten über 700 Architekten aus aller Welt. Es war keine leichte Aufgabe, aus dieser großen Anzahl die Referenten für die Foren, Workshops und Seminare auszuwählen. Damit möglichst keiner der übrigen herausragenden Beiträge verloren ging, wurden deren Verfasser gebeten, sie auf Postern dargestellt einzureichen. Einer der informativen Höhepunkte des Weltkongresses wird deshalb die Ausstellung dieser Poster im Internationalen Congress Centrum sein.

Diese Dokumentation enthält außerdem das soeben fertig gestellte UIA-Weißbuch „Die Zukunft der Architektur – Vorschläge für die gebaute Umwelt", in dem Beiträge zu möglichen Antworten auf die Fragen des Kongresses enthalten sind. Es soll der Begründung einer Architekturpolitik in Verantwortung gegenüber der Baukultur, der Gesellschaft und der Umwelt dienen und reflektiert den gegenwärtigen Stand der Diskussion innerhalb der UIA.

Wir wünschen den Teilnehmern am XXI. Architektur-Weltkongress UIA Berlin 2002 mit diesen Unterlagen anregende Einsichten für ihre eigene Arbeit im Sinne einer menschenwürdig gebauten Umwelt.

Andreas Gottlieb Hempel

RESOURCE ARCHITECTURE

Contributions, Statements, Questions by the Scientific Committee

RESSOURCE ARCHITEKTUR

Beiträge, Statements, Fragen des Wissenschaftlichen Komitees

Andreas Gottlieb Hempel Berlin

Reinhart Wustlich Hennef

Karl Ganser Breitenthal **et al**

Jewish Museum, Berlin Architect: Daniel Libeskind

Reflections on the Theme of the XXI World Congress of Architecture
Gedanken zum Thema des XXI. Architektur-Weltkongresses

Andreas Gottlieb Hempel Berlin

Treating architecture as a resource enables us to fully comprehend the wide-ranging material, spiritual and intellectual prospects it harbours. Our environment, both built and natural, is a finite, dwindling resource. Our thoughts and feelings are stirred by the aura and atmosphere of nature and architecture. Our minds, which are capable of great feats of imagination, creativity and inventiveness, constitute an infinite resource that can be harnessed for planning and building, for architecture and building culture.

These overlapping dimensions must be captured, characterised and construed. Forward-looking strategies must be devised that will map out the potential for development that resides in the material, the spiritual and the intellectual:

· a new approach to nature that renounces exploitation and destruction
· a new global peace policy that also considers mankind's spiritual needs
· an architectural policy capable of stimulating creativity, innovation and beauty.

New policies and approaches of this kind can only emerge from a dialogue between regions, social systems and cultures. This entails a dialogue between disciplines, which must join forces to shoulder a new, comprehensive responsibility for the common objectives they share. In the process, architects will need to re-examine the image they have of themselves and redefine their goals. The only way they can do this is by taking stock of the elements of which architecture is comprised.

Architecture is more than building. Architecture constitutes an added value that is attached to the built environment. While this added value is difficult to quantify, its absence is immediately recognisable. Added value is apparent not just in terms of encapsulated beauty – beauty being a term that needs very careful definition in this context – but also in the fact that architecture stands as a universal work of art in its respective surroundings, a product of cultural awareness, historical tradition and regional specificity. This work of art rests on a complex mosaic of factors that interact with one another to form a convincing whole. They include

· sparing use of natural resources
· respect for history
· knowledge of cultural relationships
· a response to local characteristics
· consideration of social conditions
· optimum exploitation of economic factors
· the development of value-based criteria
· up-to-date design.

Wendet man sich der Architektur als Ressource zu, so ergibt sich eine weiträumige Perspektive. Die Spannweite reicht von der materiellen Dimension über die seelischen Bereiche bis in die geistige Welt. Unsere natürliche und unsere gebaute Umwelt bieten endliche, sich erschöpfende Ressourcen. Unsere seelischen Empfindungen werden von Stimmung und Ausstrahlung der Natur und der Architektur genährt. Unser Geist mit seiner Phantasie, Kreativität, Vorstellungs- und Erfindungskraft stellt eine unendliche Ressource dar – auch für das Planen und Bauen, für die Architektur und Baukunst.

Diese sich überlagernden Dimensionen müssen erfasst, beschrieben und ausgedeutet werden. Mehr noch – es müssen zukunftsweisende Strategien entworfen werden, die Entwicklungsmöglichkeiten für diese drei Ebenen darlegen:

· ein neuer Umgang mit der Natur, der auf Ausbeutung und Verwüstung verzichtet
· eine neue globale Friedenspolitik – auch für die Seele der Menschheit
· eine Architekturpolitik für die Entwicklung von Kreativität, Innovation und Schönheit

Dies alles kann nur im Dialog der Regionen, der Gesellschaftssysteme und der Kulturen erfolgreich sein. Es ist ein Dialog der Disziplinen, die sich zu einer neu verstandenen ganzheitlichen Verantwortung für die genannten, gemeinsamen Ziele zusammenfinden müssen. Dabei muss auch die Disziplin der Architekten ihr Selbstverständnis und ihre Ziele neu definieren. Dies kann nur über die Inhalte geschehen, die die Architektur erfüllen. Architektur ist mehr als Bauen. Architektur bildet einen Mehrwert, der unserer gebauten Umwelt hinzugefügt wird. Dieser Mehrwert ist schwer messbar – aber er wird sofort bemerkt, wenn er nicht vorhanden ist. Man erkennt diesen Mehrwert nicht nur an der Schönheit – ein Begriff, der in diesem Zusammenhang einer sehr sorgfältigen Definition bedarf –, sondern auch daran, dass Architektur ein Gesamtkunstwerk in der Umweltgestaltung ist, hervorgegangen aus kulturellem Bewusstsein, historischen Überlieferungen und regionalen Besonderheiten. Dies basiert auf komplexen Fügungen zu ganzheitlichen Lösungen unter Einbeziehung folgender Faktoren:

· Schonung natürlicher Ressourcen
· Respekt vor der Geschichte
· Kenntnis kultureller Zusammenhänge
· Umgang mit örtlichen Besonderheiten
· Berücksichtigung gesellschaftlicher Bedingungen
· optimaler Einsatz ökonomischer Möglichkeiten
· Entwicklung von Maßstäben auf der Grundlage von Werten
· zeitgemäße Formgebung

Architecture is the product of the complex interplay between these factors. 50 to 70 per cent of the national wealth is tied up in buildings and infrastructure. One of the goals of architectural policy must, therefore, be to give these resources added economic, environmental, social and cultural value – an added value that only emerges through architecture.

The Beauty of Architectural Design

The beauty of architectural design cannot be seen in absolute terms since it is modified by the technical, functional and material dimensions involved. Nevertheless, the essence of beauty is that it should give clear expression to a product of the mind. In the course of the last century, in particular, the dichotomy between good and evil in human nature taught us to apply the concept of beauty not just to the expression of what is noble, harmonious and uplifting. However, to jettison the concept of beauty as abused, worn out and inadequate would be an inadmissible consequence. The concept of beauty needs rather to be determined by whether the form really matches the content. Throughout history the criterion of beauty has been a genuine correspondence between form and meaning. However, if the form is designed to camouflage reality the upshot is kitsch, the deceptive expression of something bogus.

If, therefore, the concept of beauty is to be used once more with reference to architecture it can only be in the context of the above definition. This entails the use of a language that is readily intelligible, because surrogate terms, such as form quality, aesthetics, design, etc., needlessly hinder communication between architects and members of the public who are interested in architecture. If architects fight shy of using the words "beauty" and "beautiful" this will only diminish the regard for a profession that is expected to treat the creation of "good design" as one of its major tasks.

Today, beauty may well manifest itself not only in the form of what is visibly "beautiful", but also as a paradigm of what is invisible. Architecture could, therefore, be interpreted as the "art of interweaving". It is not the clearly defined areas that are the problem today, but rather the joints in between them – in a figurative sense the junctures, the connections, the transitions between the visible and the invisible, between form and meaning.

Aus dieser Komplexität entsteht Architektur. Etwa 50 bis 70 Prozent des Volksvermögens sind in Bauwerken und infrastrukturellen Einrichtungen festgelegt. Architekturpolitisches Ziel muss es sein, diesen Vermögenswerten einen ökonomischen, ökologischen, sozialen und kulturellen Mehrwert zu verleihen – einen Mehrwert, der erst durch Architektur entsteht.

Schönheit architektonischer Gestaltung

Die Schönheit architektonischer Gestaltung kann nicht absolut gesehen werden, sondern wird relativiert durch ihre Bezüge auf Technik, Funktion und Material. Dennoch bleibt dieser Schönheit in unserem Sinn die Aufgabe, semantisch Inhalte einer geistigen Welt sichtbar zu machen. Das Spannungsfeld zwischen Gut und Böse in der menschlichen Natur hat uns vor allem im vergangenen Jahrhundert gelehrt, den Schönheitsbegriff nicht nur für die Gestalt des Edlen, Harmonischen und Aufbauenden anzuwenden. Deshalb aber den Begriff der Schönheit als missbraucht, als unzureichend und verbraucht über Bord zu werfen, ist unzulässig. Es geht vielmehr darum, ihn anhand des Maßes der Entsprechung von Gestalt und Inhalt neu zu definieren. So war eine Entsprechung von Form und Bedeutung zu allen Zeiten ein Kriterium für Schönheit. Täuscht Gestaltung über die wirklichen Verhältnisse hinweg, entsteht hingegen Kitsch als schöner Schein der Lebenslüge.

Wenn also der Begriff der Schönheit in Zusammenhang mit Architektur wiederverwendet wird, dann in dieser differenziert anzuwendenden Definition. Zudem soll damit eine allgemein verständliche Sprache verbunden werden, denn mit Ausdrücken wie Gestaltqualität, Formgebung, Ästhetik, Design als sprachliches Surrogat wird die Verständigung zwischen den Architekten und der baukulturell interessierten Öffentlichkeit unnötig erschwert. Wenn im Vokabular der Architekten die Worte „Schönheit" und „schön" vermieden werden, schwindet das Verständnis für einen Berufsstand, von dem erwartet wird, dass er „schöne Gestaltung" als wesentliche Aufgabe ansieht.

Schönheit tritt heute vielleicht nicht allein als sichtbares „Schönes", sondern auch als Paradigma des Unsichtbaren auf. Architektur ließe sich demnach – in der Aufnahme eines berühmten Wortspiels – als „Kunst der Fuge" verstehen. Nicht die Flächen sind heute das Problem der Architektur, sondern die Fügungen zwischen den Flächen – im übertragenen Sinne die Fügungen, die Zusammenhänge, die Übergänge zwischen Sichtbarem und Unsichtbarem oder zwischen Form und Bedeutung.

Genius Loci or the Invisible

The genius loci, the spirit of a place, is invisible and yet it fills every nook and cranny with a distinct atmosphere. This atmosphere cannot be perceived by the eyes alone; it can only be grasped by all the human senses reacting together and felt by the heart. It is the response to the atmosphere of a place that determines whether we feel at home there or not. The genius loci determines the true beauty of the form. If it does not exude this spirit even the most artistic of forms will leave us cold.

Spirit, atmosphere and matter all stem from a single source, but they exist in different physical states rather like water, which takes the form of steam, liquid or ice depending on the temperature and molecular density. To that extent, the spirit which gave rise to built architecture, the atmosphere of the materials used for its construction and the material design devised by its creators – individuals, developers, architects and builders working in their time – join with one another to testify to the nature of its invisible inner world.

Anyone walking through buildings or open spaces with an open mind will soon notice that the atmosphere of the space says much more about the essence of a place than its surface design. Those who resort to images and spaces to comprehend the spirit of a place, its history and its culture will be able to comprehend its essential, invisible aspects because they manifest themselves in the imagination of the soul.

Before the First World War, Berlin was one of the four largest industrial cities in the world. The years of rapid industrial expansion in Germany in the latter part of the 19th century were accompanied by land and building speculation. This resulted in the development of very densely built-up areas and the construction of crowded tenement blocks with unbearable living conditions for the industrial workers housed in them. These inhumane living conditions led to the development of alternative plans in the 1920s that envisaged the construction of low-cost housing "in leafy surroundings", which would be subsidised by the state or local authorities. Visitors to these housing estates today never fail to be impressed by the classical Modernist architecture and the structural ideas deriving from the "New Building" concept.

What remains in the memory, however, is not so much the design of the buildings and facades produced by Taut, Mies van der Rohe, Gropius, Scharoun and others as the new, improved living conditions for the lower social classes: the little gardens full of blooms in front of the terraced houses, the sunlight falling through the gently swaying tips of the Scots pines between the rows of white houses with their bright-coloured window frames, and the quiet, paved side-streets with children playing in them. The images conjured up in one's mind are of social peace and

Genius Loci oder das Unsichtbare

Der Geist, das Wesen oder die Stimmung eines Ortes sind unsichtbar und erfüllen doch jeden Winkel mit Atmosphäre. Atmosphäre kann nicht allein mit den Augen, sondern nur mit allen Sinnen, mit dem gesamten Empfinden, das dem Menschen möglich ist, und mit dem Herzen wahrgenommen werden. Die Empfindung für die Atmosphäre entscheidet darüber, ob wir uns an einem Ort wohl fühlen. Dieser Geist bestimmt auch die Schönheit der Gestalt. Strahlt sie ihn nicht aus, so lassen uns auch die kunstvollsten Formen kalt.

Geist, Atmosphäre und Materie sind einer Herkunft – nur in verschiedenen Aggregatzuständen, vergleichbar etwa mit dem Wasser, dessen Erscheinung je nach Temperatur und molekularer Dichte als Dampf, Flüssigkeit oder Eis auftritt. Insofern gibt die gebaute Architektur durch den Geist, aus dem sie entstand, durch die Atmosphäre ihrer Gestalt und durch die ihr übertragenen Absichten der materiellen Ausgestaltung ihrer Schöpfer – Menschen in ihrer Zeit, Bauherr, Architekt und Bauleute – Zeugnis vom Wesen ihrer unsichtbaren Innenwelt.

Wer mit offenem Empfinden durch Stadträume oder Gebäude geht, wird sehr schnell wahrnehmen, dass die Atmosphäre des Raumes mehr über das Wesentliche des Ortes aussagt als die Gestaltung der Oberflächen. Wer über Bilder und Räume den Geist von Ort, Geschichte und Kultur zu verstehen sucht, wird das unsichtbare Wesentliche deshalb begreifen, weil es in der Imagination der Seele wieder innere Gestalt annimmt.

Berlin war vor dem Ersten Weltkrieg eine der vier größten Industriestädte der Welt. Der rasche Aufbau in den Gründerjahren am Ende des 19. Jahrhunderts hatte mit seiner Boden- und Bauspekulation, der übermäßig verdichteten Baublocks mit den so genannten Mietskasernen zu unerträglichen und unsozialen Wohnzuständen für die Industriearbeiter geführt. Wegen dieser unmenschlichen Wohnverhältnisse entstand in den zwanziger Jahren als Gegenentwurf der soziale Wohnungsbau „im Grünen". Wer heute die Bauten dieser Siedlungen besucht, ist von der Architektur der klassischen Moderne und den gestalterischen Ideen beeindruckt, die dem Gedanken des „Neuen Bauens" folgen.

Stärker jedoch als die Baukörper- und Fassadengestaltungen der Architekten Taut, Mies van der Rohe, Gropius, Scharoun und anderer bleibt die Atmosphäre der neuen, besseren Lebensbedingungen für die unteren sozialen Schichten im Gedächtnis. Die kleinen Gärten der Reihenhäuser, das Sonnenlicht, das durch die Wipfel der Föhren zwischen den weißen Bauzeilen mit den bunten Fensterrahmen fällt, die gepflasterten Seitenstraßen mit spielenden Kindern. Es bleiben Gedanken an sozialen Frieden und an einen Zustand, der zu Unrecht „das kleine Glück" genannt wird.

the kind of pleasure that for some strange reason is unjustly referred to as "humble".

Many other examples could be added. What it is important to understand is that the invisible essence – the genius loci – can be built around and thus made visible. Perhaps the spirit of a place is always there before it is given structural shape. The henges in England were surely only built after the priests, sun watchers and other inhabitants sensitive to their surroundings had established that they were places with a very special atmosphere. Not only did the sun's rays fall symbolically on certain days and make cosmic relations visible. People could also sense the special aura of such places and the concentration of spiritual powers there. They, therefore, marked them with megaliths to give material expression to these powers. Many important buildings have been constructed in places with an intense atmosphere or beautiful surroundings to deliberately enhance the magic inherent in them. Such buildings include the theatre in Epidaurus, the temple at Paestum and the Temple Mount in Jerusalem. The Statue of Liberty in New York and the Eiffel Tower in Paris are likewise symbols of social and technical prowess. The destruction of the twin towers of the World Trade Center in New York is a powerful warning to us to engage in a dialogue between religions and civilisations and to exercise tolerance in the creation of a more just and peaceful world order. There is an astonishing interaction between the essential, which can only be perceived by the whole range of human senses and by the heart, and its materialisation in stone, which makes it possible to "grasp" the invisible in the literal sense of the word.

The Spirit of the Times and Globalisation

The most important factor is the time, i.e. the historical period, in which a building is constructed, thus giving "timely" expression to an inner awareness or feelings. This period of development cannot be repeated or emulated at some later date because attitudes to spiritual relations and the way they are interpreted change considerably over time. The capacity to grasp the essential meaning of things and the ability to give them expression are both influenced by the time in which people live.

This is the point at which criticism of historicist architecture must set in. What may be very convincing as an original expression of people's inner response to their outer world at a certain period in time induces no more than boredom at best in the case of a historicist reproduction for the simple reason that there is no correspondence between the external form and the inner truth. In other words, the tautness that springs from the relationship between the building and the time it was constructed is missing. A characteristic feature of historicism is its fanatical attention to detail combined with a penchant for the elaborate embellishment

Die Reihe lässt sich fortsetzen. Wichtig ist, zu verstehen, dass das unsichtbare Wesentliche – der Genius Loci – umbaut werden, sichtbar gefasst werden kann. Vielleicht besteht der Geist eines Ortes schon, bevor ihm bauliche Gestalt verliehen wird: Sicherlich sind die Henges in England erst entstanden, nachdem Priester, Sonnenbeobachter und Sensitive festgestellt hatten, dass es sich dort in ihrem Verständnis um einen Ort von besonderer Atmosphäre handelte. Nicht nur die Sonnenstrahlen fielen an bestimmten Tagen zeichenhaft ein und machten kosmische Verhältnisse sichtbar – auch die besondere Ausstrahlung, die Verdichtung geistiger Kräfte war den Menschen spürbar. Sie markierten mit Megalithen den Ort, um dieser Kraft dinglichen Ausdruck zu verleihen. Ganz bestimmt sind viele bedeutende Architekturen aus den Vorgaben eines Platzes mit atmosphärischer Dichte oder auch landschaftlicher Besonderheit in der Absicht entstanden, einen „magischen" Ort zu überhöhen. Das Theater in Epidaurus, der Tempel in Paestum ebenso wie der Tempelberg in Jerusalem. Aber auch die Freiheitsstatue in New York oder der Eiffelturm als Symbole gesellschaftlicher oder technischer Errungenschaften. Nicht zuletzt erscheint die Zerstörung der Twin-Towers in New York wie eine signalhafte Mahnung, für eine gerechtere Weltordnung im Dialog der Religionen und der Kulturen zu arbeiten und in Toleranz eine gerechtere und friedlichere Weltordnung zu entwerfen. Die Wechselwirkungen vom Wesentlichen, das nur mit der Vielfalt der menschlichen Empfindungen, mit dem Herzen gesehen werden kann, und der steingewordenen Materialisation sind erstaunlich: Das Unsichtbare wird im wahren Sinne des Wortes „begreifbar".

Zeitgeist und Globalisierung

Wohl der wichtigste Faktor ist dabei die Zeit – im Sinne von Geschichte –, in der ein Bau verwirklicht wurde und zu der er den inneren Erkenntnissen oder Empfindungen „zeitgerechte" Gestalt verlieh. Dieser Entstehungszeitraum kann zu keinem späteren Zeitpunkt wiederholt oder nachempfunden werden. Zu stark verändert sich im Ablauf der Zeiten die Empfindung für geistige Verhältnisse und deren Interpretation. Aber nicht nur das Ausdrucksvermögen, sondern auch die Fähigkeit zum originären Verständnis ist dem Zeitpunkt des Entstehens unterworfen.

Hier muss eine Kritik jedweder historisierenden Architektur einsetzen. Was als originaler Ausdruck des Zeitempfindens zur inneren Wirklichkeit überzeugen kann, ruft als historisierender Nachbau bestenfalls Langeweile hervor, da keine Entsprechung der äußeren Gestalt zur inneren Wahrheit besteht, also die Spannung vom Geist der Zeit zum baulichen Ausdruck fehlt. Allem historisierenden Nachempfinden zu Eigen ist ein detailversessener Hang zum Dekor, zum Schmuck der Oberfläche mit Emblemen der jeweiligen Epoche. Da die Strukturen geistiger Wirklichkeiten, die der noch nicht materialisierten Innenwelt entsprechen, künstlerisch nicht zeitgemäß interpretiert, sondern

of facades with emblems of the respective period. Since the structures of spiritual realities corresponding to the as yet non-materialised inner world are not expressed artistically in a manner appropriate to the time, but are merely reflected in a historicist style, the outcome is devoid of all fascination. This is always the case when no serious attempt is made to address the task at hand; the spiritual pole does not fit with the material pole and thus no flow of power can be generated. Whenever the spiritual content cannot be given adequate formal expression, the substitute form that is found is empty or replaced by formulas. What is missing, however, is the semantic quality.

Contemporary means must be employed to supply a contemporary interpretation of content, function, representative aspirations and feeling for space. However, this does not mean that the existing fabric should be dispensed with. On the contrary, it is an invaluable resource. Developers, architects and craftsmen should count themselves lucky if they can make use of existing buildings or parts of buildings of historical value or take them as a point of departure for new developments. Such buildings incorporate aspects of memory, self-identification, awareness of tradition and the flow of consciousness, which themselves constitute an often invaluable resource for the genius loci.

Atmosphere, local identification and memory, sometimes only at the subconscious level, are a veritable gift for a building project in which the existing fabric is to be incorporated in an innovative manner. This is of the utmost importance at a time in which designer fashions are being exploited in desperate attempts to establish a distinctive corporate identity. Building materials obtainable locally without any major transport costs, forms of agriculture, processing procedures, climatic conditions and traditional methods of construction employed by master builders, including the use of details that had proved their technical and economic worth over the centuries, were the reasons for the many different types of houses that were built in many different surroundings.

It is interesting to note that very detailed forms of decoration developed only very gradually. Absolute priority was accorded to the proper functioning of buildings, to economic aspects and durability. The types of houses that strike us today as being minor works of art with their well-balanced proportions, properly placed openings and aesthetic use of materials were the product of structural necessity and the need for a harmony of scale, which is still readily apparent today. An adherence to scale applies not just to the dimensions and proportions but, above all, to a proper balance between needs and costs.

Architecture that makes the genius loci visible and perceptible can only emerge if serious thought is given to the place, the people and the existing structures, of which architectural tradition forms a part. The genius loci is suffocated by too many "toos":

nur historisierend nachempfunden werden, entsteht Spannungslosigkeit. Sie tritt immer dann auf, wenn keine wirkliche Lösung der Aufgabe versucht wird, der geistige Pol dem materiellen Pol nicht entspricht und somit kein Kraftfluss entstehen kann. Was an geistigem Gehalt nicht qualitativ in entsprechender Gestalt wiedergegeben werden kann, wird nur durch hohle Form, durch Formeln ersetzt. Ihnen fehlt aber die Qualität des Semantischen.

Die zeitgenössische Interpretation von Inhalten, Funktionen, Repräsentationsansprüchen, räumlichem Empfinden muss mit zeitgenössischen Mitteln erfolgen. Das bedeutet allerdings nicht, dass auf den vorhandenen Baubestand verzichtet werden könnte. Ganz im Gegenteil – das Gebaute ist eine unschätzbare Ressource. Bauherren, Architekten und Handwerker sollten sich glücklich schätzen, wenn bestehende Gebäude oder Bauteile mit historischem Wert weiterverwendet werden können oder Anlass für eine Neuorientierung geben. In der konstruktiven Auseinandersetzung mit dem Bestand eröffnen sich Dimensionen der Rückerinnerung, der Selbstidentifikation, der Tradition und der Kontinuität des Bewusstseins dafür – eine Ressource für den Genius Loci von oft unschätzbarem Wert.

Atmosphäre, Ortsverbundenheit und Rückerinnerung – manchmal nur im Unterbewussten – werden für die Bauaufgabe, die mit der Verwendung des Bestandes innovativ arbeitet, geradezu zum Geschenk. Dies ist besonders wichtig in einer Zeit, in der gestalterische Unverwechselbarkeit als „corporate identity" nach Designermode meist krampfhaft gesucht wird. Örtlich ohne große Transportkosten erhältliche Baumaterialien, Formen der Landwirtschaft, Funktionsabläufe in der Verarbeitung, Klimaverhältnisse und baumeisterliche Ausführungstraditionen mit Details, die sich seit Jahrhunderten technisch und ökonomisch bewährt hatten, waren die Ursachen für die Vielfalt von Hausformen in einer Vielfalt von Landschaften.

Interessant zu beobachten ist, dass sich Schmuckformen im Detail nur sehr sparsam entwickelt haben. Funktion, Ökonomie und Dauerhaftigkeit hatten absoluten Vorrang. Die heute schon fast als Kunstform erscheinenden Hausstrukturen mit ihren guten Proportionen, richtig gesetzten Öffnungen und ästhetisch erscheinender Materialverwendung ergaben sich aus der konstruktiv notwendigen und in ihrer Notwendigkeit auch optisch nachvollziehbaren maßstäblichen Harmonie. Dabei ist unter Maßstäblichkeit nicht nur das Maßverhältnis zu verstehen, sondern vor allem die Angemessenheit von Notwendigkeit zu Aufwand.

Baukunst, die den Genius Loci fühlbar oder sichtbar macht, ist nur in der ernsthaften Auseinandersetzung mit dem Ort, den Menschen und den vorhandenen Strukturen – zu denen auch die Bautradition gehört – möglich. Der Genius Loci wird vor allem durch ein „Zuviel" zerstört: zu hoch, zu breit, zu luxuriös, zu exotisch, zu reich, zu bunt, zu aufgeregt, zu individualistisch und doch

too tall, too wide, too luxurious, too exotic, too rich, too colourful, too restless, too individualistic and yet impersonal, too expensive, too short-lived, too alien, too ill-considered, too disproportionate, too vain, too brash, but also too meagre, too misshapen, too uncomfortable, etc.

Above all in the age of globalisation, in which banks, construction companies and planners are global players engaged in worldwide activities to supply real estate for investment purposes – mostly without any knowledge of who the future occupants will be – architects must join forces with politicians and responsible members of the public and harness their consciences to their knowledge in order to do justice to the expectations of architecture even on distant construction sites and to produce architecture that is worthy of the genius loci. This requires an ability to empathise and a capacity to sense what is specific about a certain location or situation as well as the courage to repudiate the sort of thoughtless, insensitive and purely business-minded planning throughout the length and breadth of the country that destroys for ever situations and surroundings that are truly unique. Architects should be able to take a back seat, to regard themselves as comprising just one profession among many and not to copy every passing fashion in every conceivable place without thinking and without considering the genius loci. At a time of levelling, egalitarianism, arbitrariness and the thoughtless and relentless frittering away of unique places, resources, irretrievable situations and personal friendships in all fields of human activity, it is clear that the preservation of spirit, atmosphere and form is an existential issue for mankind, a question of its identity – and dignity.

Infinite and Finite Resources

It is easy to see that the visible and the invisible are closely intertwined in architecture, just as a person's invisible psychological state is expressed in his bodily nature, facial expressions, handwriting and movements. The most fascinating resource that architecture can draw on is the interaction between the immaterial and the material, between the mind and the body, between the idea and built reality. The most valuable and durable resource architecture has is the imagination, creativity and ideas of architects. However, this resource must be shaped by criteria stemming from moral and ethical value systems. In assessing the image they have of themselves, architects do not require a discussion of professional requirements in the traditional vein, but rather a discussion of values so that they can generate viable intellectual resources. The most important characteristic of intellectual resources, however, is that they are inexhaustible, constantly renewable and full of surprises and visions for the future. People, architects included, will never run out of ideas. Ideas are the plants of the unseen world that first have to be translated into words and plans before they can grow into the trees of this world.

unpersönlich, zu teuer, zu kurzlebig, zu fremd, zu unüberlegt, zu unmaßstäblich, zu eitel, zu protzig, aber auch zu dürftig, zu ungestalt, zu unwohnlich.

Gerade im Zeitalter der Globalisierung, in dem Banken, Baufirmen und Planer als Global Players weltweit tätig werden, um – meist, ohne den künftigen Nutzer zu kennen – Immobilien als Ware für Kapitalanleger anzubieten, müssen sich die Architekten mit Politikern, verantwortungsbewussten Bürgern vor Ort, mit ihrem eigenen Gewissen, aber auch mit ihrem Wissen verbünden, um der Baukunst auch an weit entfernten Baustellen gerecht zu werden, um eine Architektur zu entwickeln, die dem Geist des Ortes gerecht wird. Dazu gehört ein intensives Einfühlen und Erspüren der jeweiligen Situation, aber auch der Mut, sich den materiellen Einflüssen zu verweigern, die mitverantwortlich sind, dass landauf, landab einmalige Situationen unwiederbringlich durch gedankenloses, unsensibles, allein geschäftstüchtiges Planen zerstört werden. Architekten sollten sich zurücknehmen können, sich als eine Stimme unter vielen verstehen und nicht jede Mode an jedem nur denkbaren Platz unreflektiert und ohne Rücksicht auf den Genius Loci nach- oder vormachen. In Zeiten der Nivellierung, der Gleichmacherei, der Beliebigkeiten und des gedankenlosen und gnadenlosen „Verbrauchs" von einmaligen Orten, Ressourcen, unwiederbringlichen Situationen und persönlichen Freundschaften auf allen Gebieten des menschlichen Handelns ist das Bewahren von Geist, Atmosphäre und Gestalt ein menschliches Existenzproblem, eine Frage der Identität und der Würde.

Unendliche und endliche Ressourcen

Man kann erkennen, dass das Unsichtbare und das Sichtbare in der Architektur eng verbunden sind – so wie sich der unsichtbare Seelenzustand eines Menschen in seiner Körperlichkeit, seinem Mienenspiel, seiner Handschrift, seinen Bewegungen ausdrückt. Die spannungsvollste Ressource der Architektur ist das Wechselspiel zwischen Immateriellem und Materiellem, zwischen Geist und Körper, zwischen Idee und gebauter Wirklichkeit. Die werthaltigste Ressource der Architektur ist die Phantasie, die Kreativität, sind die Ideen der Architekten. Kanalisiert werden muss diese Ressource allerdings durch Maßstäbe, die aus moralisch und ethisch geprägten Wertvorstellungen entwickelt werden müssen. Der Berufsstand der Architekten braucht für sein Selbstverständnis keine Berufsbilddiskussion im herkömmlichen Sinn, sondern eine Wertediskussion, damit er tragfähige geistige Ressourcen entwickeln kann. Das Wichtigste an den geistigen Ressourcen ist jedoch, dass sie unerschöpflich sind, ständig erneuerbar, voller Überraschungen und voller Zukunftsperspektiven. Den Menschen, den Architekten werden die Ideen niemals ausgehen. Ideen sind Pflanzen der jenseitigen Welt, die zunächst in Worte und Pläne umgetopft werden, um sich schließlich zu Bäumen in dieser Welt zu entwickeln.

Material resources, by contrast, are finite. Two examples make this clear. Open land is being transformed into suburban areas that are euphemistically referred to as "in-between towns" or "garden towns" even. We need to find a balance between populated areas and open land if we want to go on living in and with this world. Energy from fossil fuels is limited, even though many people may close their eyes to the fact and refuse to accept it. Hence we must turn to alternative sources of power, such as water, wind and the sun, if we are to satisfy our future energy requirements in a manner that is compatible with the environment. Much the same applies to the materials we use to build houses. They must form part of material cycles if we wish to give our children a chance to build, too. The very least we should insist on is that once materials have been used they should be recycled in a different form if they cannot be returned to the cycle of nature.

In the course of human history there have been countless examples of architecture being treated and used as a valuable and long-lasting resource. One of the most outstanding examples, without a doubt, is Venice. Rising up out of the mud and swampland is a magnificent city with a rich abundance of culture. The complete lack of natural resources meant that every stone and every brick were used more than once. The pressure exerted by scarce material resources opened up the floodgates of immaterial resources. Ingenuity, the arts and inventiveness flourished in architecture as they did in politics. An awareness of the need for harmony with nature supplemented by a balance between the political forces and the stabilisation of social tensions led to the creation of the republic – a form of state organisation that has lasted almost 1,200 years, making it one of the longest-lasting in history. What this teaches us is that architecture is more than just the work of construction.

Architecture, as the art of building, is part of the very essence of humankind. Its outstanding achievements demonstrate that it is a very valuable resource for the economy and for materials cycles in harmony with nature. Moreover, it constitutes the backdrop to human co-existence, social relations and society itself. Towering above all this, however, is the cultural value of architecture as a bridge linking relations in this world with the world of the invisible. This bridge between the spirit, the intellect and the material world merits a political dimension of its own in the form of an architectural policy and political strategies for the sustainable, global development of architecture, cities and the countryside. This policy must be understood as constituting a new paradigm within the framework of a new, global peace policy for mankind and the environment.

Die materiellen Ressourcen dagegen sind endlich. Nur zwei Beispiele: Freie Landschaften verwandeln sich in Suburbia, die in der Theorie euphemistisch „Zwischenstadt" oder gar „Parkstadt" genannt werden. Wir müssen ein Gleichgewicht finden zwischen besiedelter Fläche und Freiräumen, wenn wir weiterhin auf und mit dieser Welt leben wollen. Die Energie, die aus fossilen Brennstoffen gewonnen werden kann, ist unwiderruflich beschränkt – auch wenn alle die Augen verschließen und das nicht wahrhaben wollen. Wir müssen uns Alternativen wie Wind, Sonne und Wasser zuwenden, wenn wir auch künftig unseren Energiehaushalt umweltverträglich decken wollen. Ähnlich wie mit den landschaftlichen Ressourcen und den fossilen Brennstoffen steht es um die Materialien, mit denen wir unsere Häuser bauen. Sie müssen in wohl durchdachte Materialkreisläufe überführt werden, wenn wir unseren Kindern die Möglichkeit lassen wollen, zu bauen. Zumindest sollten einmal verwendete Materialien wenigstens in anderer Form in den Kreislauf der Verwendung zurückkehren – wenn nicht gar in den Kreislauf der Natur.

In der Menschheitsgeschichte hat es zahlreiche Beispiele für Architektur als werthaltige und dauerhafte Ressource gegeben. Eins der beeindruckendsten Beispiele ist sicherlich Venedig. Aus Sumpf und Schlamm ist eine wundervolle Stadt mit einer reichen Kultur entstanden. Der völlige Mangel an natürlichen Ressourcen hat dazu geführt, dass jeder Stein wiederverwendet, jeder Ziegel zweimal umgedreht wurde. Das Leben mit den knappen materiellen Ressourcen hat die Schleusen der immateriellen Ressourcen geöffnet. Genie, Künste und Erfindungsreichtum blühten – nicht nur in der Architektur, auch in der Politik. Das Wissen um das Gleichgewicht mit der Natur hat durch die Balance der politischen Kräfte, den Ausgleich der sozialen Spannungen, eine republikanische Staatsform geschaffen, die fast 1.200 Jahre existierte und damit eine besonders dauerhafte Staatsform der Geschichte darstellt. Wir lernen daraus, dass Architektur nicht nur etwas mit Bauen zu tun hat.

Architektur als Baukunst, als Baukultur gehört zur Substanz des Menschen. Ihre Blüteformen beweisen, dass sie eine werthaltige Ressource für die Ökonomie für Materialkreisläufe im Einklang mit der Natur ist. Darüber hinaus ist sie der Rahmen für das menschliche Zusammenleben, für soziale Verhältnisse und die Gesellschaft. Am stärksten jedoch ist der kulturelle Wert der Architektur – Baukultur ist die Brücke unserer diesseitigen Verhältnisse in die Welt des Unsichtbaren. Diese Brücke zwischen Geist und Welt verdient eine eigene politische Dimension: Architekturpolitik oder politische Strategien für eine nachhaltige und globale Architektur-, Stadt- und Landschaftsentwicklung. Diese Politik muss als neues Paradigma im Kontext einer neuen und globalen Friedenspolitik für die Menschheit und die Umwelt verstanden werden.

The Image of Planet Earth
Reflections on Architecture as Part of a Global Environmental Policy
Das Bild der Erde
Überlegungen zur Architektur als Teil einer Welt-Umweltpolitik

Reinhart Wustlich Hennef

In the traditional architectural debate, global scenarios that address truly world-shaking issues within the framework of the discipline's practical orientation are still the exception. Nevertheless, the XXI World Congress of Architecture would be performing a great service if our discipline were to open up, look beyond the boundaries of its field of vision, and enter into a dialogue with other disciplines and international organisations that are dealing with problems that could – under the guise of "social progress" – affect architecture, urban development and landscape planning in the northern and southern hemispheres. Furthermore, it would be an honourable task for the disciplines involved in architecture, urban development and landscape planning to help develop an interdisciplinary concept for a global environmental policy.

In drawing an analogy between cultural diversity and biodiversity as the foundations of human development, the UNESCO Universal Declaration on Cultural Diversity emphasises the complex impact that architecture, urban development and landscape planning have on the environment. However, there is no globally accepted image capable of conveying the concept of a world community in the "horizon of the earth" and in the "horizon of the cultures", let alone one that would incorporate the fundamental role played by architecture, urban development and landscape planning in the image of the world's civilisations.

There are an increasing number of studies that see megacities and their environments in the southern regions of the world teetering on the brink of a catastrophe. A wide-ranging study commissioned by the Asian Development Bank (ADB), for instance, has shown that the rapid population growth in the region and the accelerating pace of urbanisation there could mean that, by 2015, Asia will have replaced the leading industrial nations in the OECD as the main source of the greenhouse gases that are changing the world's climate. The study also warns of the illusion that the destruction of the environment can be tolerated in the interests of economic development. "The destruction of the environment hits the poor hardest", says ADB expert, Tahir Qadri. "The struggle against poverty is inseparable from protection of the environment."[1]

Political debate is dominated by highly abstract concepts. And never before in the history of the natural sciences, which call themselves life sciences, has so little been offered to the eye of the beholder, to immediate sensorial perception and, above all, to the public. "The eye is starving in a feast of facts. The interested onlooker is becoming bored through being fed a daily dose of fresh insights into the blueprint of humanity and nature."[2] The GEO 2000 study submitted by UNEP (the United Nations Environ-

In der traditionellen Architekturdebatte sind Welt-Szenarien, die weltbewegende Probleme in den Rahmen der Handlungsorientierungen der Disziplin einbeziehen, noch ungewöhnlich. Es wäre jedoch ein Verdienst des XXI. Architektur-Weltkongresses, wenn sich die Disziplin, offen und über die Grenzen der eigenen Wahrnehmung hinaus, im Dialog mit anderen Disziplinen und globalen Organisationen mit Problemen konfrontierte, die im Gewand des so genannten „gesellschaftlichen Fortschritts" auf die Architektur-, Stadt- und Landschaftsentwicklung im Norden und Süden der Erde zukommen können. Mehr noch, für die an der Architektur-, Stadt- und Landschaftsentwicklung beteiligten Disziplinen wäre es eine ehrenvolle Verpflichtung, an der Erarbeitung einer übergreifenden Konzeption einer Welt-Umweltpolitik mitzuwirken.

Die UNESCO-Deklaration zum Wesen der Kultur hat mit der Analogie von kultureller Vielfalt und Biodiversität als Grundlagen der humanen Entwicklung die Berücksichtigung der komplexen Umweltbeziehungen von Architektur-, Stadt- und Landschaftsentwicklung bereits nahe gelegt. Dennoch gibt es kein global akzeptiertes Bild, das eine Vorstellung vom gemeinsamen Dasein einer Weltgesellschaft im „Horizont der Erde" und im „Horizont der Kulturen" repräsentieren könnte, geschweige denn ein solches, das die grundlegende Rolle der Architektur-, Stadt- und Landschaftsentwicklung in das Bild der Zivilisationen der Welt einbezöge.

Es mehren sich jedoch die Studien, die vor allem die Entwicklung der Megacitys und ihrer Umweltbeziehungen in den südlichen Gürteln der Erde an den Rand einer Katastrophe treiben sehen. So belegt die Asian Development Bank ADB mit einer umfassenden Untersuchung, dass rasantes Bevölkerungswachstum in der Region und sich beschleunigende Verstädterung dazu führten, dass Asien schon 2015 die in der OECD zusammengeschlossenen führenden Industrienationen als Hauptquelle für klimaverändernde Treibhausgase ablösen könnte. Die Studie warnt zugleich vor der Irrlehre, dass Umweltzerstörung zu Gunsten der ökonomischen Entwicklung in Kauf genommen werden könne: „Umweltvernichtung trifft zuallererst die Armen", warnt der ADB-Experte Tahir Qadri. „Armutsbekämpfung und Umweltschutz lassen sich nicht trennen."[1]

In den politischen Debatten herrscht extreme Abstraktion. Und nie zuvor in der Geschichte der Naturwissenschaften, die sich Lebenswissenschaften nennen, ist dem Auge des Betrachters, der unmittelbaren sinnlichen Erkenntnis, zumal der Öffentlichkeit, so wenig geboten worden: „Das Auge hungert beim Festmahl der Fakten. Der interessierte Zaungast langweilt sich bei der täglichen Präsentation frischer Einsichten in den Bauplan von Mensch und Natur."[2] Die von der UNEP (United Nations Environment Programme) vertretene Studie GEO 2000 „müsste Panik auslösen, aber die Mensch-

ment Programme) "ought to make people panic, but humanity simply takes everything in its stride, as if the future were ordained by fate. It is seemingly tired of being presented with ever-new scenarios of doom that, in any case, always take place in the distant future; they constitute a threat that is about as vague as the Last Judgement; and maybe it is not the will to find a solution – but the urge to repress – which is growing with the danger."[3]

Paradoxically, the development of ultra-modern technology for observing the earth from outer space has confirmed the wise words of Socrates: "If we could rise above the earth we would understand the world in which we live".[4]

However, observing the world at a distance from all the routines and ostensible certainties of everyday life is a rare and, indeed, scientific virtue that architecture, urban development and landscape planning occasionally seem to lack. If we could only rise above the earth and regard the tasks facing these disciplines in a global context, we would also be in a better position to understand the world we are changing.

The nightly satellite pictures of the earth, in particular, provide urbanists, demographers and climatologists with compelling evidence of dramatic increases in population densities and of the growing urbanisation and immense consumption of energy in many regions of the earth. In theory, it would be possible to make interdisciplinary diagnoses and gain a new understanding of the preconditions for architecture, urban development and landscape planning in a global context, as well as insights into the overall context in which the urbanisation of the earth is proceeding. This would help us to identify the interrelationships between architecture, urban development and the use of land and to gain insights into the transformation of major cultural regions. However, the consequences of this vision – a graphic picture of our common existence as a world community living within the "horizon of the earth" – persistently eludes us.

Radiant Images of our Energy-based Civilisation

The archipelagos of light glowing at night (the urban networks and infrastructures spread across the globe), which are clearly visible through the eye of the satellite, ought to have made it clear to us long ago – with one illuminating image – that our global civilisation is energy-based. This is the reason for the waste-producing trends in the old industrialised countries in the West and the under-supply in the South and the East – in the emerging megacities. All differences notwithstanding, the diverse developments in the major centres are operating synergistically to generate a single global climate with dynamics of its own. As a result, the key areas of architecture and urban/infrastructure development, with landscaping more or less following on

heit nimmt sie gefasst auf, wie ins Schicksal schon gefügt. Müde, so scheint es, von immer neuen Szenarien des Untergangs, der doch immer in ferner Zukunft spielt, eine Drohung, so wolkig wie das Jüngste Gericht, und vielleicht wächst mit der Gefahr nicht das Rettende, sondern die Kraft zur Verdrängung."[3]

Paradoxerweise hat die Entwicklung der modernsten Technik der Observation der Erde vom Weltall aus die Weisheit des Sokrates bestätigt: „Könnten wir uns über die Erde erheben, würden wir die Welt, in der wir leben, verstehen".[4] Doch die Welt aus der Distanz zu den Routinen und Scheingewissheiten des Alltags zu betrachten, ist ungewöhnlich, eine wissenschaftliche Tugend, an der es der Architektur-, der Stadt- und Landschaftsentwicklung gelegentlich zu mangeln scheint. Denn: Könnten wir deren Angelegenheiten, über die Erde erhoben, im großen Zusammenhang wahrnehmen, würden auch wir die Welt, die wir verändern, besser verstehen.

Vor allem die nächtlichen Satellitenbilder der Erde ermöglichen Urbanisten, Demographen und Klimaforschern schon lange eindrückliche Hinweise auf dramatisch ansteigende Bevölkerungsdichten, auf den wachsenden Grad der Urbanisierung und den enormen Energieverbrauch in den Regionen rund um den Globus. Übergreifende Diagnosen, ein neues Verständnis der Voraussetzungen von Architektur-, Stadt- und Landschaftsentwicklung im globalen Zusammenhang wären möglich, Einsichten in die großen Zusammenhänge der Urbanisierung der Erde, in die Zusammenhänge von Architektur-, Stadtentwicklung und Landschaftsverbrauch, in die Veränderung großer Kulturräume. Doch das Resultat dieser Perspektive, ein anschauliches Bild über das gemeinsame Dasein der Weltgesellschaft im „Horizont der Erde" will noch nicht zustande kommen.

Ein strahlendes Abbild der Energiezivilisation

Die mit dem Satellitenauge überschaubaren nächtlichen Archipele des Lichts: die verstädterten Netze und Infrastrukturen der Erde hätten uns längst mit einem einzigen, einleuchtenden Bild zeigen können, dass die globale Zivilisation eine Energiezivilisation ist. Sie begründet Tendenzen der Verschwendung in den alten, westlichen Industrienationen, und Tendenzen der Unterversorgung in den südlichen und östlichen Gürteln, in den heranwachsenden Megacitys. Bei allen Unterschieden besteht die Konsequenz, dass die unterschiedlichen Entwicklungen in den großen Zusammenhängen einer einzigen globalen Klimadynamik zusammenwirken. Und die Konsequenz, dass Architektur-, Stadt-, Infrastrukturentwicklung primär und Landschaftsentwicklung eher als Resultierende zu den zentralen Aggregaten des weltweiten Energieverbrauchs zählen (Materialien, Herstellung, Bauweisen, Betriebsweisen, Stadtklima, Kommunikation und Verkehr, regionaler Austausch etc.), folgt daraus.

behind, are among the biggest single consumers of energy worldwide (materials, manufacture, modes of construction, modes of operation, urban climate, communications and traffic, regional exchanges, etc.).

Today, global research into the effects of man-made processes on the earth as a whole, as well as on global warming, is supplying information on the consequences – with a dramatically altered empirical data frame – that regional planning and infrastructure, urban and landscape development have to deal with.

Taken together, both perspectives – the distanced, all-embracing view from above and meticulous, long-range, on-the-spot research – will help to define the altered paradigms for architecture, urban development and landscape planning within the horizon of the earth, the aim being not to preserve any outmoded conceptions held by members of our professions, but to create a new concept of practice.[5]

This also means asking whether architecture and urban development should conceive of themselves as professions that have to operate within prescribed information contexts that they do not elaborate themselves – or whether, as disciplines, they need to develop their own instruments for conducting scientific research and controlled experiments.

To underline just how valid the "large-scale" perspective is, the UN launched a world environmental study on World Environment Day in 2001. This four-year project is to compile the most comprehensive report to date on the state of the oceans and seas, the rivers and forests, and cultivated areas. The study represents a new quality in integrated research, with science operating on a "large scale". Our own disciplines – architecture, urban development and landscape planning – would be well advised to adopt its research motives and perspectives and to share in its objectives and fieldwork. After all, this inventory of the earth's destruction by human beings will raise fundamental, existential questions of concern to architecture, urban development and landscape planning.

Other disciplines are sometimes fascinated by the fact that architecture is able to rise above dialogues on everyday affairs and discussions of things that are familiar and well known. The French ethnologist, Marc Augé, says that what is of compelling interest in ethnology is the moment at which the projected future assumes form as architectural, urban and landscaped space. This applies not just to material structures, but to cultural symbols, foresight and visions as well. Aren't projections, exploration, experimentation and curiosity in the face of the unknown at the very heart of the future of architecture? Isn't the opportunity to help shape this future a powerful motivating force?[6] And aren't the disciplines of architecture, urban development and landscape

Heute geben die globalen Forschungsergebnisse zu den Auswirkungen der Zivilisationsprozesse auf die Erde im Ganzen und zur Erwärmung des Weltklimas über die Folgen Aufschluss – mit einem dramatisch veränderten empirischen Datenrahmen, mit dem sich Raumordnung, Infrastruktur-, Stadt- und Landschaftsentwicklung auseinander zu setzen haben. Beide Perspektiven zusammengenommen, der distanzierte, zusammenfassende Blick von oben und die akribische Langzeitrecherche vor Ort, helfen veränderte Paradigmen der Architektur-, Stadt- und Landschaftsentwicklung im Horizont der Erde zu begründen, um nicht überholte Berufsverständnisse zu prolongieren, um stattdessen neue Praxis-Verständnisse zu etablieren.[5] Und: Um zu fragen, ob Architektur und Städtebau sich als Professionen zu verstehen haben, die sich in vorgegebene Informationskontexte einordnen, ohne sie selbst zu erarbeiten – oder als Disziplinen, die eigene wissenschaftliche Instrumente der Recherche, der forschenden, kontrollierten experimentellen Anwendung zu entwickeln haben.

Um die Perspektive der „großen Maßstäbe" zu bestätigen, haben die UN zum „Tag der Umwelt" des Jahres 2001 eine Welt-Umweltstudie aufgelegt. Die auf vier Jahre angelegte Untersuchung soll die bisher umfangreichste Bestandsaufnahme von der Verfassung der Weltmeere, Flüsse, Wälder und landwirtschaftlichen Anbauflächen werden. Die Studie ist Ausdruck einer neuen Qualität ganzheitlicher Forschung, Ausdruck einer Wissenschaft der „großen Maßstäbe", deren Forschungsinteresse und Perspektive sich die Architektur-, Stadt- und Landschaftsentwicklung zu Eigen machen müssten, an deren Zielsetzungen und Recherchen unsere eigenen Disziplinen beteiligt sein müssten. Denn die Ergebnisse der Bestandsaufnahme aller durch den Menschen verursachten Störungen auf der Erde werden die Architektur-, Stadt- und Landschaftsentwicklung vor grundlegende, vor existentielle Fragen stellen.

Andere Disziplinen fasziniert – gelegentlich – an der Architektur, dass sie aus dem Dialog über den Alltag, aus dem Austausch über das Bekannte, Vertraute – aufzubrechen vermag. In der Ethnologie – bei dem französischen Anthropologen Marc Augé ist es nachzulesen – interessiert der Moment, in dem die projizierte Zukunft in Form von Architektur, städtebaulichen und landschaftlichen Räumen Gestalt gewinnt. Es geht dabei nicht nur um materielle Strukturen, sondern um kulturelle Symbole und um Voraussicht, um Visionen. Ist das Projektive, die Exploration, das Experiment, die Neugier auf das Unbekannte der Architektur nicht der Kern der Zukunft der Architektur? Bietet die Chance, an der Gestaltung dieser Zukunft teilzunehmen, nicht eine starke Motivation?[6] Und sind die Disziplinen, welche die Architektur-, Stadt- und Landschaftsentwicklung vertreten, nicht in der Lage, die Abstraktionen der Politik, der Ökonomie und der Naturwissenschaften, die sich Lebenswissenschaften nennen, in anschauliche, kulturell vermittelte, verstehbare Bilder zu übersetzen? An dieser Vermittlungs-, an dieser Übersetzungsaufgabe wird deutlich, dass Architektur-, Stadt- und

planning in a position to translate the abstractions of politics, economics and the natural sciences, which call themselves life sciences, into graphic, culturally communicable and intelligible images? This task of communication and translation makes it quite clear that architecture, urban development and landscape planning constitute a "medium", into which the future we wish to have must be projected. In making this projection, it is impossible to abstract from the social conditions (from the implantation of cities into landscapes and the environment), from the conditions of controlled innovation and the conditions of aesthetic and cultural space.

Ultimately, architecture is the material, fashioned and symbolic synthesis of social demands and requirements and their translation into space. If this is the case, can the discipline create a new paradigm[7] for itself – a paradigm of architecture, urban development and landscape planning that incorporates a global dimension? Is this discipline in a position to fire people with enthusiasm for architecture, for the town and city, and for the landscape as an idea, a vision and a utopia? Can it redefine itself in the context of a global environmental policy?

Resource Architecture – a Concept within a Global Environmental Policy

It is no accident that the Executive Director of UNEP, Klaus Töpfer, described the elaboration of a world environmental policy as a task that should form part of a global peace policy. The UNEP studies show quite clearly that the earth's temperature is rising. The consequences are visible throughout the world. Deserts are expanding, the sea level is rising, and droughts, hurricanes and cyclones are increasing in both intensity and frequency. The effects of carbon dioxide emissions are an added burden, especially for the poorest of the poor living in the developing countries. It is they who are paying the harshest price for our actions in the North.[8] In many parts of the world, desertification and water shortages are driving people from their homes. There are already 25 million environmental refugees in Africa and in parts of Eastern Europe alone. We humans are largely responsible for this, mostly as the result of the dramatic increase in the combustion of coal, mineral oil and gas and the concomitant emissions of carbon dioxide. Above all, it is we – through our behaviour as consumers – in the wealthy countries of the North who are the cause of this development. Nature is relentlessly calling us to account for our negligent behaviour[9]. The current political differences highlight the consequences that await us if we fail to develop a modern global environmental policy.

The UNEP, which reminds us that the USA is the world's biggest producer of greenhouse gases, nonetheless points out that it is up to the developed and developing countries to co-operate if

Landschaftsentwicklung „Medien" sind, in die eine gewollte Zukunft hineinprojiziert werden muss. Bei dieser Projektion kann von den gesellschaftlichen Bedingungen, der Einlagerung in Landschaft/Umwelt, von den Bedingungen einer kontrollierten Innovation und von den Bedingungen des ästhetischen, des kulturellen Raums nicht abstrahiert werden. Architektur ist ja geradezu die materielle, die gestaltete, die symbolische Synthese der gesellschaftlichen Anforderungen – und deren Transkription in den Raum.

Wenn das so ist: Kann sich die Disziplin ein neues Paradigma[7] einer um die Weltdimension erweiterten Architektur-, Stadt- und Landschaftsentwicklung geben? Kann sich die Disziplin selbst für die Architektur, für die Stadt, für die Landschaft als Idee, als Vision, als Utopie begeistern? Kann sie sich im Rahmen einer Welt-Umweltpolitik neu definieren?

Ressource Architektur – Konzept im Rahmen einer Welt-Umweltpolitik

Nicht von ungefähr bezeichnet der Executive Director der UNEP, Klaus Töpfer, die Konzeption einer Welt-Umweltpolitik als Aufgabe einer globalen Friedenspolitik. Die Studien der UNEP liefern klare Belege dafür, dass die Erdtemperatur ansteigt. Die Folgen sind überall auf der Welt spürbar: Wüsten vergrößern sich, der Meeresspiegel steigt, Dürren, Hurrikane und Zyklone nehmen an Intensität und Häufigkeit zu. Die Auswirkungen der Kohlendioxid-Emissionen sind gerade für die Ärmsten der Armen, die Menschen in den Entwicklungsländern, eine zusätzliche Last. Sie zahlen den Großteil der Zeche unseres Verhaltens im Norden[8]: Versteppung und Wasserknappheit vertreiben Menschen in vielen Teilen der Welt aus ihrer Heimat. Allein in Afrika und in Teilen Osteuropas gibt es schon heute 25 Millionen Umweltflüchtlinge.

Es sind wir Menschen, die das hauptsächlich selbst verursachen: vor allem durch die drastisch angestiegene Verbrennung von Kohle, Mineralöl und Gas, wodurch in gleichem Maß Ausstoß von Kohlendioxid bewirkt wird. Vor allem die Menschen in den Wohlstandsländern des Nordens haben durch ihr Verbraucherverhalten diese Entwicklung zu verantworten. Die Natur präsentiert uns unbarmherzig die Rechnung für unser fahrlässiges Verhalten[9]. Welche Folgen sich beim Scheitern einer modernen Welt-Umweltpolitik einstellen könnten, zeigen die aktuellen Differenzen.

Die UNEP, die daran erinnert, dass die USA der weltweit größte Emittent von Treibhausgasen sei, erklärt dennoch, dass es gerade auf die Kooperation zwischen den entwickelten und den sich entwickelnden Ländern ankomme, um die prognostizierten Folgen des Klimawandels zu vermeiden: Die fortgeschrittensten Nationen haben UNEP zufolge die Technologien, die erforderlich sind, um die Emissionen der Treibhausgase zu reduzieren. Und diese müssten

the predicted consequences of climate change are to be averted. According to UNEP, the most advanced nations already have the technologies needed to reduce emissions of greenhouse gases. These technologies must be transferred to the developing regions of the world to reduce the air pollution caused by the growth in their economies.[10] It is important to remember that the costs of climate change are considered to be just as high as the costs of slowing down climate change. The North will incur the bulk of the costs for forward-looking strategies to avoid such changes, however, while the South will bear the brunt of the costs caused by global warming.[11]

One of the determining factors in the regional application of novel development principles in architecture, urban development and landscape planning would be a new kind of architecture within the horizon of the earth. Like the concepts of sustainable urban development and landscape planning, this new kind of architecture could prove effective as a platform constituting part of a global industrial and labour market policy. It would also need to devise strategies for promoting an international exchange, i.e. strategies transcending the boundaries between individual disciplines. If we wish to address the real problems facing the world, we must focus first on some dramatic findings and insights stemming from the World Environment Conference (1992), World Population Conference (1994), World Habitat Conference (1996), World Climate Conference (1997) and the empirical findings, such as those contained in the current IPCC (Intergovernmental Panel on Climate Change) report on climate warming.

If we then wish to raise our sights higher we must state quite clearly which of the general solutions we consider to be correct and then stand by them. This takes us beyond architecture and planning in the narrow sense. What we are talking here about is a statement of general political principles befitting the world of architecture.

Dramatic environmental catastrophes are no longer the product of horror scenarios in Hollywood films. They have become everyday occurrences affecting by no means just the crisis-ridden poor countries of the South. We are now seeing pictures of floods in England and Italy, too. Climate change knows neither winners nor losers. The costs of the environmental catastrophes are rising sharply. Hitherto, architecture, urban development and landscape planning have not featured prominently in traditional political debates when it comes to pinpointing factors that are crucial to our modern energy-based civilisation and its global expansion. However, it would appear to be only a matter of time before this changes.

The lack of interdisciplinary guidelines that could commit architecture, urban development and landscape planning to address global environmental policy issues leads, with little ostensible

in die sich entwickelnden Regionen der Welt transferiert werden, um eine Reduzierung der Luftverschmutzung im Zuge des fortschreitenden Wachstums der Ökonomien dieser Länder zu erreichen.[10] Dabei werden die Kosten des Klimawandels ebenso hoch geschätzt wie die Kosten, den Klimawandel zu bremsen. Die Kosten für eine vorausschauende Vermeidung würden jedoch überwiegend im Norden anfallen, die Folgekosten der Erwärmung überwiegend im Süden.[11]

Einer der bestimmenden Faktoren der regionalen Anwendung neuartiger Entwicklungsprinzipien für die Architektur-, Stadt- und Landschaftsentwicklung wäre eine innovative Architektur im Horizont der Erde, die, wie Konzepte einer nachhaltigen Stadt- und Landschaftsentwicklung auch, als Plattform, als Teil einer modernen Welt-Umweltpolitik, als Teil einer globalen Industrie- und Arbeitsmarktpolitik wirksam werden könnte. Sie hätte gleichermaßen auf Konzeptionen für den internationalen Austausch zu zielen, mithin Konzeptionen innerhalb übergreifender Rahmenbedingungen zu entwickeln. Wenn wir die wirklichen „Weltprobleme" einbeziehen wollen, müssen am Beginn dramatische Befunde und Einsichten stehen, wie sie die Welt-Umwelt-Konferenz 1992, die Welt-Bevölkerungs-Konferenz 1994, die Welt-Habitat-Konferenz 1996 und die Welt-Klima-Konferenz 1997 und die aufregenden empirischen Befunde des Berichts der IPCC (Intergovernmental Panel on Climate Change) zur Klimaerwärmung erbracht haben. Wenn wir mehr wollen, müssen wir deutlich machen, welche der generellen Lösungsvorschläge wir für richtig halten – um uns dazu zu bekennen. Das hat nicht nur mit Architektur und Planung im engeren Sinne zu tun; das ist ein allgemeines politisches Bekenntnis, das der Architektenwelt gut ansteht.

Längst sind Berichte über dramatische Umweltkatastrophen nicht mehr Horrorrszenarien aus Hollywood-Filmen, sondern alltäglich – nicht nur in den krisengeschüttelten armen Ländern des Südens. Die Bilder zeigen auch Überschwemmungen in England und Italien. Der Klimawandel kennt keine Gewinner oder Verlierer. Die Kosten der Umweltkatastrophen steigen rasant. In den traditionellen Strategien der politischen Auseinandersetzungen spielt die Architektur-, Stadt- und Landschaftsentwicklung bei der Benennung von Faktoren, die prägend sind für die moderne Energiezivilisation und deren globale Ausbreitung, bisher noch keine zentrale Rolle. Aber das scheint lediglich eine Frage der Zeit zu sein.

Der Mangel an übergreifenden Leitlinien, die Architektur-, Stadt- und Landschaftsentwicklung auf Themen im Maßstab einer Welt-Umweltpolitik verpflichten könnten, führt – nur scheinbar begründet – zur Beliebigkeit der Handlungsmöglichkeiten, zur Architektur des anything goes, die der New Yorker Architekt Bruce S. Fowle benennt: „We have too many choices, too many responsibilities, and too few conventions from which we can

justification, to arbitrariness and an architecture of "anything goes". As the New York architect, Bruce S. Fowle, put it: "We have too many choices, too many responsibilities, and too few conventions from which we can base evaluation of our actions and our work." If this attitude were to be preserved, without any criticism being expressed, it would be so indissolubly fixed in the mainstream way of life in Germany, France, Italy and, above all, in the USA, that architecture, urban development and landscape planning would remain permanent components of our contemporary energy-based civilisation, which divides the world into North and South.

Faced with the possible abandonment of the Kyoto Protocol, Klaus Töpfer pulled off the wraps in his diagnosis of the situation by referring to the absence of a climate policy in the rich countries as a form of "ecological aggression by the North against the South".[12] To gain an appreciation of the role played by architecture, urban development and landscape planning in this "ecological aggression", it is worth recalling Töpfer's assessment that they account for the consumption of 60 to 70 per cent of the world's primary energy sources (UNEP), even though they are not actually aware of the fact.

With this information increasingly becoming public knowledge, architecture, urban development and landscape planning have no choice but to join in the elaboration of a modern global environmental policy.

Eurocentrism and the North-South Issue

In the last Club of Rome report, Ernst Ulrich von Weizsäcker, Amory B. Lovins and L. Hunter Lovins describe the "paradox" of the present situation as follows: "The least we can do is to ask whom 'sustainable development' is supposed to benefit. The North still believes that the concept of sustainable development is designed to put the development of the South on an ecological footing. The South, on the other hand, thinks that the concept is geared to maintaining the unsustainable lifestyle of the North. The Rio Declaration permits both interpretations."[13] The authors believe that the North overlooks the fact that "it is the expensive form of environmental protection focused on pollutants and the North's totally excessive consumption of resources that are causing the South to come up with an ecologically and economically unsettling response, one which aims primarily at the traditional kind of growth achieved in the industrialised countries and defers environmental protection in order to catch up". How can we all collectively visualise and understand the role of architecture, urban development and landscape planning and the way they fit into worldwide mechanisms of economics, land use and the dynamic development of the countryside, the climate and energy consumption? If we are to be capable of doing so, we in Europe

base evaluation of our actions and our work." Diese Haltung, würde sie unkritisch fortgeschrieben, wäre in den Mainstream des Way of Life in Deutschland, in Frankreich, in Italien und vor allem in den USA so unauflöslich eingelagert, dass Architektur-, Stadt- und Landschaftsentwicklung auf Dauer der unveränderliche Bestandteil der zeitgenössischen Energiezivilisation bliebe, die die Welt in den Norden und in den Süden teilt.

Klaus Töpfer hat, in der Reaktion auf die Möglichkeit des Scheiterns des Kyôto-Protokolls, seine Diagnose zur Lage verschärft und die mangelhafte Klimapolitik der reichen Länder „eine ökologische Aggression des Nordens gegen den Süden" genannt.[12] Um einschätzen zu können, welche Rolle Architektur-, Stadt- und Landschaftsentwicklung dabei spielen, sei Töpfers Einschätzung zitiert, dass dieser Bereich sechzig bis siebzig Prozent des Primärenergieaufkommens (UNEP) weltweit binde, ohne sich dessen bewusst zu sein. Architektur-, Stadt- und Landschaftsentwicklung kommt, da dieser Zusammenhang mehr und mehr Öffentlichkeit genießt, gar nicht daran vorbei, sich in den Rahmen einer modernen Welt-Umweltpolitik zu stellen.

Eurozentrismus versus Nord-Süd-Problematik

Im letzten Bericht für den Club of Rome beschreiben Ernst Ulrich von Weizsäcker, Amory B. Lovins und L. Hunter Lovins die „Paradoxie" der Ausgangslage so: „Das mindeste, was bleibt, ist die Frage, wer eigentlich der Hauptaddressat der ‚nachhaltigen Entwicklung' sein soll. Der Norden glaubt weiterhin, dass das Konzept der nachhaltigen Entwicklung dazu da ist, die Entwicklung des Südens zu ökologisieren. Der Süden denkt umgekehrt, dass der Begriff auf den nicht tragfähigen Lebensstil des Nordens gemünzt ist. Die Rio-Deklaration lässt beide Interpretationen zu."[13] Aus der Sicht der Autoren übersieht der Norden, „dass es die Form des teuren, auf Schadstoffe fixierten Umweltschutzes und die weit überhöhten Naturverbrauchsraten des Nordens sind, die den Süden zu einer ökologisch und ökonomisch beunruhigenden Haltung veranlassen, zuerst auf Wachstum im traditionellen Sinne der Industrieländer zu setzen, um aufzuholen und den Umweltschutz zurückzustellen. Wie machen wir uns alle gemeinsam auch bildlich verständlich, wie Architektur-, Stadt- und Landschaftsentwicklung im Kontext weltweiter Mechanismen (des Wirtschaftens, der Landnutzung, der Landschafts-, Klima- und Energiedynamik) verortet sind? Dazu benötigen wir in Europa eine Erweiterung der traditionellen eurozentristischen Wahrnehmung (Lehrmeinung!) sowie eine Öffnung des Verständnisses für die Wechselwirkungen zwischen Nord und Süd, zwischen alten Industriezivilisationen und nachwachsenden Entwicklungsländern: das Verständnis, dass alle Regionen der Welt eine global zusammenhängende Energiezivilisation bilden, die die Ressourcen der Erde in unterschiedlicher Weise ausbeutet. Und zwar so, dass wir die Erde viermal brauchten, wenn wir so weitermachten. Wir

need to expand our traditional Euro-centrist view (school of thought) and develop a better understanding of the reciprocal relationship between North and South, between old industrial civilisations and up-and-coming developing countries – an understanding of the fact that all the regions of the world constitute a globally interlinked, energy-based civilisation that exploits the world's resources in different ways. In fact, if we carry on as we are doing at present we are going to need four worlds. The problem is we only have one. (Thomas Herzog) "Carrying on regardless" would mean that the Western countries would not change their energy consumption course, not adapt their systems to circulatory management and the protection of resources, and not take corrective measures in architecture, urban development and landscape planning. Moreover, it would mean that the developing and threshold countries would adopt and imitate this concept of development at a high level of energy consumption.

In the North, we can interpret the bogus certainties of Western civilisation as meaning that we, who labour under the illusion that we are relatively free of the town-planning problems haunting the megacities in the South, need to consume some 60 to 70 per cent of the world's energy (source) in order to preserve what is, in effect, our urban stagnation. In doing so, we are causing problems in other regions of the earth, which we are fond of dealing with in an abstract way because we are simply incapable of understanding them.[14]

The satellite view of our global energy-based civilisation is very instructive. It counteracts cultural traditions that are focused on their own ways of life in their own parts of the world and are thus self-centred. Eurocentrism is the term for the European version of self-centredness; but there are also North and South American versions, African, Middle and Far Eastern variations, and specially defined versions in Australia and Oceania. With the discovery of the world by the seafarers, this view initially meant subjugation, which, in the nineteenth century, was of an imperialist nature. With the discovery of globalisation it has come to mean subjugation by expanding economies. The dialogue between civilisations ought to encourage an interdisciplinary perspective that situates innovative architecture within the horizon of the earth and makes sustainable urban development and landscape planning a task for a dialogue among civilisations.[15]

For all their critical foresight, the Declarations of Rio and Istanbul, of Kyoto and The Hague have one "flaw". They lack visual clarity and fail to appeal to the senses. They have neither a distanced view of the earth, nor an image of the horizon of the earth that can be grasped by the senses. The XXI World Congress of Architecture will become a truly world congress if it takes the world at its word and by its image as a world with inter-related problems.

haben aber nur eine." (Thomas Herzog). So weitermachen wie bisher hieße, dass die westlichen Länder ihre Linie der Energiezivilisation nicht korrigierten, ihre Systeme nicht auf Kreislaufwirtschaft und Ressourcenschonung umstellten, ihre Architektur-, Stadt- und Landschaftsentwicklung nicht korrigierten. Und dass die Entwicklungs- und Schwellenländer dieses Konzept der Entwicklung auf hohem Energieniveau übernehmen und nachvollziehen.

Die Scheingewissheiten der westlichen Zivilisation können wir im Norden so verstehen, dass wir, die wir in Relation zu den südlichen Megacitys kaum städtebauliche Probleme zu haben „glauben", zur Aufrechterhaltung dieser urbanen ‚Stagnation' aber 60 bis 70 Prozent des Weltenergieverbrauchs benötigen. Und damit Verursacher von Problemen in anderen Regionen der Erde sind, von denen wir gern abstrahieren, weil sie sich uns nicht vermitteln.[14]

Der Blick vom Himmel auf die globale Energiezivilisation ist ein didaktisches Hilfsmittel. Er wirkt den kulturellen Traditionen entgegen, mit denen der Blick auf die jeweils eigene Lebensweise in den Regionen der Welt zentriert ist und dadurch selbstbezogen wird. Eurozentrismus heißt die europäische Variante der Selbstbezogenheit, und es gibt sie in nord- und südamerikanischen Varianten, in afrikanischen, nah- und fernöstlichen Spielarten, ganz eigen definiert in Australien und Ozeanien. Mit der Entdeckung der Welt durch die Seefahrer war dieser Blick zunächst einer der Unterwerfung. Im 19. Jahrhundert einer des Imperialismus. Mit der Entdeckung der Globalisierung einer der expansiven Ökonomien. Der Dialog der Zivilisationen sollte eine übergreifende Perspektive ermöglichen, welche die innovative Architektur in den Horizont der Erde stellt, die nachhaltige Stadt- und Landschaftsentwicklung zur Aufgabe eines Dialogs der Zivilisationen machte.[15]

Die Deklarationen von Rio und Istanbul, von Kyôto und Den Haag haben, bei aller kritischen Voraussicht, einen „Mangel": den der fehlenden sinnlichen, den der fehlenden visuellen Anschaulichkeit. Ihnen fehlt der Blick aus der Distanz auf die Erde, das sinnlich nachvollziehbare Bild vom Horizont der Erde. Der XXI. Architektur-Weltkongress wird dann ein Weltkongress, wenn er die Welt beim Wort und beim Bild nimmt: die Welt mit ihren übergreifenden Problemen.

ENE

Development in the Horizon of the Earth

The statistical trends reflecting the development of the world's urban population speak for themselves. At the beginning of the 20th century, 10 per cent of the world's population lived in towns and cities. In the year 2000, that figure had risen to 50 per cent. It is forecast that five billion people will be living in towns and cities by the year 2025, two-thirds of them (3.4 billion) in large cities and megacities in the poor countries.[16] In 1950, only New York and London had more than eight million inhabitants. Now there are already 22 urban agglomerations, to which the term "megalopolis" is applied. It is predicted that this number will rise to 33 by the year 2015. 27 of these megacities will be in the less developed countries, 19 of them in Asia. Among the 10 largest cities in the world, only Tokyo will be rich.[17]

The 20th century experienced a population explosion and mass migrations on an unprecedented scale. Human beings have spread to every part of the inhabitable earth (by 1800 one billion, by 1930 two billion, by 1960 three billion, by 1975 four billion, by 1987 five billion, by 1999 six billion). Historically, human beings have preferred to settle on attractive, productive land with moderate temperatures, avoiding very mountainous, as well as very dry and cold regions. The unbridled growth of the megacities and large agglomerations has transformed the relationship we have with our natural surroundings. The growth of cities and infrastructures (land consumption), the intensification of agriculture, industrialisation (clearing) and climate change (growth of deserts, floods) are threatening to destroy our environment.[18]

Thinking in terms of the new/old horizon of the earth has characterised the anti-civilisation philosophy in Japan since the earthquake in Kobe and it is a typical feature of the philosophies of sustainable development in Europe, too. The German cultural analyst, Hartmut Böhme, calls for the development of cultural reflection in societies throughout the world. In his view, there will be dramatic shifts triggered by climate changes and energy consumption, by the explosive growth of the megacities, by the huge migratory movements in East Asia and Africa, by the radical innovations in biotechnology and by the devastating over-exploitation of the earth. He argues that these changes will affect both the state of our natural environment as it has developed during the history of the planet and the existence of all that has emerged to create a humanly dignified life in many of the world cultures. He suggests that we need binding agreements susceptible of development by future generations in respect of what it means for human societies to develop within the horizon of the earth. This entails recognising that humanity is terrigenus, a creature of the earth. According to Böhme, this is no longer self-evident. If the concept of "urban development within the horizon of the earth" is taken seriously, the paramount objective is to feel at home on the earth with a different architecture. A look at

Entwicklung im Horizont der Erde

Die statistischen Tendenzen der Entwicklung der urbanen Welt-Bevölkerung sprechen eine eigene Sprache: Zu Beginn des 20. Jahrhunderts lebten zehn Prozent der Welt-Bevölkerung in Städten. Im Jahr 2000 waren es 50 Prozent. Für das Jahr 2025 wird prognostiziert, dass fünf Milliarden Menschen in Städten leben werden – zwei Drittel davon (3,4 Milliarden) in großen Städten und Megacitys der ärmeren Länder.[16] 1950 hatten nur New York und London mehr als acht Millionen Einwohner. Heute gibt es bereits 22 Urbanisationen, die Megalopolis genannt werden. Für das Jahr 2015 wird die Zahl von 33 Megacitys vorausgesagt, 27 davon in den weniger entwickelten Ländern, davon 19 in Asien. Unter den zehn größten Städten der Welt wird Tokio die einzige reiche Stadt sein.[17]

Das 20. Jahrhundert hat eine Bevölkerungsexplosion und Bevölkerungswanderungen erlebt wie noch kein Jahrhundert zuvor. Menschen haben sich über die gesamte bewohnbare Erde ausgebreitet (bis 1800: eine Milliarde, bis 1930: zwei Milliarden, bis 1960: drei Milliarden, bis 1975: vier Milliarden, bis 1987: fünf Milliarden, bis 1999: sechs Milliarden). Historisch gesehen bevorzugen Menschen für ihre Siedlungen attraktives, ergiebiges Land mit gemäßigten Temperaturen und vermeiden allzu bergige, trockene oder kalte Regionen. Mit dem ungebremsten Wachstum der Megacitys und großen Agglomerationen verändert sich die Korrespondenz mit den Landschaften – bis schließlich durch das Wachstum der Städte und Infrastrukturen (Landschaftsverbrauch), durch die Intensivierung der Landwirtschaft und die Industrialisierung (Ausräumung) und durch den Klimawandel (Ausdehnung der Wüsten, Überschwemmungen) die Landschaften bedroht werden.[18]

Das Denken im neu-alten Horizont der Erde kennzeichnet die zivilisationskritische Philosophie im Japan nach dem Erdbeben von Kobe ebenso wie die Philosophien der nachhaltigen Entwicklung in Europa. Der deutsche Kulturwissenschaftler Hartmut Böhme fordert die Etablierung kultureller Reflexion in den Gesellschaften selber: Es sei davon auszugehen, dass der Bestand der natürlichen Umwelt, wie sie erdgeschichtlich gebildet ist, und der Bestand dessen, was in vielen Weltkulturen als menschenwürdiges Leben ausdifferenziert wurde, sich dramatisch ändern werde – durch die Folgen der Klima- und Energiedynamik, durch die Explosion der Megacitys, durch die großen Wanderungsbewegungen in Ostasien und Afrika, durch die umwälzenden Biotechnologien ebenso wie durch den ruinösen Raubbau an der Erde. Das erfordere verbindliche und für nachfolgende Generationen anschlussfähige Konsense darüber, was es für menschliche Gesellschaften heißen soll, sich innerhalb des Horizonts der Erde zu entwickeln. Zuerst hieße es anzuerkennen, dass die Menschheit terrigenus ist, ein Geschöpf der Erde. Dies sei nicht mehr selbstverständlich. Nimmt man das Konzept der „Stadtentwicklung im

the state of our planet reveals that we are far from achieving this goal.[19]

The radical criticism that has been expressed has helped to sharpen awareness of the fact that the objective of architecture, urban development and landscape planning in the horizon of the earth is currently geared to the worldwide preservation, development and humane structuring of urban civilisations, incorporating due consideration for the specific cultural characteristics of the world's regions. The vision of the invulnerability of cities – and of their culture, which has also been radically called into question – cannot be debated within the current confines imposed by our perception of globalisation. The invulnerability of cities and their culture is reflexive both as a thesis and an antithesis and cannot be restricted solely to the preservation of the existing fabric, which has developed over a period of centuries. On the contrary, it must be extended to include the unfinished urban development in the regions of the world. The invulnerability of cities is tantamount to the invulnerability of an ideal or utopia, which is far removed from reality in most places in the world. If the prospect for the future is that Tokyo will be the only rich city among the world's 10 largest megacities 15 years from now, the course that clearly needs to be charted is one that will lead to a balancing up of living conditions, to urban development aid, justice and equality of opportunity.

World Urban Policy

The vision of the invulnerability of cities has never arisen either in history or in the genealogy of cities in the southern hemisphere. The concept of urban culture is the concept of a "world whose borders are drawn by the boundaries of awareness. It involves a conviction that nothing can arrest or disturb the continuous flow of consciousness. The battles of life, politics and economics have been fought. All that matters now are the disassociations and distinctions that take place in the mind."[20] The notion of an unendangered cultural society no longer exists. Moreover, the events of New York have demonstrated that the concepts of culture in the regions of the world are determined by different boundaries of awareness. The global culture of architecture, urban development and landscape planning now once again incorporates the awareness that the material substructure of every form of urban culture is politically, militarily and economically vulnerable and can be called into question by external developments.

Many economists and political scientists are agreed that there is a need to radically modify the economic processes of globalisation. Attention must likewise be paid to cultural differences. "In the process of globalisation America has overlooked the reality of life in the poor countries. Yet our prosperity depends on the proper functioning of the globalised world (...). There is a growing

Horizont der Erde" ernst, dann geht es primär darum, mit einer veränderten Architektur auf der Erde heimisch zu werden. Davon sind wir, im Weltmaßstab betrachtet, weit entfernt.[19] Die weltweite Erhaltung, Entwicklung und humane Gestaltung der städtischen Zivilisationen in den kulturell spezifischen Eigenarten der Regionen der Welt ist als Ziel der Architektur-, Stadt- und Landschaftsentwicklung im Horizont der Erde gerade dadurch, dass es nun so radikal in Frage gestellt worden ist, umso stärker zu Bewusstsein kommen. Bereits die Perspektive der Unverletzlichkeit der Städte – und ihrer Kultur, die ebenso radikal in Frage gestellt worden ist, kann nicht innerhalb der herrschenden Grenzen der Wahrnehmung der Globalisierung debattiert werden. Unverletzlichkeit der Städte und ihrer Kultur ist als These wie als Antithese reflexiv, kann nicht nur Themen der Erhaltung des einmal entwickelten Bestands umfassen, der sich jahrhundertelang ausbilden konnte – sondern muss sich zugleich der globalen, der unvollendeten städtischen Entwicklung in den Regionen der Welt widmen. Unverletzlichkeit der Städte – das wäre die Unverletzlichkeit eines Ideals, einer Utopie, von der die Realität fast in der Mehrheit der Orte der Welt entfernt ist. Wenn die Weltsicht beinhaltet, dass von den zehn größten Megacitys der Welt in fünfzehn Jahren Tokio die einzige reiche Stadt sein wird, können die Bedingungen der Reflexivität beschrieben werden: als eine Perspektive des Ausgleichs der Lebensbedingungen, als eine Aufgabe der urbanen Entwicklungshilfe, der Gerechtigkeit und der Chancengleichheit.

Welt-Städtepolitik

Die Perspektive der Unverletztlichkeit der Städte hat sich weder in der Geschichte noch in der Genealogie der Städte des Südens gestellt: Die Vorstellung von Stadtkultur ist die Vorstellung einer „Welt, deren Grenzen durch die Grenzen des Bewusstseins gezogen sind. Sie ist überzeugt davon, dass nichts den kontinuierlichen Fortlauf des Bewusstseins aufhalten oder stören kann. Die Kämpfe des Lebens, der Politik, der Wirtschaft sind geschlagen; jetzt kommt es nur noch auf die Distanzierungen und Distinktionen an, die im Kopf stattfinden."[20] Das Bild der ungefährdeten Kulturgesellschaft gibt es nicht mehr. Zugleich haben die Ereignisse des 11. September verdeutlicht, dass die Vorstellungen von Kultur in den Regionen der Welt durch unterschiedliche Grenzen des Bewusstseins bestimmt sind. Nun gehört zur globalen Kultur der Architektur-, Stadt- und Landschaftsentwicklung wieder das Bewusstsein, dass der materielle Unterbau jeder Form urbaner Kultur politisch, militärisch und ökonomisch angreifbar ist, in Frage gestellt durch externe Entwicklungen.

Bei zahlreichen Ökonomen und Politikwissenschaftlern besteht Übereinstimmung, dass es darauf ankommt, die ökonomischen Prozesse der Globalisierung entscheidend zu modifizieren. Analog dazu muss die kulturelle Differenzierung mitbedacht werden. „Amerika hat die Realität armer Länder im Globalisierungsprozess

impression that the world economy is dominated by opportunism and that a large part of the world has no share in prosperity (...). Cultural convergence is difficult to achieve. But top-down solutions do not work. What is needed are closer relations in civil society."[21]

Amongst the progressive, internationally oriented groups in the USA, the New America Foundation stands out for the positions it has adopted recently. It pinpointed clear alternatives to the strategies of isolationist forces even before the events of 11 September 2001. "The United States would be well advised to think ahead. For instance, about how the International Monetary Fund (IMF) can be transformed into a consultative agency for the developing countries so that they no longer have to meekly accept their role as the manufacturers of cheap goods for US consumers, but instead can set up economic structures capable of supporting a national middle class and long-term domestic growth. It is high time Washington began working together with these regions of the world as well as with non-governmental organisations (NGOs) active on their behalf. In other words, the USA should approach the poor regions of the world with understanding rather than confronting them with its ideology and assertion of power."[22]

Saskia Sassen reminds us that differentiated assessments and an open dialogue between cultures entails seeing development in the different regions of the world in context. As she says, "the fact that poverty and inequality are steadily increasing and that the national debt of countries is rising, which means that they have less and less money available for development programmes, produce the sort of conditions that, in my view, generate helplessness and anger. The attacks of 11 September could have set the alarm bells ringing." Saskia Sassen sees development programmes along the lines of the Marshall Plan as being the smallest conceivable common denominator for global development and the improvement of the situation in the world's largest cities. Programmes of this kind would make it possible for developing countries facing debt-servicing charges of the kind encountered in the period after the Second World War to create completely new structural conditions in urban and rural areas. This would require far-reaching innovation. "The world is moving ever closer together; its regions are now more closely intertwined than ever before. This has given new meaning to old misunderstandings. And it has produced new assymmetries. Increasing debt, poverty and epidemics in the southern hemisphere are now spreading to the rich countries of the North. Today, we can no longer afford to turn our backs on the many forms of misery, as we have done in the past. We may find it hard to address these problems seriously on humanitarian grounds. But even where there is a lack of altruism, self-interest should provide adequate motivation. (...) Will these events not trigger the learning process that is so overdue?"[23]

vernachlässigt. Unser Wohlstand hängt aber davon ab, dass die globalisierte Welt angemessen funktioniert (...). Der Eindruck wächst, dass die Weltwirtschaft von Opportunismus dominiert wird und ein großer Teil der Welt nicht am Wohlstand teilhat (...). Kulturelle Annäherung ist zwar schwer. Aber Lösungen von oben funktionieren nicht. Notwendig sind tiefere Beziehungen in der Zivilgesellschaft."[21]

Bei den progressiven, internationalistisch denkenden Gruppierungen in den USA sind in letzter Zeit Positionen der New America Foundation aufgefallen, die bereits vor den Ereignissen des 11. September 2001 gegenüber isolationistischen Kräften klare Alternativen aufzeigen: „In den Vereinigten Staaten sollte man besser nach vorn denken. Zum Beispiel darüber, wie sich der Internationale Währungsfonds (IWF) in eine Beratungsinstanz für die Entwicklungsländer verwandeln ließe, damit diese sich nicht einfach in ihre Rolle von billigen Produzenten für die US-amerikanischen Verbraucher fügen, sondern stattdessen ökonomische Strukturen aufbauen, die eine eigene Mittelklasse und ein langfristiges Binnenwachstum tragen können. In Washington muss man also endlich mit diesen Teilen der Welt zusammenarbeiten und auch mit den Nicht-Regierungs-Organisationen (NGO), die sich für sie engagieren. Mit anderen Worten: Die USA müssten den armen Teilen der Welt mit Verständnis gegenübertreten, statt wie bisher nur mit ihrer Ideologie und ihrem Herrschaftsanspruch."[22]

Zu den differenzierten Einschätzungen, zu einem offenen Dialog der Kulturen zählt, die Entwicklung in den Weltregionen im Zusammenhang zu sehen, wie es Saskia Sassen anmahnt: „Dass Armut und Ungleichheit stetig anwachsen, dass die Verschuldung der Staaten zunimmt, so dass sie immer weniger Mittel für Entwicklungsprogramme zur Verfügung haben – das alles schafft nach meiner Einschätzung die Bedingungen, unter denen Wut und Hoffnungslosigkeit gedeihen. Die Attentate vom 11. September hätten ein Weckruf sein können." Als denkbar kleinsten gemeinsamen Nenner der Weltentwicklung und der Entwicklung der Situation in den großen Städten der Welt nennt Sassen Entwicklungsprogramme ähnlich dem Marshall-Plan, der es den betroffenen Entwicklungsländern, bei einem der Zeit nach dem Zweiten Weltkrieg vergleichbaren Schuldendienst, ermöglicht, in den Städten wie in den ländlichen Räumen völlig neue strukturelle Bedingungen zu schaffen. Das erfordere grundlegende Innovationen. „Die Welt rückt immer näher zusammen, ihre Teile sind heute enger miteinander verbunden als je zuvor. Das hat alten Missverhältnissen eine neue Bedeutung verliehen. Und es hat neue Asymmetrien geschaffen. Steigende Verschuldung, Armut und Seuchen auf der Südhalbkugel – das alles greift inzwischen auch auf die reichen Staaten des Nordens über. Heute können wir es uns nicht mehr leisten, wie in der Vergangenheit den vielfältigen Formen des Elends den Rücken zuzukehren. Es mag uns schwer fallen, diese Probleme aus humanitären Gründen

Thoughtful people in the different regions of the world are pinning their hopes in the nascent debate on a better balance in living conditions, on changed, small-scale economic structures that are compatible with urban life in all parts of the world. "In shaping a New World order after its triumph over communism, capitalism neglected to cultivate the ground to aerate the fields for small and medium size enterprises to grow. This includes caring for the environment and resolving unjust situations. This situation is now the vulnerable underbelly of global capitalism. In the urban centres and in the natural environment, dispossession can be seen in the homogenisation and coarsening of the urban fabric, the life within and the degradation of the environment. In architecture, corporate taste is for uniform steel and glass towers, and big brutish blocks stamp out the fine grain of humanised environment. Small-scale areas and cultural areas in cities are destroyed for new blockish developments. Kitsch replaces and displaces authentic aesthetic values. Kitsch is the narcotic of the middle classes. Cultural politics of the nation-state and civilisations is the anaesthetic of a truly global consciousness. A few individuals through hijack, kidnap and through the possession of weapons of mass destruction can wreak havoc."[24]

The Pressing Need for Dialogue

The statements taken from different institutions and disciplines on current world developments and the role of architecture, urban development and landscape planning show that a global environment policy must be pursued in the interests of peace (Klaus Töpfer). This is supported by the conviction that a new kind of global urban policy must also be a policy for peace.

Counteracting this crucial realisation are some hard political facts that, for the moment at least, threaten to restrict freedom of choice in respect of complex human development concepts. The Indian writer, Arundhati Roy, has raised the very relevant question for future visions as to who can force the regions to choose between two antithetical positions that are at loggerheads with one another, when basically there are so many other existential matters at stake that are capable of being perceived very differently in the different regions of the world.[25] She raised this question explicitly from her own perspective of the Third World, but basically it also valid for the global perspective of the discussion of architecture, urban development and landscape planning in the horizon of the earth.

The concept of the worldwide preservation, development and humane structuring of urban civilisations, incorporating due consideration for the specific cultural characteristics in the regions of the world, as the objective of architecture, urban planning and landscape development in the horizon of the earth has been called radically into question and thus exposed as a concept for

ernst zu nehmen. Doch selbst wo es an Altruismus mangelt, müssten handfeste Eigeninteressen als Motivation ausreichen. (...) Sollte dieses Ereignis nicht endlich den fälligen Erkenntnisprozess in Gang bringen?"[23]

Die nachdenklichen Stimmen in den unterschiedlichen Regionen der Welt setzen in der aufkommenden Debatte auf einen besseren Ausgleich der Lebensbedingungen, auf veränderte, stadtverträgliche, kleinteilige Wirtschaftsformen in allen Teilen der Welt.

"In shaping a New World order after it's triumph over Communism, Capitalism neglected to cultivate the ground to create the fields for small and medium size enterprises to grow. This includes caring for the environment and resolving unjust situations. This situation is now the vulnerable underbelly of global capitalism. In the urban centres and in the natural environment, the depossession can be seen in the homogenisation and coarsening of the urban fabric, the life within and the degradation of the environment. In architecture, corporate taste is for uniform steel and glass towers and big brutish blocks stamp out the fine grain of humanised environment. Small-scale areas and cultural areas in cities are destroyed for new blockish developments. Kitsch replaces and displaces authentic aesthetic values. Kitsch is the narcotic of the middle classes. Cultural politics of nation-states and civilisations is the anaesthetic of a truly global consciousness. A few individuals through hijack, kidnap and through the possession of weapons of mass destruction can wreak havoc."[24]

Dialoge als Gebote der Stunde

Die aus verschiedenen Institutionen und Disziplinen entlehnten Statements zur Entwicklung der aktuellen Weltlage und der Einbettung der Architektur-, Stadt- und Landschaftsentwicklung zeigen, dass Welt-Umweltpolitik Friedenspolitik sein müsse (Klaus Töpfer). Hinzu kommt die Überzeugung, dass eine globale Stadtpolitik neuen Zuschnitts gleichfalls Friedenspolitik sein müsse.

Neben diesen Kern ist mit unmissverständlicher Härte eine politische Dimension getreten, die die Wahlfreiheit im Rahmen komplexer humaner Entwicklungsvorstellungen und -alternativen vorerst einzuschränken droht. Die indische Schriftstellerin Arundhati Roy hat die für zukünftige Perspektiven wesentliche Frage gestellt, wer die Regionen der Welt zwingen könne, zwischen zwei antithetischen, miteinander im Krieg befindlichen Positionen zu entscheiden, wo es doch im Kern um so viele andere existentielle Fragestellungen gehe, die in den unterschiedlichen Regionen der Welt sehr unterschiedlich wahrgenommen werden könnten.[25] Sie stellte diese Frage ausdrücklich unter dem Vorbehalt einer eigenen Perspektive der Dritten Welt, grundsätzlich gilt sie aber auch für eine globale Perspektive der Erörterungen der Architektur-, Stadt- und Landschaftsentwicklung im Horizont der Erde. Die Vorstellung

the safeguarding of culture. If it is accepted that the conditions for urban development all over the world are bound up with one another, then culture, urban culture and urban building culture could assume a new (old) role that first emerged in the time of the Greek polis – a role in the reconstitution of political publicness in global development.

At the same time, it has been shown and confirmed in truly apocalyptic manner that the concerns of global architecture, urban development and landscape planning are an expression of conflicting social and cultural concepts, the subject of contradictory forces for modernisation and inertia and an expression of the complex developments and global inequalities that architecture cannot turn its back on. If there ever were "simple answers for architecture" they no longer exist now.

Everywhere there is talk of the need for a shift in paradigms, now that an outdated view of the world, which for a long time was rationalised, immunised, justified and protected against competition from other paradigms, has finally collapsed.

einer weltweiten Erhaltung, Entwicklung und humanen Gestaltung der städtischen Zivilisationen in den kulturell spezifischen Eigenarten der Regionen der Welt als Ziel der Architektur-, Stadt- und Landschaftsentwicklung im Horizont der Erde, ist dadurch, dass sie radikal in Frage gestellt worden ist, als Vorstellung einer gesicherten Kultur bloßgestellt worden. Wenn zugelassen wird, dass die Entwicklungsbedingungen der Städte im Weltmaßstab reflexiv miteinander verbunden sind, könnte der Kultur, der Stadtkultur, der Stadt-Bau-Kultur eine neue (alte) Rolle zuwachsen, die bereits in der griechischen Polis begründet war, eine Rolle bei der Neukonstituierung der politischen Öffentlichkeit in der globalen Entwicklung.

Zugleich ist aber aufgezeigt und auf apokalyptische Weise bekräftigt worden, dass die Angelegenheiten der globalen Architektur-, Stadt- und Landschaftsentwicklung Ausdruck konfligierender Gesellschafts- und Kulturkonzepte, Gegenstand widersprüchlicher Modernisierungs- und Beharrungskräfte und Ausdruck komplexer Entwicklungen und globaler Ungleichheiten ist, denen sich die Architektur nicht entziehen kann. Wenn es je „einfache Antworten der Architektur" gegeben haben sollte, jetzt gibt es sie nicht mehr.

Allenthalben ist vom Erfordernis eines Paradigmenwechsels die Rede, da eine überkomme Weltsicht, die lange rationalisiert, immunisiert, gerechtfertigt und gegen die Konkurrenz anderer Paradigmen geschützt worden ist, schließlich in sich zusammenbrach. In dieser Konstellation eröffnet sich die Krise als Chance.

1 Asian Development Bank Institute, Asian Environment Outlook 2001, Manila 2001; Am Rand einer Katastrophe, Bericht mit Wachstumsprognose von acht ausgewählten asiatischen Megacitys, in: Der Spiegel magazine 27/2001, p. 113

2 Durs Grünbein, Mit Darwins Augen, in: FAZ Frankfurter Allgemeine Zeitung newspaper, Bilder und Zeiten, 23 June 2001

3 Ullrich Fichtner, Mr. Tapfer, Der Retter der Welt, in: Der Spiegel magazine 24/2001, p. 76

4 Hans Blumenberg, Das Lachen der Thrakerin. Eine Urgeschichte der Theorie, 1987

5 According to Thomas Herzog

6 Idem

7 Thomas S. Kuhn, The Structure of Scientific Revolutions, General Philosophy of Science

8 According to Klaus Töpfer

9 Klaus Töpfer – Info: Global Warming Report Details – www.unep.org/Documents

10 Klaus Töpfer – Info: Statement on Climate Change by Klaus Töpfer, 15 March 2001 www.unep.org/Documents

11 The 1997 Kyoto Protocol – www.wri.org/wri/wr2000; Info: World Resources Institute/New report on global health – www.wri.org/wri/press

12 Klaus Töpfer – Info: Statement on Climate Change by Klaus Töpfer, 15 March 2001 www.unep.org/Documents

13 Ernst Ulrich von Weizsäcker, Amory B. Lovins, L. Hunter Lovins, Faktor vier. Doppelter Wohlstand – halbierter Naturverbrauch [Doubling Wealth Halving Resource Use] – Der neue Bericht an den Club of Rome, München 1995, p. 242

14 Donata Valentien, propositions for discussion by the Scientific Committee of 19 March 2001

15 Dialogue Among Civilizations – www.unep.org/Documents – United Nations Year of Dialogue Among Civilizations

16 Source: Global Urban Observatory

17 Source: Mutations, exhibition catalogue, arc-en-rêve, Bordeaux 2000

18 Lester C. Thurow, The Future of Capitalism, London1997, Chapter 5, Plate Three: Demography – Growing, Moving, Getting Older

19 Hartmut Böhme, Wer sagt, was Leben ist? Die Zeit newspaper No. 49 of 30 November 2000; Info: SDIS Sustainable Development Information Service, Resources at Risk – www.wri.org/wri/trends;
Info: Demographic Dynamics and Sustainability –www.unep.org/Documents

20 Mark Siemons, Jenseits des Bewußtseins. In Kampfzeiten: Steht der Kultur eine neue Politisierung bevor? FAZ Frankfurter Allgemeine Zeitung newspaper of 16 October 2001

21 Jeffrey Sachs, Amerika muß die Augen öffnen, Die Zeit weekly of 27 September 2001

22 New America Foundation

23 Saskia Sassen

24 Tay Kheng Soon, Dimensions in New Urban Form and Architecture Post 11 September 2001, working paper of 27 October 2001

25 Arundhati Roy, television interview, Magazin Monitor (ARD) of 18 October 2001

10 Questions by the Scientific Committee to the XXI World Congress of Architecture UIA Berlin 2002

10 Fragen des Wissenschaftlichen Komitees an den XXI. Architektur-Weltkongress UIA Berlin 2002

Question 1
On what moral values does the responsibility of those substantially involved in planning and building rest?

Question 2
Which keystones for a new global peace order can architects offer with their architectures?

Question 3
How can the ecological costs for the built environment be internalised in economic efficiency more than has previously been done while observing their global effects?

Question 4
How can innovations in architecture build on traditions and the history of construction?

Question 5
How can regional identity in architecture, economy and society be perceived as an added value and how can it be continued in a modern guise?

Question 6
How can beauty in architecture correspond to contemporary content and assume a timeless shape?

Question 7
How can the social value of planning as a holistic way of thinking in relation to independent decisions and individual buildings be increased?

Question 8
How can sustainable, resource-saving construction contribute to a greater degree of social justice?

Question 9
In what ways can planning and building protect the material resources and increase the spiritual resources of beauty and identity?

Question 10
How are architects to take a stand when political changes are required, but when the realities of planning and building take the same old form?

Frage 1
Welche moralischen Werte begründen die Verantwortung der maßgeblich am Planen und Bauen Beteiligten?

Frage 2
Welche Bausteine zu einer neuen globalen Friedensordnung können Architekten mit ihren Architekturen anbieten?

Frage 3
Wie können die ökologischen Kosten der gebauten Umwelt mehr als bisher in Wirtschaftlichkeit internalisiert werden unter Beachtung der globalen Auswirkungen?

Frage 4
Wie können Innovationen in der Architektur auf Traditionen und (Bau-)Geschichte aufbauen?

Frage 5
Wie kann regionale Identität in Architektur, Wirtschaft und Gesellschaft als Mehrwert begriffen und modern fortgeführt werden?

Frage 6
Wie kann Schönheit in der Architektur zeitgemäßen Inhalten entsprechen und zeitlose Gestalt annehmen?

Frage 7
Wie kann der gesellschaftliche Wert von Planung als ganzheitliche Denkweise im Verhältnis zu Einzelentscheidung und Einzelbauwerk gestärkt werden?

Frage 8
Wie kann nachhaltiges, ressourcensparendes Bauen zu mehr sozialer Gerechtigkeit beitragen?

Frage 9
Wie können Planen und Bauen die materiellen Ressourcen schonen und die geistigen Ressourcen der Schönheit und der Identität stärken?

Frage 10
Wie sollen Architekten Position beziehen, wenn politische Veränderungen notwendig sind, die Realitäten des Planens und Bauens aber alten Mustern folgen?

Prologue
Prolog

Of all the arts architecture bears the greatest social responsibility and has the greatest impact on society. Architects cannot build heaven on earth, but they can design homes and cities that are fit for people to live in; cities that are more than just land use plans cast in concrete.

Dr. h. c. Johannes Rau
President of the Federal Republic of Germany
Patron of the XXI UIA World Congress of Architecture Berlin 2002

Die Architektur ist diejenige von allen Künsten mit der größten gesellschaftlichen Verantwortung und der größten gesellschaftlichen Wirkung. Architektur kann nicht den Himmel auf Erden schaffen, aber Architekten können menschliche Städte bauen, menschliche Häuser bauen, Städte, die etwas anderes sind als in Beton gesetzte Flächennutzungspläne.

Dr. h.c. Johannes Rau
Bundespräsident der Bundesrepublik Deutschland
Schirmherr des XXI. Architektur-Weltkongresses Berlin 2002

"Resource Architecture", the main theme of the Berlin Congress, is of fundamental importance for the future development of the world's cultures and for our profession. We see our task as being to establish a more sustainable relationship with the natural and built environment and to work for a more equitable world, in which greater attention is paid to the needs of minorities and the homeless and respect is shown for the diversity of architecture.

Vassilis Sgoutas
President of the Union Internationale des Architectes (UIA)

„Ressource Architektur" – das Thema unseres Berliner Kongresses ist von fundamentaler Bedeutung für die künftige Entwicklung der Weltkulturen und für unseren Berufsstand. Wir haben uns zur Aufgabe gemacht, auf einen nachhaltigeren Umgang mit natürlicher und gebauter Umwelt und auch auf eine gerechtere Welt hinzuarbeiten, die Obdachlosen und Minderheiten mehr Beachtung schenkt und die die Vielfalt der Architektur respektiert.

Vassilis Sgoutas
Präsident der Union Internationale des Architectes (UIA)

The XXI World Congress of Architecture of the Union Internationale des Architectes (UIA), which is being held in Berlin in 2002, will continue the international dialogue on how living standards in cities and regions throughout the world can be maintained and enhanced. Two years ago, the capital of the Federal Republic of Germany hosted the Urban 21 Global Conference on the Urban Future. Architects and architecture bear a great responsibility for the quality of the built environment.

Kurt Bodewig
Federal Minister of Transport, Construction and Housing

Mit dem XXI. Architektur-Weltkongress der Union Internationale des Architectes (UIA) 2002 in Berlin wird der weltweite Dialog über die Sicherung und Verbesserung der Lebensbedingungen in unseren Städten und Regionen an exponierter Stelle fortgesetzt. Die Hauptstadt der Bundesrepublik Deutschland war schon zwei Jahre zuvor Gastgeber der Weltkonferenz Urban 21, auf der über die Zukunft der Städte diskutiert wurde. Architekten und Architektur tragen wesentliche Mitverantwortung für die Qualität unserer gebauten Umwelt.

Kurt Bodewig
Bundesminister für Verkehr, Bau- und Wohnungswesen der Bundesrepublik Deutschland

Renewed efforts are required to achieve a global balance between economic interests and environmental concerns and the needs of different societies and cultures. Regional traditions must be preserved and linked with innovations that are capable of providing a better future. We are confident that the XXI World Congress of Architecture will be an important forum for debate and the presentation of new ideas in this respect.

Andreas Gottlieb Hempel
President of the XXI World Congress of Architecture Berlin 2002

Neue Anstrengungen sind nötig, um weltweit Ökonomie und Ökologie, Gesellschaft und Kultur im Gleichgewicht zu halten und regionale Traditionen mit zukunftsfähigen Innovationen fortzuführen. Wir sind davon überzeugt, dass der XXI. Architektur-Weltkongress ein bedeutendes Forum zur Diskussion und Präsentation neuer Ideen dafür sein kann.

Andreas Gottlieb Hempel
Präsident des XXI. Architektur-Weltkongresses Berlin 2002

1
Responsibility for Global Problems
Verantwortung für die Weltprobleme

The XXI World Congress of Architecture UIA Berlin 2002, which is being held early in the 21st century, intends to remind architects, engineers and planners in all countries and on all continents that they have a joint responsibility to help resolve the global problems mankind will face in the new century.

The 21st century marks the beginning of the end of the industrial age as we have known it so far, as the success and progress it has achieved are being undermined by severe conflicts[1]:

· the domination of capital over labour,
· the pursuit of growth and prosperity to the detriment of the environment,
· the purely rationalistic approach to science and technical progress,
· and the inequitable distribution of wealth

have proved to be inhumane and prone to crisis. They are about to founder on their own inherent contradictions.

The time has come to do away with these contradictions and bid farewell to one-sided Modernist paradigms. A dialogue between cultures, incorporating due respect for their long-standing traditions, must lead to the elaboration of new principles that treat Man not as the ruler of the cosmos, but as part of it.

The approach to planning and the architecture of Modernism reflect the structural forms of this era that will soon be replaced by another.

Architecture as a resource is a concept that forges a relationship with history and nurtures the hope that mankind can live in harmony with nature.

The participants in the XXI World Congress of Architecture are called upon to make a stand, to demonstrate a political commitment and to produce innovative solutions. Essentially, this is a slight rephrasing of the much-used slogan "Think globally, act locally".

Der XXI. Architektur-Weltkongress UIA Berlin 2002 am Beginn des 21. Jahrhunderts will die Architekten, Ingenieure und anderen Planer in allen Ländern und Erdteilen zur Mitverantwortung für die Lösung der Weltprobleme in diesem neuen Jahrhundert aufrufen.

Das 21. Jahrhundert leitet die Ablösung vom Industriezeitalter bisheriger Prägung ein, dessen Erfolge und Fortschritte von großen Konflikten begleitet sind[1]:

· die Dominanz des Kapitals gegenüber der Arbeit,
· Wachstum und Wohlstand zu Lasten der Umwelt,
· das nur rationalistische Verständnis von Wissenschaft und technischem Fortschritt
· und die ungerechte Verteilung des Wohlstandes

erweisen sich als inhuman und krisenschaffend, sind dabei, an ihren eigenen inneren Widersprüchen zu zerbrechen.

Die Zeit ist längst gekommen, diese Prinzipien abzulösen und von den einseitigen Interpretationen der Paradigmen der Moderne Abschied zunehmen. Im Dialog der Kulturen und im Respekt vor ihren tausendjährigen Traditionen sind neue Prinzipien zu erarbeiten, die den Menschen aus seiner beherrschenden Stellung im Verhältnis zum Kosmos zurücknimmt und ihn als einen Teil des Kosmos begreift.

Das Planungsverständnis und die Architektur der Moderne spiegeln die baulichen Ausdrucksweisen dieser Zeit wider, die jetzt an der Schwelle zu ihrer Ablösung steht.

Die Architektur als Ressource ist eine Begrifflichkeit, die einen Bezug zur Geschichte und eine Hoffnung für ein Leben der Menschen im Einklang mit der Natur herstellt.

Die Teilnehmerinnen und Teilnehmer des XXI. Architektur-Weltkongresses werden aufgerufen, Stellung zu beziehen, sich politisch zu engagieren und sich praktisch mit innovativen Lösungen zu betätigen. Dieses ist die etwas andere Umschreibung des inzwischen viel strapazierten Slogans „Global denken und lokal handeln".

Question 1
On what moral values does the responsibility of those substantially involved in planning and building rest?

Frage 1
Welche moralischen Werte begründen die Verantwortung der maßgeblich am Planen und Bauen Beteiligten?

1 Clarifying remark: The majority of people in the developing countries is turning from a society characterised by manual production to a society with industrial production. Even in highly industrialised countries, industrial productions evolve, modified by electronic control. We are talking here about the turn towards new design principles of industrial production.

1 Nachbemerkung zur Klarstellung: Die Mehrzahl der Menschen in den Entwicklungsländern (Developing Countries) ist auf dem Weg von einer durch handwerkliche Produktion charakterisierten Gesellschaft zur Gesellschaft mit Industrieproduktion. Auch bei den hoch industrialisierten Ländern entwickeln sich die industriellen Produktionen weiter, modifiziert durch elektronische Steuerung. Es geht also um die Hinwendung zu neuen Gestaltungsprinzipien der industriellen Produktion.

2
In Accord with the United Nations
Im Einklang mit den Vereinten Nationen

The XXI World Congress of Architecture UIA Berlin 2002, following in the footsteps of the major conferences held by the United Nations, will propose that the resources of architecture should be used to meet the demands raised by these conferences and by UNESCO, in particular at its General Assembly of November 2001. With this in mind, it will call for a dialogue between
· cultures,
· civilisations,
· disciplines.

and will refer to the
· Environment Conference in Rio de Janeiro (1992),
· World Population Conference in Kairo (1994),
· Conference on Human Settlements in Istanbul (1996),
· demands made by the World Heritage Centre.

The UIA World Congress will provide an opportunity to continue the debate on cities in the 21st century, which began with the Urban 21 Global Conference on the Urban Future that was held in Berlin in the year 2000.

The XXI World Congress of Architecture UIA Berlin 2002 would like to contribute a number of building blocks to the new global peace policy long since called for by the Executive Director of UNEP, Klaus Töpfer, for the benefit of
· humankind,
· the environment
· and cultures.

Der XXI. Architektur-Weltkongress UIA Berlin 2002 reiht sich in die großen Konferenzen der Vereinten Nationen ein, will dazu aufrufen, durch Mittel der Architektur die Forderungen dieser Konferenzen und der UNESCO, insbesondere der UNESCO-Generalversammlung vom November 2001, mit dem Aufruf zum Dialog der
· Kulturen,
· Zivilisationen,
· Disziplinen auszufüllen.

Der UIA-Weltkongress nimmt Bezug
· zur Umweltkonferenz von Rio de Janeiro (1992),
· zur Weltbevölkerungskonferenz von Kairo (1994),
· zur Habitat-Konferenz von Istanbul (1996),
· zu den Forderungen des World Heritage Centre

und sieht sich auch als Fortführung der Diskussion über die Stadt im 21. Jahrhundert, der Konferenz URBAN 21 vor zwei Jahren in Berlin.

Der XXI. Architektur-Weltkongress möchte Bausteine liefern zu einer – schon vom Direktor des United Nations Environmental Programme (UNEP), Klaus Töpfer, – seit langem geforderten neuen globalen Friedenspolitik gegenüber
· der Menschheit,
· der Umwelt
· und den Kulturen.

Question 2
Which keystones for a new global peace order can architects offer with their architectures?

Frage 2
Welche Bausteine zu einer neuen globalen Friedensordnung können Architekten mit ihren Architekturen anbieten?

3
Sustainability as the Foundation
Nachhaltigkeit als Fundament

The need to adopt a sustainable approach to the use of natural and social resources has become a fundamental political and moral principle. All over the world, however, economies continue to be propelled by an outdated logic of development, which operates to the detriment of the environment and the economically weak.

The reports issued by the United Nations and the measurements recorded by well-known research institutes make it quite clear that

- forests are continuing to die as a result of "acid rain" and the clearing of tropical forests;
- global warming is proceeding unabated as a result of the unchecked consumption of fossil energy supplies;
- the denaturing of water cycles in the dry and humid regions of the world, which has been brought about by traditional systems of water supply and drainage, the use of large reservoirs and "state-of-the-art" re-routing of rivers, has not changed;
- 30 million people are fleeing war and hunger. Third World debt is increasing relentlessly.

The highly industrialised states, in particular, are reluctant to set a good example by putting their economic activities on a sustainable footing and ensuring a socially more equitable balance. Clear evidence of this was provided by the tough negotiations on implementation of the Kyoto Protocol to reduce greenhouse gases and the attitude adopted by the United States.

Every man-made structure is an intrusion on the natural environment. Keeping such intrusions to a minimum would be one way in which architecture could contribute to resolving the global environmental problems that will result from development in the future. Every time a building is demolished stored energies are destroyed, material flows are stepped up and waste is inevitably generated. Hence there are two sides to "Resource Architecture". Given the urgent need for sustainability and the maintenance of national wealth, the preservation of existing buildings and urban facilities ranks on a par with the construction of new buildings that are in harmony with the natural environment.

But what is the situation in real life? How many of the new buildings erected in developed and highly developed countries satisfy the criteria of sustainability? The existing fabric is being ruthlessly demolished and devalued in order to make room for the construction of new buildings. Building projects involving the sparing use of energy and materials are still a rarity and many high-rise buildings held up to be environmentally compatible may turn out to have a poor eco-balance when examined more closely.

Even among experts little distinction is made between integration and application. Rather than applying an integrated, resource-

Die Verpflichtung zum nachhaltigen Umgang mit den natürlichen und gesellschaftlichen Ressourcen ist inzwischen ein hochrangiges politisch-moralisches Postulat. Die real-ökonomische Situation in den Regionen der Erde vollzieht sich aber immer noch nach der alten Entwicklungslogik zu Lasten der Umwelt und der ökonomisch Schwachen.

Die Berichte der Vereinten Nationen und die Messungen renommierter Forschungsinstitute zeugen davon:

- Das Waldsterben als Folge des „sauren Regens" und der Rodung des Tropenwaldes geht weiter.
- Die Erderwärmung als Folge des noch immer wachsenden Verbrauchs fossiler Energievorräte schreitet fort.
- Die Denaturierung der Wasserkreisläufe in den trockenen und in den feuchten Regionen der Welt durch herkömmliche Formen der Wasserversorgung und Entwässerung, durch Großspeicher und Flussumleitungen „nach dem Stand der Technik" hat sich nicht geändert.
- 30 Millionen Menschen sind auf der Flucht vor Krieg und Hunger.
- Die Verschuldung der Dritten Welt steigt unaufhörlich.

Vor allem die hoch entwickelten Industriestaaten zeigen sich zögerlich, wenn sie bei nachhaltiger Wirtschaftsweise und sozial gerechterem Ausgleich mit gutem Beispiel vorangehen sollen. Die zähen Verhandlungen über die Umsetzung des „Kyôto-Protokolls" zur Minderung der Treibhausgase, nicht zuletzt das Verhalten der Vereinigten Staaten, sind dafür ein Beleg.

Mit jedem Bauwerk ist ein Eingriff in den Naturhaushalt verbunden. Diesen Eingriff so klein wie irgend möglich zu halten, wäre ein Lösungsbeitrag der Architektur zu den globalen ökologischen Problemen der künftigen Entwicklung. Mit jedem Abbruch eines Bauwerkes werden gespeicherte Energien vernichtet, Materialströme erhöht und Abfälle zwangsläufig erzeugt. Die „Ressource Architektur" hat insoweit ein doppeltes Gesicht: Der Erhalt bestehender Gebäude und Stadtanlagen ist unter dem Gebot der Nachhaltigkeit und des Erhalts von Volksvermögen zumindest ebenso wichtig wie das neue Bauen im Einklang mit der Natur.

Aber wie sieht es in der Realität aus? Wie viel der erstellten Neubaumasse in den hoch entwickelten Ländern und in den entwickelten Ländern wird den Maßstäben der Nachhaltigkeit gerecht? Um Neubauten Platz zu machen, wird in rücksichtsloser Weise Bestehendes abgerissen und entwertet. Noch sind die Vorhaben des energieschonenden und materialsparenden Bauens eine Seltenheit und so manches in Großwerbung propagierte ökologische Hochhaus hat bei Licht besehen eine miserable Öko-Bilanz.

Selbst in der Fachwelt wird kaum unterschieden zwischen Integration und Applikation. In der Mehrzahl der Vorhaben wird

saving design from the outset, the majority of projects go to great technical lengths to integrate energy-saving features in conventionally constructed buildings. This makes them complicated, increases the outlay for construction and maintenance and creates no more than the appearance of sustainability. The term "environmentally compatible" is compromised in the process and turned into an "increased cost factor". Seen in global terms, that is very short-sighted.

There is no justification for claiming that the conditions for sustainable methods of building are poor, energy too cheap, rules and regulations too lax, clients not sensitive and an awareness of the issues non-existent. The good examples provided by dedicated architects and intelligent clients all over the world demonstrate that there is a great deal of freedom and room for manoeuvre.

If we really want improved conditions for sustainable building we need to establish first whether we can envisage such conditions and whether we are prepared to wholeheartedly embrace them in policy terms. These principles apply first and foremost to highly developed countries and regions with a stagnating population and declining growth figures. In countries and regions with continuing strong economic growth, priorities need to be set that encourage the sparing use of resources.

1. Every building is given energy coefficients as the norm for its life cycle. Ideally this will permit the consumption of environmental energy only and allow a diminishing ratio between fossil energy and environmental energy for a transitional period. These measured values contain the overall energy balance for preparatory demolition, preparation of the subsoil, construction of the building and subsequent operation.

2. As regards drainage systems, the principle should apply that the run-off from a property/area of land should not be any worse after preparation/building than it was before in terms of the ratio between rainwater infiltration, evaporation and storage.

3. New buildings should preferably be erected on areas that have already been used for construction purposes. This is a general recycling precept, while a return to free space may also be sensible if a better ecological balance can be obtained by building on a different site.

4. Every building must be disposed of after the expiry of its life cycle at the expense of its owners.

mit gewaltsamem technischen Aufwand Energiesparen in ein konventionell erstelltes Gebäude addiert, anstatt eine ressourcensparende ganzheitlich begriffene Entwurfslogik anzuwenden. Das macht ein Gebäude kompliziert, erhöht den Bau- und Erhaltungsaufwand und hängt sich nur den Anschein der Nachhaltigkeit um. Auf diese Weise wurde der Begriff „ökologisch" desavouiert und zu einem „erhöhten Kostenfaktor" – wie kurzsichtig bei globaler Betrachtung!

Man sage nicht, die Rahmenbedingungen für nachhaltige Bauweise seien zu schlecht, Energie zu billig, Vorschriften zu lasch, Bauherren nicht sensibel, Bewusstsein nicht vorhanden. Die guten Beispiele aus der Hand engagierter Architekten und kluger Bauherren aus allen Regionen der Welt zeigen, dass viel Freiheit und Entfaltungsspielräume bestehen.

Aber sollten wir denn wirklich bessere Rahmenbedingungen für das nachhaltige Bauen wünschen, dann haben wir zu prüfen, ob wir uns solche vorstellen können und ob wir mit unserer Persönlichkeit politisch dafür eintreten.

Diese Grundsätze gelten zuerst für die hoch entwickelten Länder und Regionen mit stagnierender Bevölkerung und abnehmenden Wachstumskennziffern. In Ländern und Regionen mit noch starkem Wachstum sind darin Prioritäten zu sehen, die zu einem sparsamen Umgang mit den Ressourcen anhalten.

1. Jedes Bauwerk erhält Energiekennzahlen als Norm vorgegeben für seine Lebensdauer, die im Idealfall nur den Verbrauch von Umweltenergie zulassen, für eine Übergangszeit ein degressives Verhältnis von fossiler Energie und Umweltenergie eröffnen. In diesen Maßzahlen ist die energetische Gesamtbilanz für vorbereitende Abrisse und Baugrundaufbereitungen, Bauwerkserstellung und späteren Betrieb enthalten.

2. Für die Entwässerungssysteme hat der Grundsatz zu gelten, dass die Abflussverhältnisse auf einem Grundstück/einem Areal nach einer Bebauung/Aufbereitung zumindest nicht schlechter sein dürfen als die vorangegangene Situation, was das Verhältnis von Regenwasserversickerung, -verdunstung und -speicherung anbelangt.

3. Vorrangig sind nur Flächen für eine neue Bebauung zugelassen, die schon einmal bebaut waren. Das ist ein generelles Wiederaufbereitungs-Gebot, wobei auch eine Rückführung in Freifläche sinnvoll sein kann, um stattdessen mit besserer ökologischer Bilanz an anderer Stelle zu bauen.

4. Jeder Bau muss auf Kosten seiner Eigentümer nach Ablauf seiner Lebensdauer entsorgt werden.

In this way the prevailing system of consumption would be replaced by a system of cycles. The polluter-pays principle will have a key role to play in this respect. How are we to respond to such demands? Shall we list all the practical constraints to show that this approach is mere "wishful thinking"; or shall we rise to the innovative challenge by demonstrating its feasibility?

The internalisation of environmental costs has long been a standard theorem in environmental economics and a recognised economic principle. In the course of this century product liability and the obligation to take back products will become a normal feature of production and corporate responsibility. A start has already been made with automobiles and other product areas will follow. Why has this not long been the case in the construction industry, where large amounts of energy are stored and where around 80 per cent of the waste are "produced"?

The question we have to ask ourselves, therefore, is whether and when there will be a take-back obligation for buildings and their parts and how we can make the necessary preparations. Here again there are two regulatory mechanisms we can use to assess our attitude and approach.

1. In future every building permit will incorporate the need for a demolition description documenting how the new building can be safely reintegrated into the energy and materials cycle.

2. The granting of a demolition permit will entail the setting up of a reserve fund (guaranteed by a bank and secured against bankruptcy), which can be used to pay for the demolition costs even if no-one is responsible any more for unoccupied buildings, abandoned properties and vanished companies.

The fact that design and the architecture that flows from it could take on very different forms as a result of the demolition commitment should stimulate our imagination:
· no destruction of materials
· flexible handling of construction, completion and installation
· the passing on of rooms and facilities that are no longer required to those in need etc.

Question 3
How can the ecological costs for the built environment be internalised in economic efficiency more than has previously been done while observing their global effects?

Auf diese Weise würde das gegenwärtig herrschende Verbrauchssystem ins System der Kreisläufe überführt werden. Das Verursacherprinzip spielt dabei eine grundsätzliche Rolle.

Wie stellen wir uns nun vor oder hinter diese Forderungen? Tragen wir alle Sachzwänge zusammen, um den Nachweis zu führen, dass es sich hier um „idealistisches Denken" handelt, oder reizen wir unsere innovativen Fähigkeiten aus, um die Machbarkeit zu belegen?

Längst ist die Internalisierung der Umweltkosten gängiges Theorem der Umweltökonomie und anerkanntes Prinzip der Volkswirtschaft. Produkthaftung und Produktrücknahme werden im begonnenen Jahrhundert zur Regel in der Produktionsweise und der Unternehmensverantwortung. Beim Automobil sind die Anfänge gemacht, andere Produktbereiche werden folgen. Warum nicht längst beim Bau, wo man viel Energie speichert und ca. 80 % des Abfalls „produziert"?

Also müssen wir uns fragen, ob und ab wann es auch eine Rücknahmeverpflichtung für Bauwerke und ihre Teile geben wird und wie wir uns darauf vorbereiten?

Auch dazu zwei Regelungsmechanismen zur Prüfung unserer Einschätzung und Einstellung:

1. Jede Baugenehmigung benötigt in Zukunft auch eine Rückbaubeschreibung, in der der Nachweis geführt wird, wie das neue Bauwerk schadlos in den Energie- und Materialkreislauf zurückzuführen ist.

2. Mit der Rückbaugenehmigung ist eine bankengesicherte und konkursfeste Rücklage zu bilden, aus der die Rückbaukosten finanziert werden; auch dann, wenn für leer stehende Gebäude, verlassene Grundstücke und verschwundene Unternehmen niemand mehr verantwortlich ist.

Dass eine Entwurfslogik und eine daraus entstehende Architektur aus der Rückbauverpflichtung grundlegend andere Formen annehmen könnte, sollte unsere Phantasie beflügeln:
· keine Zerstörung von Material,
· flexible Nutzung von Konstruktion, Ausbau und Einrichtung,
· Abgabe nicht mehr benötigter Räume und Einrichtungen an Bedürftige usw.

Frage 3
Wie können die ökologischen Kosten der gebauten Umwelt mehr als bisher in die Wirtschaftlichkeit internalisiert werden unter Beachtung der globalen Auswirkungen?

4
Respect for History
Respekt vor der Geschichte

"Resource Architecture" is more than just a stored material resource; it is also a repository of history, cultural awareness and identity.

Cultural values are vanishing at a breathtaking pace in the wake of worldwide modernisation and globalisation. They include landscapes, buildings, urban facilities, cultural monuments and traditions, languages and national heritage. Aggravating the situation is the destruction caused by seemingly archaic wars fought with ultra-modern weapons.

This phenomenon dates from the onset of industrialisation. Nature conservation and the preservation of historical monuments both date from the middle of the 19th century.

The challenge the design of buildings, cities and landscape faces in the 21st century is to resolve the contradiction between conservation and modernisation. Our task is to build upon existing cultural values, i.e. to lay a new cultural layer on top of the old and not destroy it in the process. Modernisation can only be accepted if it forms a relationship with the existing fabric.

If in certain cases the new should eliminate the old, strict criteria must be applied requiring proof that what is new embodies an "added value". Hence nature conservation and monument preservation demands should not be perceived as difficulties, but as quality requirements for conversion.

Yet how many architects are there who attempt to create an image for themselves in theory and practice by protesting vehemently against the restrictions that are imposed in protection provisions for conservation purposes?

If we do not forge a link between new buildings and the past and fail to ensure that new buildings form a part of future sustainable development in the world, more and more people will turn away from architecture and devote their energies to renaturalisation, restoration and the reconstruction of buildings from the past. "Progress without a historical context" will find itself increasingly at odds with public opinion and community action groups. Planning and building propelled by an outdated logic of development will become more conflict-prone, time-consuming, cost-intensive, incalculable and, as a result, uneconomic in overall terms.

The reports and memoranda of the UNESCO World Heritage Centre illustrate just how dramatic the destruction of pre-modern cultures is.

By no means all countries have sophisticated systems for the conservation of nature and historic monuments and even if they did the question arises as to how they should be handled. The

Die „Ressource Architektur" ist nicht nur gespeicherte materielle Ressource, sondern auch ein Speicher der Geschichte, des kulturellen Bewusstseins und der Identität.

Begleitend zur weltweiten Modernisierung und zur Globalisierung gehen in atemberaubendem Tempo kulturelle Werte verloren: Landschaften, Bauwerke, Stadtanlagen, Kulturdenkmäler, kulturelle Traditionen, Sprachen und Heimat. Dazu kommt die Zerstörung durch archaisch anmutende Kriege mit modernsten Waffen.

Diese Erfahrung wurde schon zu Beginn der Industrialisierung gemacht. Naturschutz und Denkmalschutz gehen auf die Mitte des 19. Jahrhunderts zurück.

Die Bau-, Stadt- und Landschaftsgestaltung im 21. Jahrhundert ist aufgerufen, den Gegensatz von Konservieren und Modernisieren aufzuheben. Der Auftrag lautet, auf den kulturellen Werten aufzubauen, die jüngste Kulturschicht also über die alten zu legen und diese nicht zu zerstören, Modernisierung nur im Kontext zum Bestand zuzulassen.

Sollte im Einzelfall das Neue Bestehendes beseitigen, dann ist eine hohe Hürde für den Nachweis anzulegen, dass das Neue einen „Mehrwert" aufweist. Daher sind die Hürden des Naturschutzes und des Denkmalschutzes für das Neubauen nicht als Erschwernis, sondern als Qualitätsanforderung der Umnutzung zu begreifen.

Aber fragen wir uns, wie viele Architekten sich in Theorie und Praxis mit lauter Stimme gegen die Beengungen von konservierenden Schutzbestimmungen profilieren?

Wenn wir das neue Bauen nicht rückbinden in die Vergangenheit und vorausbinden in eine nachhaltige Weltentwicklung, dann werden immer größere Teile der Öffentlichkeit sich von der Architektur abwenden und ihre Emotionen in die Renaturierung, Restaurierung, Rekonstruktion und in den Wiederaufbau vergangener Bauwerke investieren. Der „geschichtslose Fortschritt" wird in immer stärkeren Gegensatz zur öffentlichen Meinung und zu Bürgerbewegungen geraten. Planen und Bauen in der alten Entwicklungslogik wird konfliktreicher, zeitraubender, kostentreibender, unkalkulierbarer und damit – ganzheitlich gesehen – unwirtschaftlich.

Die Berichte und die Denkschriften der UNESCO World Heritage zeugen von der Dramatik der Zerstörung vormoderner Kulturen.

Längst nicht alle Länder verfügen über ein ausgeformtes System für Denkmalpflege und Naturschutz, und selbst wenn solche Systeme etabliert werden, müssten wir uns fragen, wie wir mit ihnen umgehen. Am Beispiel der Bundesrepublik Deutschland lässt sich zeigen, dass ein ausgeformtes Schutzsystem dann kontra-

example of Germany shows that an advanced system of protection and conservation can be counterproductive if it is perceived and criticised as being an obstacle to future development, the modernisation of the economy, the creation of new jobs and the activities of global investors. We, therefore, need to ask what our attitude is and whether we can devise innovative solutions that will make it possible to map out a future based on "architectural as a historical resource".

produktiv wird, wenn es als Hemmnis für die künftige Entwicklung, für die wirtschaftliche Modernisierung, für die Schaffung neuer Arbeitsplätze, für die Tätigkeit globaler Investoren diffamiert wird. Also müssen wir uns fragen, wo steht unser Bewusstsein und wo sind unsere innovativen Lösungen, um aus der „Geschichtsressource Architektur" zukunftsfähige Lösungen zu entwickeln?

Question 4
How can innovations in architecture build on traditions and the history of construction?

Frage 4
Wie können Innovationen in der Architektur auf Traditionen und (Bau-)Geschichte aufbauen?

Exposition Demolition

On the handling of the most recent period in architectural history and the method of building employed in industrialised societies

- on the elimination of entire industrial landscapes as part of the process of de-industrialisation
- on the demolition of housing developments erected as part of modern urban planning (incorrectly limited to buildings made from prefabricated slabs)
- on the return of nature to towns and cities

To what extent is there an historical obligation in this demolition; how many resources are wasted in the levelling that is usually carried out? What can we learn for future urban planning from the experience of demolition, bearing in mind that megacities throughout the world, and especially in the Far East and China, develop in accordance with the same logic?

Exkurs Rückbau

Über den Umgang mit der jüngsten Epoche der Baugeschichte und der Bauweise der Industriegesellschaft

- zur Entsorgung ganzer Industrielandschaften im Rahmen der De-Industrialisierung
- zum Rückbau der Siedlungskomplexe des modernen Städtebaus (fälschlicherweise reduziert auf den Begriff „Plattenbauten")
- und zur Rückkehr der Natur in die Städte

Wie viel geschichtliche Verpflichtung besteht bei diesem Rückbau, wie viele Ressourcen werden bei der vorherrschenden Plattmacherei verschwendet? Was lernen wir für den künftigen Städtebau aus den Rückbauerfahrungen, wenn weltweit nach derselben Logik sich die Megacitys entwickeln, vor allem in Asien, Lateinamerika und Afrika?

5
Regionalism versus Global Uniformity
Regionalität gegen globale Uniform

Sustainability and historicity are of equal value as the foundations on which regional architecture develops. "Regional building" is generally seen as the antithesis to the increasingly uniform nature of cities around the world. Indeed, international investors currently seem to have precious little interest in the regional building sector. The result is that products of planning and building can end up standing next to each other even though they are completely different in terms of their location, building volume, investors, financing, developers, architecture and appearance and thus in their atmosphere and identity.

Ideally, you (still) have the client, the town surveyor and the building contractor involved in construction at the local level, where houses are built by small or medium-size construction companies. At the international level, however, matters are in the hands of developers/investors with an anonymous design department in the background and occasionally an international star architect in the foreground who, together with corporate construction groups, are interested in producing properties that will be attractive in the financial market place. The difference is that, while buildings at the local level are generally erected for owner occupation or a specific occupant, at the international level they are produced for capital investment. This pair of opposites is inappropriate to future developments, however. Clinging to them will simply mean that the international level will override the local level.

Regionalism is a general aspiration that opposes the uniformity produced by globalisation by preserving the criteria of sustainability and by respecting the demand for historicity. Together with sustainability, the new interpretation of structural traditions generates regional diversity and economic benefits.

A major prerequisite for the accomplishment of this goal is that architecture should not be dependent on investors, but should be able to hold its own in the push and pull between short-term return on investment and long-term profitability. This again depends on a global, integrated consideration of – and responsibility for – the costs involved.

The best prospect of such architecture emerging is through independent architects from independent organisations, who are sufficiently well versed in the principles governing global investment that they are capable of understanding, mastering and handling the relevant processes, rules and regulations and can realise buildings which function reliably, use durable materials and establish an identity.

Nachhaltigkeit und Geschichtlichkeit sind die beiden gleichwertigen Fundamente, auf denen sich regionale Baukultur entwickelt. „Regionales Bauen" wird landläufig als Gegenpol zur zunehmenden globalen Uniformierung der Städte dargestellt. In der Tat dreht sich die Scheibe des internationalen Investments an der des regionalen Bausegments zur Zeit vorbei. So stehen unterschiedliche Produkte von Planen und Bauen unverbunden nebeneinander, die sich nach Standortwahl, Baumasse, Investor, Finanzierung, Bauträger, Architektur und Aussehen und dadurch in Atmosphäre und Identität als wesensverschieden darstellen.

Idealistisch dargestellt bewegen sich auf der örtlichen Scheibe (noch) der Bauherr, der Stadtbaurat, der Baumeister um ein Haus, das von mittelständischen Betrieben des Baugewerbes gebaut wird. Auf der internationalen Scheibe dagegen herrscht der Developer/Investor mit einer anonymen Entwurfsabteilung im Hintergrund, gelegentlich dekoriert durch einen internationalen Star-Architekten, über eine finanzmarktfähige Immobilie, verbunden mit global tätigen Baukonzernen. In der Regel sind beide Bereiche dadurch gekennzeichnet, dass der Erste für den Eigenbedarf oder einen definierten Nutzer baut, der Zweite für die Kapitalverwertung.

Aber dieses Gegensatzpaar ist nicht tauglich für künftige Entwicklungen. In diesem Gegensatz zu verharren, würde nur bedeuten, dass die große Scheibe die kleine überdeckt.

Regionalität ist ein allgemeiner Anspruch, der sich gegen die durch Globalisierung verursachte Uniformität stellt, indem die Kriterien der Nachhaltigkeit Geltung erhalten und das Postulat der Geschichtlichkeit geachtet wird. Die neue Interpretation baulicher Traditionen erzeugt zusammen mit der Nachhaltigkeit regionale Vielfalt und ökonomische Vorteile.

Eine wichtige Voraussetzung, um dieses Ziel zu erreichen, ist eine investorenunabhängige Architektur, die sich im Spannungsverhältnis von kurzfristigem Renditekalkül und nachhaltiger Ertragslage zu behaupten weiß. Grundlage dafür ist eine globale und ganzheitliche Kostenbetrachtung und -verantwortung.

Das geschieht am ehesten durch unabhängige Architekten in unabhängigen Organisationsformen, die aber mit den internen Prinzipien des globalen Investments so vertraut sind, dass sie eben damit verbundene Prozesse und Regelhaftigkeiten verstehen, beherrschen und zu gestalten wissen und damit in der Lage sind, Bauwerke zu realisieren, die sich als funktional verlässlich, als materiell dauerhaft und als identitätsstiftend erweisen.

If regionalism were to reject these challenging demands, "regional building" would run the risk of withering away in traditionalism and historical decor or of being overwhelmed by global anonymity.

The mood is favourable for regionalism because the need for "regional identity" is increasingly being expressed not just by traditionalists, but also by forward-looking politicians and key companies. However, regional identity can only emerge out of a respect for history and from the bold shaping of the future.

Würde sich Regionalität diesen unbequemen Anforderungen verweigern, würde „regionales Bauen" große Gefahr laufen, in Traditionalismus und im historischen Dekor zu verkümmern oder von globaler Anonymität überrollt zu werden.

Die Stimmung ist gut für Regionalität, denn immer mehr wird der Bedarf an „regionale Identität" nicht nur aus Kreisen der Traditionalisten, sondern auch aus der fortschrittsorientierten Politik und von beherrschenden Unternehmen formuliert. Regionale Identität entsteht aber nur aus Beachtung der Geschichte und der mutigen Gestaltung der Zukunft.

Question 5
How can regional identity in architecture, economy and society be perceived as an added value and how can it be continued in a modern guise?

Frage 5
Wie kann regionale Identität in Architektur, Wirtschaft und Gesellschaft als Mehrwert begriffen und modern fortgeführt werden?

6
Beauty as a Principle
Schönheit als Prinzip

"Beauty as a principle" is a formulation that accords a high intrinsic value to the pleasing shape, to the aesthetic quality of architecture.

Beauty of form is intimately related to the aforementioned principles – the harmony of sustainability, historicity and regionalism and a genuine correspondence with the content.

This opens up a host of opportunities for design that do not continue in the vein of those forms which originally resulted from different technological interrelations. The issue is not to develop a kind of new style that is generally acceptable.

To that extent beauty is the product of reason. Over and above this, however, beauty is credited with the capacity of rising above reason, of being more than mere rationality and of enhancing rationality in an unexpected manner in the interests of good form while also respecting functionality and economic efficiency.

This added value, which is reflected in unorthodox internal and external representation and is expressed in a "roundabout return" with regard to economic efficiency, is often rationalised away at present and maligned as "useless architecture" even. In fact, what we are increasingly seeing is a dubious division, in which more and more building volume is produced for anonymous purpose-built structures and a "protected area" is set aside for building culture enabling "architecture" to find expression in cultural and representational buildings.

At the same time, however, there is a growing need for "fashionable attire", for superficial forms and entertainment architecture. The question arises here as to whether such architecture merely has a fancy dress function with a very short life span, which flies in the face of the principles outlined above, or whether it is designed to foster the entertainment aspect.

„Schönheit als Prinzip" ist eine Formulierung, die der guten Form, der ästhetischen Qualität, der Architektur einen hohen Eigenwert zubilligt, eben ein Prinzip.

Die Schönheit der Form steht in inniger Beziehung zu den vorher dargestellten Prinzipien des Zusammenklangs von Nachhaltigkeit, Geschichtlichkeit und Regionalität und der wahrhaftigen Entsprechung der Inhalte.

Dabei eröffnet sich eine Vielfalt an Gestaltungsmöglichkeiten, ohne die Formen fortzuschreiben, die sich ursprünglich aus anderen technischen Zusammenhängen ergaben. Dabei geht es auch nicht darum, eine Art von neuem Stil mit allgemeiner Verbindlichkeit zu entwickeln.

Insoweit ist die Schönheit auch Resultat der Vernunft. Darüber hinaus wird der Schönheit aber zugedacht, sich über die Vernunft zu erheben, mehr als Rationalität zu sein und diese im Interesse der guten Form in ungeahnter Weise unter Berücksichtigung von Funktionalität und Wirtschaftlichkeit zu überhöhen.

Dieser Mehrwert, der sich in ungewöhnlicher Repräsentation nach innen und außen und in „Umwegrendite", bezogen auf die Wirtschaftlichkeit, niederschlägt, wird gegenwärtig häufig wegrationalisiert, ja sogar diffamiert als „unbrauchbare Architektur".

In der Realität breitet sich daher immer mehr eine fragwürdige Zweiteilung aus, in der immer mehr Baumasse als anonymer Zweckbau erstellt wird und der Baukultur ein Reservat für „Baukunst" bei Kultur- und Repräsentationsbauten zugewiesen wird.

Gleichzeitig aber steigt auch der Bedarf an „modischen Kleidern", an inszenierter Form, am Entertainment Architektur, die sich befragen lassen muss, ob sie lediglich modische Verkleidung mit schnellem Verfallsdatum im Dissens mit den hier dargestellten Prinzipien bewegt oder die Kultivierung des Unterhaltsamen im Sinn hat.

Question 6
How can beauty in architecture correspond to contemporary content and assume a timeless form?

Frage 6
Wie kann Schönheit in der Architektur zeitgemäßen Inhalten entsprechen und zeitlose Gestalt annehmen?

Exposition on Engineering Structures

No special mention needs to be made of the fact that the principles dealt with so far also apply to "technical buildings" and the technical infra-structure, i.e. for buildings that are generally constructed today by engineers without the assistance of architects, such as bridges, workshops, manufacturing plants, multi-storey car parks, hydraulic engineering, traffic facilities, radio towers, power stations and waste disposal facilities. These buildings and the design and manufacturing processes associated with them have increasingly divested themselves of any responsibility for the shaping of the structural environment, often preferring instead to follow the rationalist principle of technical efficiency and cost saving. Calls for their integration into urban and landscape planning and for quality of design are dismissed as being "irrelevant" almost as a matter of principle. From the start of industrialisation right on into the early 20th century, by contrast, it was taken for granted that at least the aspiration to efficiency in the use of materials, elegance and harmony ranked among the supreme design principles for technical structures along with their integration into urban and landscape planning. These principles must still apply today, all the more so given the considerable extension of creative freedom resulting from advanced and innovative methods of design and construction. The UIA World Congress 2002 will thus call upon architects and engineers, whose training and practical work have drifted far apart, to elaborate common values and design principles. Those commissioning technical buildings must come to see themselves more as responsible clients and developers. Quality of structure must play a leading role in the integrative design of a technical facility, with due consideration being given from the outset to its incorporation into urban and landscape planning.
This call contrasts sharply with widespread current practice, in which large-scale technical facilities are "dumped in towns and the countryside" without any thought being given in the technical and economic planning process to aspects of design. In Germany, e.g., the new sections built by Deutsche Bahn are engineering works that leave their mark on the landscape, even though they unfortunately do not aspire to a pronounced design principle. However, if extensive thought is given to structure and form from the very beginning of the design process, it is possible that a technically and economically attractive project will ultimately be rejected for the simple reason that it is deemed inappropriate in urban and landscape planning terms. The design process can then be terminated without any major costs being incurred up front.

Exkurs Ingenieurbauten

Es sollte nicht der gesonderten Erwähnung bedürfen, dass die bisher genannten Prinzipien auch für „technische Bauwerke" und die technische Infrastruktur gelten, also für Bauwerke, die in heutiger Zeit in der Regel von Ingenieuren ohne Architekten entstehen: die Brücken, die Werkshallen, die Fabrikanlagen, die Parkhäuser, der Wasserbau, die Verkehrsanlagen, die Sendetürme, die Kraftwerke und die Einrichtungen der Abfallbeseitigung. Diese Bauwerke und die zugehörigen Entwurfs- und Herstellungsprozesse wurden immer mehr aus der Verantwortung für die Gestaltung der baulichen Umwelt entfernt, sind oft nur noch einem rationalistischen Prinzip der technischen Effizienz und der Kostenersparnis unterworfen. Ansprüche an städtebauliche und landschaftliche Integration sowie an Gestaltqualität werden beinahe prinzipiell als „sachfremd" abgewiesen. Zu Beginn der Industrialisierung bis in die Anfänge des 20. Jahrhunderts hinein war es dagegen selbstverständlich, dass Materialeffizienz, Eleganz und Harmonie sowie städtebauliche und landschaftliche Einbindung zumindest als Anspruch zu den obersten Konstruktionsprinzipien für technische Bauwerke gehörten. Diese Prinzipien müssen aber auch heute noch gelten, zumal mit den inzwischen fortentwickelten innovativen Konstruktions- und Bauweisen der Grad der gestalterischen Freiheiten deutlich zugenommen hat. Also ruft der UIA-Weltkongress 2002 die in Ausbildung und Praxis weit voneinander abgewanderten Architekten und Ingenieure zu gemeinsamen Werthaltungen und Entwurfsprinzipien auf. Die Auftraggeber technischer Bauwerke müssen sich wieder mehr als verantwortliche Bauherren verstehen. Die Gestaltqualität muss als führende Größe im integrativen Design einer technischen Anlage stehen, wobei die städtebauliche und landschaftliche Einpassung von Anfang an zu bedenken ist.

Diese Forderung steht im deutlichen Gegensatz zur heute weit verbreiteten Praxis, dass vor allem große technische Anlagen im technischen und wirtschaftlichen Planungsprozess abseits gestalterischer Überlegungen mit Gewalt in Städte und Landschaften hineingepaukt werden. In Deutschland z. B. sind die Neubaustrecken der Deutschen Bahn landschaftsprägende Ingenieurbauwerke, die leider ohne ausgeprägten Gestaltungsanspruch entstehen. Wird dagegen die umfassende Gestaltvorsorge schon zu Beginn in den Entwurfsprozess einbezogen, dann kann sich eine technisch-wirtschaftlich verführerische Lösung als „stadt- und landschaftsuntauglich" erweisen, wobei dann ein Entwurfsprozess ohne nennenswerte Anlaufkosten abgebrochen werden kann.

7
Planning Reigns Supreme
Über allem ist Planung

The quality of a particular building depends on the quality of the building next to it and on the quality of the public spaces and the setting in which it stands. This is a truism that has traditionally contributed to the intrinsic value of rural and urban landscapes in all cultures. This requires plans that express the relationship between buildings in the form of rules, which are binding for building projects. Rules of this kind can only be altered after lengthy periods of time have elapsed so that the principles inherent in them can be properly developed. Many major architects of the past were always outstanding urban and landscape architects, too. In the 18th and 19th centuries, in particular, large parks paved the way for town expansions and "ideal towns" were mapped out for rapidly growing areas.

Nowadays – in the highly developed countries at least – there is a confusion of different kinds of specialised planning, which do not align themselves to a clear overall picture and which are often more of a hindrance than a help. This explains the present-day "aversion" to planning. Planning is not popular and in some cases it is rejected outright. There is a feeling that individual interests and projects should be able to be pursued with little or no concern for the needs of others under the guise of the "freedom to build". The argument put forward by the apologists of a supposed modern society is that it is impossible to apply planning regulations to today's variety of lifestyles and plurality of interests, accompanied as they are by rapid changes in fashion-related requirements. The prospect this holds out is one of cities in chaos and the countryside as a land reserve to be exploited at random. With this in mind, the UIA World Congress at the start of the 21st century should address the issue of how planning can re-position itself between over-regulation and under-regulation.

On the one hand, there is a need for more "protective planning", which – in the interests of sustainability and historicity – will erect barriers against arbitrariness and non-compliance with the rules for purely financial reasons. On the other hand, the deregulation of detailed planning could generate scope for innovation and individual solutions, provided those involved in the respective regions and buildings are aware of their responsibility for principles and standards and can reach an agreement accordingly. In such situations detailed planning, the binding nature of which may make it obstructive, can be replaced by intelligent interaction.

Question 7
How can the social value of planning as a holistic way of thinking in relation to independent decisions and individual buildings be increased?

Die Qualität eines einzelnen Bauwerkes ist von der Qualität des benachbarten abhängig und von der Qualität der öffentlichen Räume und der Landschaften, in dem dieses steht. Dies ist eine Binsenweisheit, die von alters her den Wert der Agrar- und Stadtlandschaften in allen Kulturen hervorgebracht hat.

Dafür sind Pläne notwendig, die das Zueinander der Bauwerke in Regeln fassen, die für die einzelnen Bauvorhaben einen verbindlichen Kanon darstellen. Solche Regeln können sich nur in längeren Zeiträumen ändern und damit sich die beabsichtigten Prinzipien auch ausbilden können. Viele große Architekten der Vergangenheit waren immer auch große Stadtbaumeister und Landschaftsarchitekten. Gerade im 18. und 19. Jahrhundert wurden Stadterweiterungen durch große Parkanlagen vorbereitet und für die rasch wachsenden Bebauungen „Idealstädte" entworfen.

Heute besteht zumindest in den hoch entwickelten Staaten ein Gewirr von spezialisierten Planungen, die sich nicht nach einem Gesamtbild ausrichten und häufig mehr verhindern als gestalten. Der „Überdruss" an Planung ist insoweit verständlich. Planung ist nicht populär, zum Teil sogar desavouiert. Einzelinteressen und Einzelprojekte sollen sich unter der Ideologie der „Baufreiheit" rücksichtsfrei durchsetzen dürfen. Die Vielfalt der Lebensstile und die Pluralität der Interessen, überwunden mit einem raschen Wandel der Moden an Anforderungen, würden sich einer Regelung durch Planung entziehen, so die Apologeten der vermeintlich modernen Gesellschaft. Stadt als Chaos und Landschaft als beliebige Landreserve! In dieser Situation soll sich der UIA-Weltkongress zu Beginn eines neuen Jahrhunderts mit der Frage auseinander setzen, wie Planung zwischen Überregelung und Unterregelung einen neuen Stellenwert erhalten kann.

Wir brauchen auf der einen Seite im Interesse von Nachhaltigkeit und Geschichtlichkeit ein Mehr an „Schutzplanungen", die hohe Hürden gegenüber Beliebigkeit und Abweichung aus banalem Kalkül aufrichten. Als Gegenleistung dafür könnte eine Deregulierung bei Detailplanungen mit Freiheit für Innovationen und individuelle Lösungen angeboten werden, dies allerdings unter der Bedingung, dass in den einzelnen Regionen und am jeweiligen Bauwerk Akteure mit Verantwortung für Prinzipien und mit Maßstäben zu einem Konsens finden. Hier wird dann verbindliche und behindernde Detailplanung durch kluge Moderation ersetzt.

Frage 7
Wie kann der gesellschaftliche Wert von Planung als ganzheitliche Denkweise im Verhältnis zu Einzelentscheidung und Einzelbauwerk gestärkt werden?

8
Justice as a Consequence
Gerechtigkeit als Folge

The World Economic Forum has descended from Davos to Manhattan to demonstrate that the world economy refuses to be blown off course by the terrorist attacks of 11 September 2001. Taking place at the same time was the World Social Forum in Porto Alegre, which was staged as an "alternative event". As an alternative to what?

We are talking here not merely about a functioning world economy that produces income from investments, but about a just world economy. Media coverage was devoted almost exclusively to the Economic Forum in New York with very little time or space being given to events in Porto Alegre.

This is a clear indication of the true balance of power despite the political demands that have been raised for a dialogue between cultures and for a just world economic order. In architecture and town planning this division leads to a socially unjust normality and a multitude of schemes that are designed to mitigate the effects of social segregation and promote the integration of minorities.

The survival of the still-growing megacities in the world depends on whether this division can be healed and a mode of development found that will pave the way for social justice and allow people to live together on an equal footing even though they may have very different lifestyles.

The principles of sustainability, historicity, regionalism and beauty must also lend support to social justice.

Question 8
How can sustainable, resource-saving construction contribute to a greater degree of social justice?

Das Weltwirtschaftsforum ist von Davos nach Manhattan hinabgestiegen, um ein Zeichen zu setzen, dass sich die Weltwirtschaft durch den Terroranschlag vom 11. September nicht beeindrucken lässt. Gleichzeitig fand in Porto Allegre das Weltsozialforum statt, das sich als „Gegenveranstaltung" verstand. Gegen was?

Es geht um mehr als eine funktionierende und Kapitalerträge abwerfende Weltwirtschaft, es geht um eine gerechte Weltwirtschaft.

In den Medien wurde fast nur über das Wirtschaftsforum in New York berichtet, Porto Allegre wurde so gut wie ganz verschwiegen. Das ist mehr als ein Indiz für die wahren Machtverhältnisse trotz der politischen Postulate, die den Dialog der Kulturen einfordern und nach einer gerechteren Weltwirtschaftsordnung verlangen.

In Kategorien der Architektur und der Stadtplanung führt diese Zweiteilung zu einer sozial ungerechten Normalität und einer Vielfalt von Initiativen, die sich um die Abmilderung von sozialer Spaltung und um die Förderung von Integration der Minderheiten bemühen.

Gerade für die weiter wachsenden Megacitys auf der Welt wird es zur Überlebensfrage, ob sich diese Spaltung aufheben lässt und ein Entwicklungsmodus gefunden wird, der soziale Gerechtigkeit und ein gleichberechtigtes Nebeneinander unterschiedlicher Lebensstile zulässt.

Die Prinzipien der Nachhaltigkeit, der Geschichtlichkeit, der Regionalität und der Schönheit müssen auch soziale Gerechtigkeit im Gepäck haben.

Frage 8
Wie kann nachhaltiges, ressourcensparendes Bauen zu mehr sozialer Gerechtigkeit beitragen?

9
Architecture as a Resource
Architektur als Ressource

"Architecture as a resource" forms part of the cycles of nature. A repository of history and a vehicle for regional culture, it is committed to beautiful design. It is thus a habitation for social, peaceful human coexistence and a prerequisite for harmony between mankind and nature.

„Architektur als Ressource" ist Teil der Naturkreisläufe, Speicher der Geschichte, Kulturträgerin der Regionen und der schönen Gestalt verpflichtet. Sie ist auf diese Weise Behausung für das soziale, das friedvolle Zusammenleben der Menschen und eine Bedingung für den Frieden zwischen Menschheit und Natur.

Question 9
In what ways can planning and building protect the material resources and increase the spiritual resources of beauty and identity?

Frage 9
Wie können Planen und Bauen die materiellen Ressourcen schonen und die geistigen Ressourcen der Schönheit und der Identität stärken?

10
Convictions or Excuses?
Reinreden oder rausreden?

At the end of these observations, a sense of reality and modesty compels us to address the power and impotence of architecture.

Modesty obliges us not to accept impotence as an excuse so that we can carry on with business as usual because we are loath to change or have given up in resignation. Modesty requires that we recognise, describe and analyse the social facts of life and the political relations of power and keep a look-out out for possible areas of reform, be they large or small.

"Small-scale reforms" exploit the potential that exists in every work of construction in order to find a better solution under the given conditions and contractual relations and to work hard to implement that solution.

Architects, i.e. architects as a community, who have come together under the roof of the International Union of Architects for its XXI World Congress, can raise their voices and join other world organisations, in particular the United Nations, in fostering development that preserves values. This presumes that we have the courage of our own convictions and a determination not to resort to the customary "excuses".

"If I don't take on the job somebody else will!"
"If it's not built in my town it'll be built down the road!"
"If we agree on higher standards, the producers and markets will simply move into neighbouring economies!"

What is at stake, however, in both our personal attitudes and in scientific theory is a reversal of the scientific principle and a renunciation of the supposed rationality inherent in Modernism. As long as there was no scientifically based, empirical evidence for the harmfulness of a product, project or intervention it could be treated as harmless. That way of thinking must now be replaced by the principle of risk avoidance.

The participants in the 21st World Congress of Architecture will return home and reflect on what it would mean if the purpose of planning and building were to be the avoidance of risks for the material and spiritual world.

Question 10
How are architects to take a stand when political changes are required, but when the realities of planning and building take the same old form?

Mit Realitätssinn und Bescheidenheit wird am Ende dieser Betrachtungen die Rede von der Macht und der Ohnmacht von Architektur sein müssen. Die Bescheidenheit fordert dazu auf, nicht die Ohnmacht als Ausrede zu nehmen, um sich bequem oder entmutigt in den bestehenden Verhältnissen einzurichten. Bescheidenheit heißt, die realen gesellschaftlichen Verhältnisse und die politischen Machtkonstellationen zu kennen, zu beschreiben, zu bewerten und nach kleinen und großen Reformspielräumen Ausschau zu halten. Die „kleinen Reformen" nutzen den Freiraum, der dem Grunde nach bei jeder Bauaufgabe gegeben ist, unter den bestehenden Rahmenbedingungen und Auftragsverhältnissen nach einer besseren Lösung zu suchen und diese mit Engagement zu vertreten. Für die „großen Reformen" können die Architekten, und gemeint sind hier die Architekten als Gemeinschaft, die sich unter dem Dach der Internationalen Architekten-Union (UIA) zum XXI. Weltkongress versammelt haben, ihre Stimme erheben, um im Einklang mit anderen Weltorganisationen, allen voran den Vereinten Nationen, eine werthaltige Entwicklung zu befördern.

Das setzt den Mut zur Einrede voraus und sich zu versagen, die allseits gegenwärtigen „Ausreden" zu bemühen:
„Wenn ich den Auftrag nicht übernehme, dann macht ihn ein anderer!"
„Wenn das in meiner Stadt nicht gebaut wird, dann wird es nebenan entstehen!"
„Wenn wir uns höhere Normen zumuten, dann wandern die Produktionen und Märkte in benachbarte Volkswirtschaften ab!"

Vor allem aber geht es in der persönlichen Grundhaltung wie in der wissenschaftlichen Theorie um eine Umkehrung des wissenschaftlichen Prinzips, um eine Abkehr von der vermeintlichen Rationalität der Moderne: Solange kein empirischer Nachweis mit den Regeln der Wissenschaft über die Schädlichkeit eines Produktes, eines Vorhabens, eines Eingriffes vorliegt, konnte gelten, dies sei unbedenklich. An die Stelle dieser Denkweise muss das Prinzip der Risikovermeidung treten.

Die Teilnehmerinnen und Teilnehmer des XXI. Architektur-Weltkongresses Berlin 2002 werden die Frage mit nach Hause nehmen, was es bedeutet, wenn Planen und Bauen auf das Vermeiden von Risiken für die materielle und geistige Welt ausgerichtet werden.

Frage 10
Wie sollen Architekten Position beziehen, wenn politische Veränderungen notwendig sind, die Realitäten des Planens und Bauens aber alten Mustern folgen?

Members of the Scientific Committee
Mitglieder des Wissenschaftlichen Komitees

Karl Ganser Spokesman Sprecher
Breitenthal

Gunnel Adlercreutz
Helsinki

Achraf Bahri-Meddeb
Tunis

Joaquim Guedes
São Paulo

Thomas Herzog
Munich

Engelbert Lütke Daldrup
Leipzig

Yannis M. Michail
Athens

Matthias Sauerbruch
Berlin/London

Jörg Schlaich
Stuttgart

Tay Kheng Soon
Singapore

Donata Valentien
Wessling

Andreas Gottlieb Hempel Moderation Moderation
Berlin

Reinhart Wustlich Secretary Sekretär
Hennef

Commentary
Kommentare

Tay Kheng Soon Singapore

Views on the 'Questions'

The idea of questions-for-the-congress is excellent.

The congress's economic assumption is silent on Globalisation's growth paradigm remaining unquestioned. This needs to be at the centre of the discussion.

Architects cannot remain ignorant of the implications of the economic growth paradigm presently underpinning Globalisation. To premise our work solely on the materiality of architectural sensations is utterly foolish and ego-centric.

I urge the Congress to examine the Zero Growth, eco-regional economy for 'good life' paradigm of Herman E Daly and others like him. We need to go in that direction to achieve social and environmental justice with an architecture that takes seriously the job of saving on energy and materials and heals the soil, protects water and restores biota. The architect is an important facilitator of a process where communities are at the centre of effort in making a new economy based on of healing social rifts the wounded earth.

I agree that 'Beauty' must be our aim but premised on above. There can be no concensus on 'beauty' without healthy communities who share common values. The paranoia of individuals and cultures presently generates a wild appetite and desperation for difference. Much 'expressive' architecture today is wasteful of energy, materials and careless of nature. It is for ego.

Matthias Sauerbruch Berlin/London

Comments

"I like Mr. Ganser's text for its simple clarity. I do think that we need a text of this quality; however, I wonder in what respect the questions asked (and answers given) are really different from what we have heard since 1970 again and again. The issues of ecology and the protection of resources are of course important (despite the fact that they have been around for some time), but I find it difficult to talk about them in such a general way. Of course there is need for change on a global or at least a continental level, but are Architects powerful enough to change world politics?

To introduce this ambition into the Congress will make anything which has actually been happening in the last two decades look compromised and insignificant. I think we should focus on our (more limited) radius of action. Of course we shouldn't loose sight of the overall picture and of course we should challenge decision makers and their preconceptions, but in the end this is a Congress of architects and an exchange of their insights and experiences. Mr. Ganser's text – by the way – sounds terribly anti-innovation, anti-spatial or formal expression, anti-anything that makes architecture fun. For me, the whole idea of construction is about a departure into the unknown and it is exactly this quality which makes architects love their profession. To want to eliminate all risks in this process is like wanting to take the mountains out of downhill-skiing.

This congress is also an opportunity to articulate and advertise the values which are being created through architecture and I find the generally defensive tone of Gansers text quite unsuitable in this respect. Yes, of course there are many mistakes being made, but there is also a lot of potential which is being realised (and will continue to be realised) all over the world. Why not concentrate on the positive aspects of the task ahead? Let's bring out and support the potential of architecture rather than dwelling on its (many) shortcomings."

Library at Luisenbad, Berlin Architects: Rebecca Chestnutt and Robert Niess

Retrospective

The XVIII UIA World Congress of Architects Chicago 1993

Declaration of Interdependence for a Sustainable Future (Excerpt)

In recognition that:

· A sustainable society restores, preserves, and enhances nature and culture for the benefit of all life, present and future; a diverse and healthy environment is intrinsically valuable and essential to a healthy society; today's society is seriously degrading the environment and is not sustainable;
· We are ecologically interdependent with the whole natural environment; we are socially, culturally, and economically interdependent with all of humanity; sustainability, in the context of this interdependence, requires partnership, equity, and balance among all parties;
· Buildings and the built environment play a major role in the human impact on the natural environment and on the quality of life; sustainable design integrates consideration of resource and energy efficiency, healthy buildings and materials, ecologically and socially sensitive land-use, and an aesthetic sensitivity that inspires, affirms, and ennobles; sustainable design can significantly reduce adverse human impacts on the natural environment while simultaneously improving quality of life and economic well being;

We commit ourselves, as members of the world's architectural and building-design professions, individually and through our professional organisations, to:

· Place environmental and social sustainability at the core of our practices and professional responsibilities
· Develop and continually improve practices, procedures, products, curricula, services, and standards that will enable the implementation of sustainable design
· Educate our fellow professionals, the building industry, clients, students, and the general public about the critical importance and substantial opportunities of sustainable design
· Establish policies, regulations, and practices in government and business that ensure sustainable design becomes normal practice
· Bring all existing and future elements of the built environment – in their design, production, use, and eventual reuse – up to sustainable design standards.

Olufemi Majekodunmi
President International Union of Architects

Susan A. Maxman
President American Institute of Architects

The XX UIA World Congress of Architects Beijing 1999

UIA Beijing Charter (Excerpt)

Coming to Terms with our Centuries
The 20th Century: Unparalleled Construction and Destruction

The 20th century has seen unprecedented magnificence and progress, and also incomparable calamity and confusion. The 20th century has enriched the history of architecture in its unique manner: architects have played an admirable role in the reconstruction that followed the two world wars; technical and artistic innovations on a massive scale have introduced fine examples of design to the populous like never before. But, this is not to deny that much of the built environment is still in a deeply unsatisfactory state. The very survival of humankind is under threat amidst squandering of the world's natural and cultural heritage. In affluent regions, redevelopment was often to become destruction by construction; in the poorer areas, pauperised masses are struggling to build their own cities of tomorrow. Over the past century, the world has turned into a very different place. Yet one thing remains the same: we architects are again at a cross roads as a world profession.

The 21st Century: a Turning Point

The diversity and complexity of the world has created much confusion; yet it is but part of the eternal process of change. The present century has seen remarkable reform and development in politics, economics, technology, and society, and the resurgence of human ideas. In the coming century, the pace of transformation is expected to accelerate, though its direction may be even harder to tell. In the coming century, the coexistence of globalisation and pluralism will bring to a head the conflicts and the contradictions that characterise our age. On the one hand, modern means of communication have brought into close contact diverse cultures and traditions; global integration of production, finance and technology continues to dominate decision making. On the other hand, the gap between the rich and the poor is widening at an alarming rate; regional strife and financial uncertainties cast a sinister shadow upon the human habitat. Whilst we should not take on tasks outside our professional remit, it would be both irresponsible and foolish to ignore the torrent of social and cultural change that is redefining the scope of the architectural profession. A conscious reconsideration of the role of 21st century architecture calls for our enthusiasm, strength and courage.

Wu Liangyong
School of Architecture Tsinghua University, Beijing

URBAN SOCIETIES
DIALOGUE BETWEEN CIVILISATIONS

Introduction and Abstracts

STADT UND GESELLSCHAFT
DIALOG DER ZIVILISATIONEN

Einführung und Abstracts

Daniel Biau Nairobi

Elena Rodina New York

Richard S. Levine Lexington

Crispino C. Ochieng Nairobi

Unmesh Kirtikar Calcutta

Ruzica Bozovic-Stamenovic Singapore

Anne Stenros Helsinki

Silvia Arango Bogotá

Regula Lüscher Gmür Zurich

Bart Goldhoorn Moscow

Bernd Zimmermann New York

Harald Bodenschatz Berlin

Introduction

Einführung

Urban culture and social change raise questions about values in relation to architecture, urban development and landscape planning. A considerable part of society's wealth is tied up in architecture, cities and infrastructure. This does not mean that they can only be judged by applying material criteria. Towns and cities, as the expression of a desired future, reflect architecture's role as an intellectual and cultural resource. Architecture, the city and the urban environment, and landscaped and natural spaces are defined as values, to which all people are entitled. By means of architecture, urban development and landscape planning, urban societies – societies in all regions around the world – produce sustainable living conditions for human beings, thereby creating, and responding to, demands for public space by social communities. The core competence of architects and other planners, who render their services to society in trust, lies in defining, planning and translating their ideas into the "language" of buildings, urban structures and infrastructures.

Context

The sociological theory of society mistrusts the Western "utopias" of "forms of life that are pre-conceived" by planning (Jürgen Habermas), not least because the planners proceed from a "hopeless underestimation of the diversity, complexity and changeability of the modern worlds in which people live". But would a "self-regulating city of projects" concerned solely with the occasional pursuit of power and economic goals be any less at the mercy of chance? Architecture must take up the problems of the majority of the world's population. Sustainable urban development and resource-saving construction can only be brought about through foresighted planning.

Sustainable Urban Development

Societies in regions all around the world are changing into urban societies. Despite all the development problems, there is no real alternative to the ambivalent "success story of the city". Nevertheless, the expression crisis of the city certainly applies to the uncontrolled process of urbanisation on all continents, to the blotting out of landscapes and the disappearance of nature. For all their cultural differences, the regional towns and cities – the local centres – are exposed to the levelling tendencies of global economies. The United Nations declared 2001 a year of dialogue between civilisations. This dialogue between civilisations was intended to show that the present globalisation process not only involves economic, financial and technological factors, but also touches the very foundations of human cultures, the spiritual dimensions of societies and their transformation – as well as the reciprocal relationship between these aspects – all over the world. Architecture, urban development and landscape planning are

Stadtkultur und gesellschaftlicher Wandel provozieren Wertefragen für die Architektur-, Stadt- und Landschaftsentwicklung. Große Anteile des gesellschaftlichen Vermögens sind in Architektur, Stadt und Infrastruktur gebunden. Das bedeutet nicht, diesen Bestand nur nach materiellen Kriterien zu bewerten: Die Stadt als Ausdruck einer gewollten Zukunft spiegelt die kulturelle Rolle der Architektur als geistige Ressource wider. Architektur, Stadt und urbane Umwelt, Landschaft und Naturräume sind als Werte definiert, auf die jedermann ein Recht hat. Urban Societies, die Gesellschaften in den Regionen der Welt, schaffen mit Architektur-, Stadt- und Landschaftsentwicklung nachhaltige Lebensgrundlagen für Menschen, begründen und beantworten mit ihnen öffentliche Raumansprüche sozialer Gemeinschaften. Die planerische Definition und Übersetzung (Transkription) in die „Sprache" der Gebäude, der städtebaulichen Strukturen und Infrastrukturen ist Kernkompetenz der Architekten und anderen Planer, deren Leistungen treuhänderisch für die Gesellschaft erbracht werden.

Kontext

Die sozialwissenschaftliche Gesellschaftstheorie misstraut den westlichen „Utopien" einer durch Planung „vorgedachten Lebensform" (Jürgen Habermas), nicht zuletzt, weil die Planer von einer „hoffnungslosen Unterschätzung der Vielfalt, Komplexität und Veränderlichkeit moderner Lebenswelten" ausgehen. Doch wäre die Behauptung einer ‚selbstregulierenden Stadt der Projekte', die nur Einzelfällen von Macht- oder Wirtschaftsansprüchen folgte, weniger dem Zufall ausgeliefert? Die Architektur hat sich mit den Problemen des größten Teils der Weltbevölkerung auseinander zu setzen. Nachhaltige Stadtentwicklung und ressourcensparendes Bauen sind nur durch vorausschauende Planung zu erreichen.

Nachhaltige Stadtentwicklung

Die Gesellschaften in den Regionen der Welt wandeln sich zu städtischen Gesellschaften. Allen Entwicklungsproblemen zum Trotz ist das ambivalente ‚Erfolgsmodell Stadt' ohne wirkliche Alternative. Dennoch trifft die Bezeichnung Krise der Stadt auf die unkontrollierte Verstädterung aller Kontinente, auf die Überformung der Landschaften und das Verschwinden der Natur zu. Die Städte der Regionen, die lokalen Zentren, so unterschiedlich ihre Kultur ist, sind kulturellen Nivellierungstendenzen – verursacht durch globale Ökonomien, ausgesetzt. Das Jahr 2001 wurde von den Vereinten Nationen zum Jahr des Dialogs der Zivilisationen der Welt ausgerufen. Der Dialog der Zivilisationen soll zeigen, dass der gegenwärtige Prozess der Globalisierung nicht nur ökonomische, finanzielle und technologische Faktoren meint, sondern zugleich weltweit die Grundlagen menschlicher Kulturen, die spirituellen Dimensionen der Gesellschaften und ihres Wandels betrifft – und deren Wechselwirkung. Zu den hervorgehobenen Querschnittsbereichen, in denen

among the areas in which the effects of development overlap. The city of today is a city of internationalisation and universalisation, the development of which goes hand in hand with the levelling of local structures and other phenomena. The city of locally invested capital is being transformed into the city of anonymous capital transfers, and the city of local cultural traditions into the theme-city of an international event economy. At the same time, the scarcity of vital resources for two-thirds of the world's population means that people in urbanised regions are having to put up with inadequate housing, poverty and starvation and are exposed to "enormous differences in development and justice" (Hartmut Böhme). The structural transformation of cities is accompanied by the dislocation of urban cultures, social inequality and segregation. "The city", Rem Koolhaas says, "the city – the way that it is – is all we have." But what is meant by "city"? The "planned" city? The "self-regulating" city? The city committed to regional culture, or the international city?

Responses

The city develops by breaking down outmoded forms and boundaries. Edge cities are created and old infrastructures transformed. The modernisation and concentration of historical city centres go hand in hand with the privatisation of public space, whilst the transformation of residential districts is accompanied by the exclusion of gated communities ("new urbanism"). The downside of this development is evident in the rapidly growing megacities of Asia and Latin America, whose economies clearly illustrate the problems of social inequality, while at the same time giving grounds for hope that there may be a reduction in poverty and income differentials. Particular attention needs to be paid to planning and environmental problems in the urbanised regions of these countries.

Proposition

The future of urban societies entails provident care for living conditions in society, for planning and urban culture. This includes the preservation of local and cultural characteristics in the various regions of the world. Factors jeopardising the future development of cities must be defined with foresight, and counterbalanced by positive reinterpretation. Processes leading to the reduction and impoverishment of urban living conditions must be stopped. Any potential contained in modernisation for endowing life with meaning must be fostered in a way that enables a distinction to be made between modernisation and modernity. Only in this way can architecture make an important contribution to the quality of the urban environment.

sich die Auswirkungen der Entwicklungsfaktoren überlagern, gehören Architektur-, Stadt- und Landschaftsentwicklung. Die Stadt der Gegenwart ist eine Stadt der Internationalisierung und der Universalisierung, deren Entwicklung mit einer Nivellierung lokaler Strukturen und Erscheinungsformen einhergeht. Die Stadt der örtlichen Kapitalbindung wandelt sich zur City des anonymen Kapitaltransfers, die Stadt der lokalen kulturellen Traditionen zur Themenstadt der internationalen Eventökonomie. Zugleich werden Menschen in den verstädterten Regionen aufgrund des Mangels an lebensnotwendigen Ressourcen für zwei Drittel der Weltbevölkerung mit unzureichenden Behausungen, mit Armut, Hunger, mit „Entwicklungs- und Gerechtigkeitsgefällen in großem Maßstab" (Hartmut Böhme) konfrontiert. Dislozierung urbaner Kultur, soziale Ungleichheit und Segregation prägen den Strukturwandel der Städte. „Die Stadt – so, wie sie ist –", stellt Rem Koolhaas dennoch fest, „ist alles, was wir haben." Doch welche „Stadt" ist gemeint? Die „geplante" Stadt? Die „selbstregulierende" Stadt? Die regionaler Kultur verpflichtete oder die internationale Stadt?

Antworten

Die Stadt entwickelt sich, indem sie überkommene Formen und Grenzen aufbricht. Edge cities werden ausgebildet, alte Infrastrukturen umgebaut. Die Modernisierung und Verdichtung historischer Stadtzentren geht einher mit einer Privatisierung des öffentlichen Raumes, während der Umbau der Wohngebiete mit der Abgrenzung von Gated Communities verbunden ist (‚New Urbanism'). Die Kehrseite dieser Entwicklung findet Raum in den rasch wachsenden Megacitys Asiens und Lateinamerikas, deren Ökonomie die Probleme sozialer Ungleichheit verdeutlicht und zugleich die Hoffnung auf eine Verringerung der Armut und der Einkommensgefälle begründet. Den Planungs- und Umweltproblemen der verstädterten Regionen dieser Länder gilt besondere Aufmerksamkeit.

Thesis

Die Zukunft der Urban Societies bedarf der reflexiven Vorsorge für die Lebenswelt der Gesellschaft, der reflexiven Planung, der reflexiven Stadtkultur. Sie ist bezogen auf die Erhaltung der lokalen und kulturellen Eigenarten der Regionen der Welt. Gefährdende Faktoren der Zukunftsentwicklung der Städte sind vorausschauend zu definieren und durch positive Umdeutung aufzufangen. Prozesse der Reduktion und Verarmung der Lebensbedingungen in der Stadt sind aufzuhalten. Ein sinnstiftender Umgang mit den Möglichkeiten der Modernisierung ist so zu fördern, dass eine Unterscheidung zwischen Modernisierung und Modernität möglich wird. Nur so wird Architektur zu einem bedeutenden Faktor der Qualität des Urban Environment.

Reinhart Wustlich Hennef

The Challenges of Sustainable Urbanisation

Daniel Biau Nairobi

1. Today, 40 per cent of the population of developing countries lives in cities. By 2020, that figure will rise to 51 per cent. Many countries in Latin America and the Caribbean already have a 75 per cent urban population. The urban population of developing countries grew from 700 million in 1970 to 1.4 billion in 1990 and 2 billion in 2000. It will reach 3.2 billion by 2020. The greatest challenge will be in Africa and Asia, where a major urban transition is taking place. By 2015, 153 of the world's 358 cities with more than one million inhabitants will be in Asia. Of the 27 megacities with more than 10 million inhabitants, 15 will be in Asia. Megacities with 20 or 30 million inhabitants, urban agglomerations of a size never known before in human history, are in the making.

2. Currently, three quarters of global population growth occurs in the urban areas of developing countries, causing hyper-growth in the cities least capable of adapting to that change. The current average annual population increase in developing country cities is estimated to be 64 million, a rate of 175,000 persons per day. Half of this increase is caused by natural population growth within these cities. This urban transition is accompanied by a significant increase in urban poverty, disproportionately affecting women and children; ethnic and racial conflicts, homelessness; environmental deterioration; and marginalization of the poor, often expressed in social exclusion, intolerable living standards and spatial segregation. All of the above contribute directly or indirectly to increased social unrest and urban violence.

3. Policies designed to reduce rural-to-urban migration have consistently failed over the last 30 years in the developing world. Not only is urbanization inevitable but its pace cannot be curbed – in democratic societies – by voluntarist policies. In fact the carrying capacity of rural areas is reaching its limits in many parts of the world and agricultural improvements are creating a surplus population (particularly in Africa and Asia) which has to move to towns and cities to find better employment opportunities. International organizations, which have invested billions of dollars in rural projects, are only starting to understand this reality: sustainable development requires first and foremost sustainable urbanization policies designed to facilitate – and not to hinder – the current urban transition.

4. Cities are often viewed as a product to be marketed at the regional and global scales. Information and communication technologies allow for internationalization of footloose investments, resulting in vast increases in the volume and speed of international capital flows of all types, ranging from foreign direct investment to short-term banking activities. In this environment, cities often have no option but to compete for these investments, exploiting whatever comparative advantages they may have, at times even to their own detriment.

5. Globalization has placed towns and cities in a highly competitive framework of inter-city linkages and networks within a geographical context limited only by planetary boundaries. This new constellation of globally linked cities is sometimes referred to as the urban archipelago. It implies that a city may have more relations with some faraway place than with its hinterland and that such cities act as energy nodes in a global force field.

6. Although the paradigm of the urban archipelago reflects real urban evolution at the global level, the relationships between rural and urban areas still include factors that do not necessarily depend on the international level. There remain considerable local economic factors (the symbiosis of rural primary production and urban secondary and tertiary activities), local demographic issues (migration and absorption of rural population surpluses), local environmental consequences (water and air) and local infrastructure matters (transport, energy and telecommunication) that still bind cities to their rural hinterland. In addition, the international mobility of labour still lags far behind the mobility of capital, and traditional urban-rural linkages remain crucial in a host of mutually supportive functions.

7. Globalization is causing the end of territorialism: the condition whereby socio-economic and political space is reducible to territorial coordinates. It creates an apparent paradox, whereby the conditions of social development are simultaneously becoming more global and more local. In many localities, people are overwhelmed by changes in their traditional cultural, spiritual and social values and norms and by the introduction of a cult of consumerism intrinsic to the process of globalization. Many localities have reacted by stressing their own identity, their own roots, their own culture and values and the importance of their own neighbourhood, area or town. In political terms, this has translated into demands, on the one hand, for political decentralization, and on the other, to filter out undesirable effects of globalization. Civil society has become important in recasting national and local politics as a third-sector actor, distinct from state and market, shaping policies, norms and values.

8. Despite these positive developments, the future of large and small settlements in the developing world looks grim if they continue to develop in the unstructured and unstable fashion that has characterized many of them until now. Cities in the developed world also face unprecedented problems of urban decay, ageing populations, widening gaps between poor and rich, layers of vested interests and large disparities between knowledge-based elites and under-educated urban populations. These problems determine the need for a new, inclusive approach to local governance and present a great challenge to central and local government. On a more positive note, the world is no longer only a community of states but also an increasingly borderless network of interconnected cities where power is being shared more evenly and where governance is becoming more democratic. Promising partnerships are evolving between the public sector, the private sector and civil society. There is a growing awareness of the needs and rights of women, the indivisibility of human rights, the need for participation in urban decision-making and the urgency for wise social, economic and environmental stewardship. To address the "glocalization" paradox that has distinctly emerged since the City Summit held in Istanbul in June 1996, Governments and their partners will have to address the untenable present course of the global urban environment. To do so, the highly articulated goals and objectives of the extraordinary series of United Nations global conferences of the 1990s, culminating in the Millennium Declaration of 8 September 2000, must remain the overriding development agenda of the international community.

9. Within cities in all regions of the world, disparities are growing between the affluent and the dispossessed, exemplified by the coexistence of thriving business districts, affluent neighbourhoods and slums (in the case of most developing countries) or distressed neighbourhoods and derelict quarters (in more affluent countries). This is the most visible trait of the divided city. According to UN-HABITAT data, nearly one third of the world's urban population (837 million in 2001) lives in slums and informal settlements.

10. The globalization of lifestyles is another aspect of the phenomenon and one that is most evident in cities, both in the cultural and consumption attitudes of the affluent and in the homogenization of the physical image of large and smaller cities. They are all increasingly alike in their modern and post-modern architecture, shopping malls, gentrified historical neighbourhoods and gated suburban housing estates. They are also similar in their derelict brownsites, their degraded and sub-standard residential districts or informal settlements, where the invisible other inhabitants of the divided city survive.

11. Cities in the developing world are faced with an enormous challenge caused by continuous population growth, a population that needs better shelter and services but whose income is insufficient and a diminishing ratio of public resources per inhabitant, compounded by a weakened support role of the Government, an inadequate revenue base and poor institutional capacity for urban planning and management. The contrast with the buoyant image of e-regions and prosperous cities of the North could not be sharper. The dilemma for cities in developing countries is that they are faced with explosive problems that require immediate action but for which no sustainable solution is in sight in the conventional terms of public housing, subsidized housing finance systems and social programmes that worked well in the industrialized countries of the North half a century ago.

12. The entry point for addressing this dilemma has been identified in the Habitat Agenda: the provision of secure tenure within comprehensive slum and squatter-settlement upgrading programmes. Various forms of tenure exist, not all of them linked to home-ownership, that can guarantee physical, social and economic security to the threatened half of the divided city. It is only on this basis that the poor can acquire dignity, improve their health conditions, invest in the improvement of their own shelter and communities and gradually earn a role as full-fledged citizens. This has been done and is being done with success. The challenge for the partners of the Habitat Agenda is to reproduce these successes on a scale unheard of before. World leaders acknowledged this challenge by including the goals of large-scale slum upgrading and good governance in the Millennium Declaration.

13. A different reality is indeed beginning to emerge from within some of the world's slums – a reality that places the poor as active participants in development and as true poverty experts. Where banks do not lend to them, they save and lend to each other. Where no housing is available, they build their own shelter. Where no education is provided, they teach each other. The poor are currently the single largest producers of shelter and builders of cities in the world. In most cases, women are taking the lead in devising survival strategies that can create the conditions for development.

14. It is clear that a number of city leaders want to meet the challenge of urban poverty by actively engaging with the poor and the marginalized. Recent progress made in participatory local democracy is providing fertile ground for innovation in the way that demands are articulated and satisfied. It is precisely these initiatives that provide the best prospects for sustainable urbanization strategies, and which international development institutions and experts are being invited to support.

15. On 29 April - 3 May 2002, UN-HABITAT organized in Nairobi the first World Urban Forum to discuss the issues of sustainable urbanization in preparation for the World Summit on Sustainable Development. More than 1000 participants agreed that the key to making cities more inclusive, more productive and environmentally sound, can be found in improved urban governance based on participatory democracy and strong partnerships between all levels of government, between the public and private sectors and between institutional actors and civil society. This is probably the major message from UN-HABITAT: rapid urbanization is manageable and necessary resources can be mobilized within the cities themselves, provided proper urban policy and related capacities are developed. We are absolutely convinced that good urban governance can bring about sustainable urbanization during the 21st century.

Typology and Conformity to Natural Laws of "Chaotic City" Development (Tokyo, New York)

Elena Rodina New York **Vladimir Rodin** Voronezh

Attractor – the term is taken from contemporary physics and used here to denote distinct elements of structural stability in the urban complex dynamic system.

Introduction

In 1980 B. Mandelbrot developed Fractal Geometry. Fractals were found in nature: coastlines, clouds, snowflakes, trees, the functioning of human lungs, blood circulation, etc... (Feder, 1988). Chaos theory, which emerged on the basis of fractal geometry, states that Chaos is the principle that ultimately reintegrates and refreshes all existing forms of life. "Chaotic behavior" is an integral principle of functioning within nature (Coping with Chaos, 1994). Archetypal examples of "chaotic" behavior are the weather and the stock market. Principle characteristics of a chaotic natural system are existence of complexity (unpredictability), self-similarity (not self-sameness), self-organization around points of attraction and the existence of Fractal Nature (fractal dimension) (Feder, 1998).

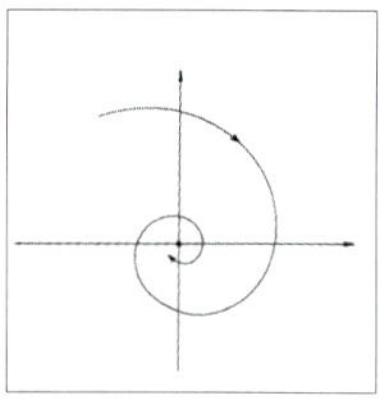

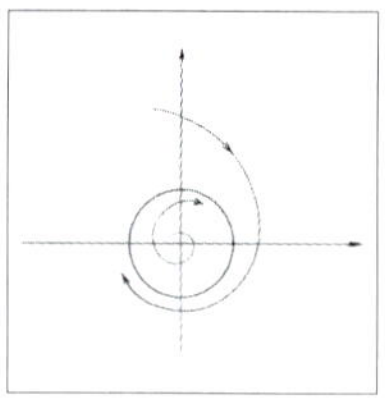

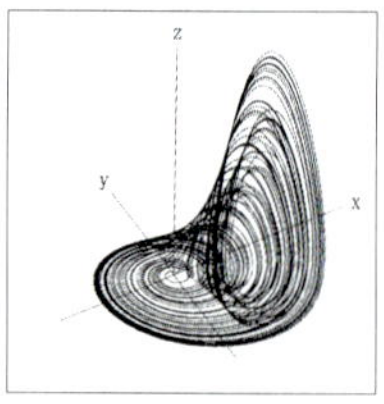

At the initial point of our investigation of the development of "Chaotic Cities" we were interested in finding out the logic hidden in the all too chaotic layout of Tokyo. We established that Tokyo's visually chaotic development conforms to the Natural Laws. We can compare Tokyo's layout with the development of natural systems in which the presence of chance, plans, fates and solutions reflect a fractal dimension (V. Rodin, E. Rodina, 2000). If the city is functioning, its development should have natural "chaotic" characteristics. In the second stage of our research, we decided to find out if "Chaos" exists in the development of New York, especially in Manhattan the part of the city famous for its regular urban pattern. During the research project, we made use of generalization of spatial urban systems, group analysis and the analytical method.

Tokyo

Conditionally we can divide Tokyo into a zone of so-called "stable chaos" or 2D "chaos" within the plane of the street pattern, and a 3D "chaos" of building mass. The rate of modification of the latter is much higher than that of the former. In fact, the 3D of urban space appeared to be "invisible". For visitors and inhabitants, 3D urban space appeared as 2D planes with advertisement boards. Similar to a natural system, Tokyo's chaotic environment contains a self-similar element of repeating advertisements and functions. The "virtual" structural organization of Tokyo's "invisible" 3D building mass shows the presence of the second main characteristic of Natural System – self-organization around 3D attractors. 3D urban attractors organize the city mass and help inhabitants to orientate themselves in the city (Fig.1). A virtual matrix of the city contains zones of attraction, attraction zones containing attractors inside; those attractors in turn contain smaller attractors, etc... Each person has a different "internalized route" containing a "virtual" itinerary within the urban mass to the urban attractors. The urban structure of Tokyo's attraction zones can be compared to the Russian toy "Matreshka" doll that contains within it similar smaller dolls, but in different costumes. Generally attractors have topologically distinct forms of dynamics: stable fixed points, limited-circle attractors, semi-periodical attractor zones, etc... (Feder, 1998). Within Tokyo's urban development, the main types of attractors distinguished were:

1. Stable fixed-point attractors (e.g. Imperial Palace). 2. Limited-circle attractors (e.g. temples and shrines). 3. Chaotic or Strange Attractors (e.g. train stations, commercial zones, entertainment zones Kabuki-cho). 4. "Frozen" or Dead Attractors – zones that were intended by the urban planning committee to be attractors, but which in fact became abundant zones (e.g. Metropolitan Area in Shinjuku, Daiba).

It was interesting to discover that the place of attractors in the city has hardly changed with time. For example, we discovered that the old city of Tokyo (Edo) lay almost entirely within the circle created by the present-day Yamanote line – the main attraction zone in Tokyo. Edo, historically attractive to surrounding villages, turned into the main commercial and entertainment attractor on the Yamanote train line. The Yamanote train line is the only border within the city that can be clearly distinguished; borders between the attraction zones in Tokyo are unclear. With changes in world-outlook, the various types of attractors changed their meaning for people. It is important to emphasize that different social groups have different types of attractors. An intersection of interests in attraction zones creates main Attractors within the city.

New York/Manhattan

Similar to the Yamanote Circle zone, Manhattan itself is a big limited attractor within New York City. It is hard to find "Chaos" in Manhattan's "Standard regular" urban structure, but it is "chaotic" in a virtual sense, in its spirit. "Chaos" we can observe

in self-organization and migration of different cultural communities. Natural Laws in the scenario of the city development are represented by self-similarity in service functions and in the organization of cultural communities. The structure of attractors within Manhattan is different from that of Tokyo. Whereas in Tokyo, the principle attractors are train stations incorporated into the big commercial zone, in New York the main attractors are streets themselves (e.g. Fifth Avenue, Broadway). There is no need (as in Tokyo) to create a virtual route to attractors; routes are attractors themselves in Manhattan and they in turn contain smaller attractors, etc... The difference in attractor structure comes from the fundamental difference in thinking about space. The Japanese approach has more affinity with area (hence the importance of "totem mat" and the floor of the building and the "machi" as an area unit of organization of the city). Western approach has more with the line (the sequential ordering of buildings along city streets) (Yoshinobu, 1987). Within main zones of attraction in Manhattan (Mid-town, East side, Downtown) we can distinguish the following types of architectural attractors: Stable fixed-point attractors (e.g. Times Square, Fifth's Avenue, Broadway), Limited-circle attractors (e.g. Central Park), Chaotic or Strange Attractors (e.g. various cultural zones in Downtown, commercial zones of Mid-town).

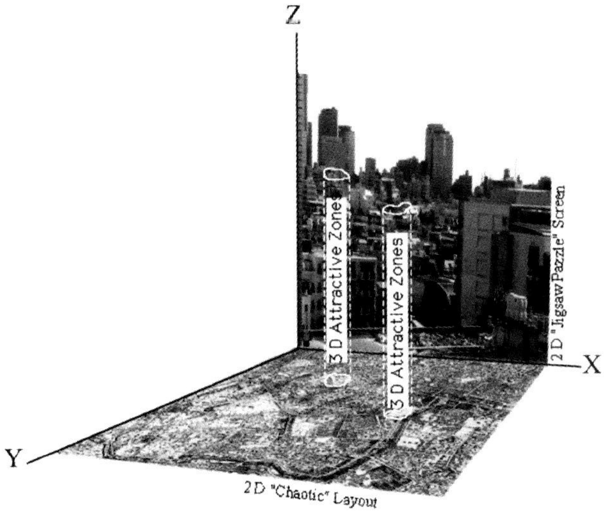

Fig. 1

It is hard to distinguish borders between zones of attraction in Downtown (Fig. 2). Tourists search for Little Italy in Tribeca and for Soho in Greenwich Village, while local habitants try to answer the typical question of exactly where Little Italy starts. There is no precise sign or architectural element (except the Arch on Washington square), that distinguishes the entrance to a district. It might be not a bad idea to create some types of

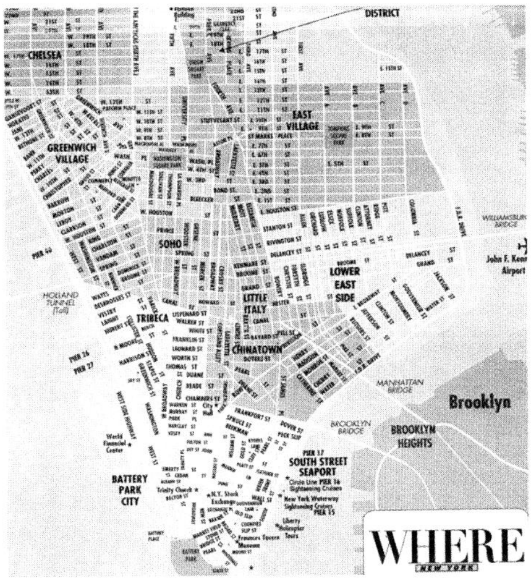

Fig. 2

architectural attractors, which will also have the meaning of a gate to a district. It is important to mention that a city in general can be represented as a multi-layer structure which develops along the time axis, the social axis, and the architectural axis. In New York, the axis of cultural community distribution and development does not correlate with the axis of architectural style distribution. Most of the buildings were built in the beginning of the 20th century, but in course of the century communities "migrated" within Manhattan and New York in general; this phenomenon results in the cultural difference between the community that originally built the buildings and the community which now occupies it (e.g. Harlem). In Japan population and architectural style are congruent, except those attraction zones in the International Global Style. In both two cities, New York and Tokyo, we can observe self-similarity and self-organization around points of attraction. In Tokyo it is self-organization of urban mass around fixed-point attraction zones, in New York it is self-organization of the social mass within stable urban structure. Indeed social complexity contributes to the penetration of "chaos" (consequently of life) in to the "standard regular" urban structure of Manhattan. Both cities have complexity: Tokyo in its urban structure, New York in its social structure. It is a paradox that the social order of Japanese society exists within the "chaotic" urban pattern, and the social "chaos" of New York exists within the strict order of an urban grid-work pattern. We found the existence of the fractal dimension within Tokyo's street pattern; the fractal nature of the development of the social communities in New York is still a question for us.

The Discovery of Fractal Dimension in Tokyo's 2D Urban Structure

In our analysis of the zone of "stable chaos" within the 2D chaotic plane of Tokyo's street pattern, we used a formula obtained by Hack during his study of 400 rivers in Virginia and Maryland (Feder,1998). He received the following ratio between the length of the longest river above a given location and its drainage area (1) where D is fractional and ranges from1<D<2.5.

D is an index of fractal dimension. The D fractional can characterize an irregular pattern. For the "standard regular" urban structure (Midtown Manhattan, Fig. 3) a well-known formula of fractal dependence of the line from the area (2) is true with D-integer, for example D=2.

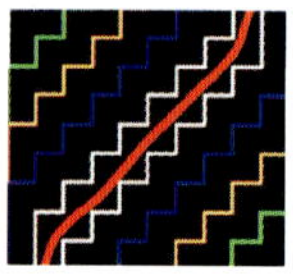

Fig. 3

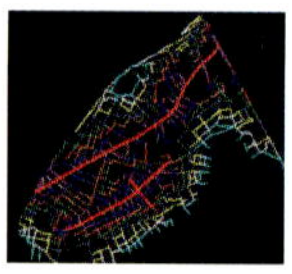

Fig. 4

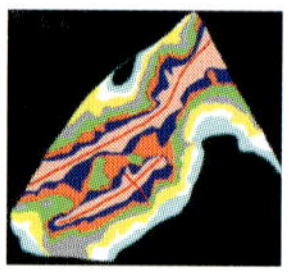

Fig. 5

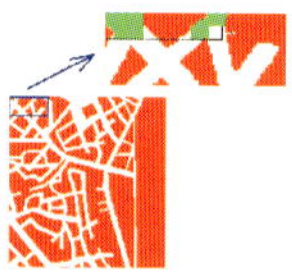

Fig. 6

We applied formula (1) to an estimate of the ratio between the longest street in the limited "chaotic" urban zone and the area that it occupies (Fig. 5, 6). Unexpectedly, we found the existence of the fractal dimension during the investigation of the 6 "most chaotic" zones in Tokyo: Shinjuku D=3.24 Shibuya D=2.34 (Fig. 5), Ikebukuro D=2.54, Ueno D=2.4, Nakano D=1.66, Aoto D=2.8 (V. Rodin, E. Rodina, 2000). Shibuya, Nakano and Ueno show the same index of the fractal dimension as rivers in Hack's study. Of course, our investigation is entirely qualitative rather than quantitative. Coefficients were calculated with a fairly large margin of error because all calculations were done manually. It is possible to create a computer program to provide a quick and precise result of estimate of the fractal dimension of city streets.

The Creation of a Computer Program

At the present time we have created a test version of a computer program that allows processing the city map and defining areas with a high index of fractal dimension, and of recommending them for prime re-planning and reconstruction. Such areas are considered negative because they are extremely difficult to navigate and escape from in case of emergency. In most cities these territories show a high rate of crime as a result of bad illumination, lack of people, distance from the main roads and police stations. The work of the program performed as follows: for dynamic graphic programming the Sprite system is used. At the initial stage, on the map of the city we mark out the Background (the static part of the map): all dwellings, industrial enterprises, squares, parks, etc... A special color is assigned to them. All basic lines of communication (roads, highways, foot and bicycle routes) are uncolored. Then the dynamic Sprite in the form of a square (after the program development we transformed it into the form of a snowflake) follows communication lines on the map (Fig. 6), distinguishing the street order and calculating the area occupied by the streets within each order.

References

1 Coping with chaos: analysis of chaotic data and exploitation of chaotic systems (1994), New York: John Wiley & Sons, Inc.

2 Feder, J. (1988), Fractals, New York: Plenum Press.

3 Vladimir A. Rodin, Elena V. Rodina, "The Fractal Dimension of Tokyo's streets", in Fractals, Vol. 8, No. 4, December 2000, pp. 413-418.

4 A. Yoshinobu (1987), Chaos and Order in the Japanese City. Japan Echo, Vol. XIV, pp. 64-68.

Sustainable Cities: A Strategy for a Post-Terrorized World

Richard S. Levine **Ernest J. Yanarella** **Taghi Radmard** Lexington **Heidi Dumreicher** Vienna

Abstract

When and if the current terrorism crisis abates, the root causes will remain. The world is awash in a growing equity crisis to which there are no global solutions. However, this paper posits that there are local solutions – solutions whose individual successes can serve as models to be emulated globally. Such solutions will require a change in the way we think about the relationship between ecology, economy and equity, but it will also involve a change in our thinking and practice concerning the role of architecture and urban design.

Introduction

Sustainable city-regions are places having a sufficiently high degree of diversity and self determination that their citizen stakeholders are able to negotiate amongst themselves their own sustainable way of life whose only limits are the limits of nature, appropriate technology and their own traditions and creativity. Although long term unsustainability is often measured on the national or global scales, (i.e. climate change) all contributions to national or global unsustainability are produced locally. This observation gives rise to the realization that if a city were able to take itself out of this global equation of unsustainability and still demonstrate a vibrant local economy and quality of life, then this would be a model that cities around the globe could emulate, in their own way and through their own cultural patterns.

The problem with global capitalism is that it is designed to fail. A healthy natural environment is the basis and precondition for any society, and a healthy society should be the basis and precondition for any economy. Economic systems should exist for the purpose of sustaining a desirable way of life and not the other way around. Because the globalization paradigm is effectively designed to marginalize social, cultural and environmental values, it also destroys both the underlying basis for its own existence as well as it's own rationale for being. Because it has no embedded structures able to consider social, cultural or environmental effects and concerns, the growing domination of the globalist economic system systematically degrades the very basis of any economy. Moreover because it robs people of their local culture by substituting western (American?) profit-driven values (or lack of values) for the traditional values of their own cultures – it robs peoples of their identity. Systematic marginalization of culturally embedded religious values is a dangerous pursuit for it robs people not only of their identity in this life but in their afterlives as well. As we have seen, this perceived cultural genocide is so threatening that it is seen to be worth sacrificing one's own life in an act of defense. In the events of September 11, 2001 we have witnessed the first major battle in the wars against globalization. These wars will be unlike any previous wars. Like globalization itself they will not be about territory, and they will not be waged by nation states. They will be waged by diverse individuals and groups driven by perceived threats against their traditional ways of life under the emerging realization that they have in effect, a common enemy.

Although it is still very much a minority view, there is a growing realization that the dominance of globalization constitutes the most pernicious force in the world today. But even the great majority of those who decry it acknowledge Margaret Thatcher's famous pronouncement, "There is no alternative," counseling that all we can do is to try to resist the juggernaut as best we can and try to moderate its most destructive tendencies. At best these strategies can only buy a little time and if that time is not used to pursue a transition to another way of operating, the ultimate outcome will be the same and any positive efforts will have been wasted.

Sustainable Area Budget

If issues of equity and human rights are to be respected, every individual is entitled to his/her share of the Earth's bounty on a regenerative basis and its capacity to absorb offences. The current authors have developed this concept as the Sustainable Area Budget (SAB). This means that each individual is entitled to one six billionth of the Earth's regenerative capacity interpreted as land area. A town's or city's working budget is thus the aggregated SAB of its citizens.

Working with these principles and within this budget the research team has developed a participatory design approach, which generates new urban models and architectural scenarios for Sustainable Cities. These principles were initially applied to the evolution of the Sustainable City-as-a-Hill project at the Westbahnhof in Vienna – a computer aided design process using a modular, participatory approach. The successful implementation of such Sustainable City Implantations and the environmental and cultural economy they reflect, provides a powerful alternative model to the now dominant globalist paradigm.

The Sustainable City Game

The proposed program works from an extended definition of the architectural endeavor. A city is not something that is first designed and then built. Its design, construction, maintenance and reconstruction is an ongoing process. When such a process has historically been an organic process, responding to the needs of people and institutions and modifying previous interventions that didn't work or are no longer appropriate, it becomes a process that continually increases the living quality of the environment.

If the process of city development is one that responds most directly to increasing speculation responding to globalist forces, then the quality of its urban structure and public life will most certainly deteriorate.

In the Sustainable City Game works with a city-region or a large bounded site within an existing city (a brownfield site as an example) together with its rural partner region. Gameplayers who may be potential stakeholders are assembled and from them a program of needs is developed. Initially it is a shopping list that, because it represents the ideas and desires of diverse interest groups, contains many inherent conflicts and contradictions. Initially a coherent set of specifications for the nature of the future Sustainable Urban Implantation proves to be impossible to put together because of these conflicts. Instead of seeing the design process bog down amidst a flurry of single issue confrontations, the process is separated into several steps. In the first step it is encouraged that any legitimate needs and ideas be placed on the table. In the second step, varied teams of stakeholders together with designers and other professionals attempt to assemble design scenarios that represent a full spectrum of urban facilities and services, within the initial framework conditions and within chosen urban design strategies. Thus the design and development of the city becomes an empowerment process, engaging citizen stakeholders in the shaping of their common sustainable destiny.

The Sustainability Engine™

These scenarios are then modeled as both physical designs and energy and material flow models using the Sustainability Engine™, a utility under development in the US that combines some of the attributes of CAD, Facilities management and GIS software together with systems to become the principal feedback, design and management tool in the negotiation of sustainable city-regions. The practice of architecture in recent years has increasingly gravitated toward the delivery of contract documents in a CAD and/or Facilities Management format. Through embedded databases these formats provide the capability of extracting many sorts of useful information about the virtual building as it has been constructed within the computer program. Material takeoffs of virtually every nut and bolt together with their locations and specifications are easily charted. Maintenance and replacement schedules can be developed and recorded. Changes made in material, size and energy performance are automatically projected through the building model and its database and the reverberations of those changes can be displayed almost instantly.

It is but a small conceptual step from the design and management of conventional buildings to the design and management of sustainable cities. One difference is that in the case of sustainable cities, a lot more information is attached to the components, systems and building blocks that make up the city implantation. Within the Sustainability Engine™ are module libraries of components and building blocks containing many attributes including such things as embodied energy, distance from source, cost, availability within the SAB, labor requirements, recyclability, u-value, land use implications, energy and material flow connections to other regenerative systems and the various inputs and outputs involved in the functioning of the module within the city-system. These modules function as plug-in, "free body" objects that provide inputs and outputs when attached to a larger Sustainable Urban Implantation scenario model. Similarly, the Engine™ contains libraries of regenerative energy modules, agricultural modules and other modules of means and technologies that are particularly supportive of ecological living.

In the playing of the Sustainable City Game, stakeholders together with architects and technicians attempt to assemble a Sustainable City Implantation, drawing on the existing building blocks that most closely meet their needs and desires already developed within the libraries. If no building blocks are suitable, existing blocks are modified or completely new ones developed by the architects. The architects also provide overall urban design strategies, either as modifications of pre-existing ones taken from the library or new urban configurations crafted to suite the particular site conditions.

Because any urban design that represents the needs or interests of only one stakeholder or group of stakeholders will not contain the diversity or complexity of a real town, such a limited model when run on the Sustainability Engine™ will appear in its first trial run as a city-system that is grossly out of balance. The feedback of this gross imbalance becomes an important moment for the stakeholder-players. It indicates to them that in spite of the fact that their immediate needs may have been well satisfied by their preferred urban proposal, still because their interests represent only a portion of the city-system, many other needs must be met in order to be approaching equilibrium. This feedback then supports a significant operational principal of the sustainability endeavor, to wit: any proposition may be put on the table, but in order to be carried forward in subsequent iterations of the Game, the overall city/system scenario in which the proposition is imbedded must be approaching equilibrium. Very quickly it is seen that no matter how beneficial a given proposition may appear to be in the first instance, it must still attach itself to a network of mutually supportive propositions to form a larger, well balanced scenario in order to remain viable as the Game progresses.

The Game is played through many iterations and at each successive step the scenarios become more sophisticated and more complex. In a similar fashion the game itself and its module libraries would be learning ecologies, becoming more elaborated and accumulating more options and able to provide more sophisticated feedback as the Game learns through its own playing. Because of its growing successes the Game and the city models its playing generates, becomes an attractor of people and interests who are in a position to act upon what they have negotiated to be their preferred form and structure of a locally determined Sustainable City Implantation. As the game becomes sufficiently serious that construction is planned and carried out and people live in the city, the same stakeholder process that has generated the city form and structure using the Sustainability Engine™ continues to be employed as the process by which the city continues its development, modification, maintenance and governance.

Conclusion

The "War on Terrorism" as currently formulated by the Bush administration is a war that in its own terms cannot be won. Because its enemy is the very enemy it is continuing to create by its policies of marginalizing local cultures, the war can only continue indefinitely, eroding traditional democratic values, alienating ever larger numbers of peoples by increasing the inequitable and oppressive dominance of the globalist system. Thus it is a war that creates its own growing enemy. Because there is now only one global super power and only one dominant economic system, such a war cannot be successfully confronted by alternative forces. There is however, a local alternative. A bumper sticker in the 1960's read, "What if they held a war and no one came?" Perhaps the only way to end this war is to opt out of it. At the individual level this is not an option. As individuals, and interest groups within sustainable city-regions, operating in balance with our ecological budgets, we can create regions that remove themselves from the global unsustainabilty equation. Such an alternative that builds within local cultures and local conditions, exploiting neither its own territory nor lands far away; exploiting neither its own people nor people in distant places – such an alternative is both necessary and possible.

Reference

"Generating Models of Urban Sustainability: Vienna's Westbahnhof Sustainable Hilltown", H. Dumreicher, R. S. Levine, E. J. Yanarella and T. Radmard, book chapter in Achieving Sustainable Urban Form, Jenks, Burton and Williams ed. London: Routledge, Ltd. 2000

Layering of a Spontaneous City over a Planned One
Case of Nairobi's Central Business District (CBD)

Crispino C. Ochieng Nairobi

Open-air market in pedestrian shopping street with foot traffic being forced into main road

Abstract

Most cities have experienced layering of planned city by a spontaneous one. In most cities, for example Nairobi, within the central Business District (CBD), this has taken the form of open-air markets. In most sectors of the CBD, activities of this occupy shopping as well as pedestrian streets. Within these, foot traffic is forced to fight for the only remaining space with passing motorcars.

There appears to be a strong connection between preferred choice of open-air market settings and human traffic flow. They are likely to be found along and/or next to those motorways with a reasonable amount of human traffic. These would be motorways that lead directly to some of the popular residential and/or transit points.

From available data it would appear that in the development of the open-air market, re-developing of open places in the CBD and some of the motorways to accommodate this development could only enhance its activities within the CBD.

Introductory Background

To a great extent, physical planning of Nairobi has been informed by ideological concepts (Nevanlinna: 96). This has resulted in two cities that exist side by side. There is the squalid one to the east, and a wealthy centre and selected well-kept suburbs to the west. In Nairobi, modern technology exists side by side with primitive living conditions and in complete contrast.

Nairobi's Central Business District has three sectors. There is the government one with all the public and administrative offices; the retail one to the east of Tom Mboya Street; and the business sector. Motorways of different grades link the CBD to the several suburbs around the city. According to Nairobi City Council sources, more than fifty percent of original renters in the CBD has relocated to the quieter and relatively secure suburbs.

An interesting development along some of the motorways within the CBD is the high level of open-air markets that continue to take place along the walkways and/or shopping streets. From as early as nine o'clock in the morning, both the pavements and shopping streets become vibrant with traders and buyers. With evening approaching streets become more vibrant. Goods that are sold are no longer the informal but cheap industrial goods from the Far East and China.

Without a concrete solution, the open-air market will continue to haunt this city. As early as 1961, attempts by individuals in city hall to solve the problem were never fully supported (Nevanllina: 96). No legislation has been put in place since.

The 1948 proposal by Thornton, Silberman and Anderson that would have contributed to a change in the landscape of Nairobi was never implemented (Thornton, White, Silberman: 48, Nevanllina: 96). Lack of fully implementing the plan has contributed to the city problem. In the 1948 proposal the improvement of motorways and articulation of open settings to supplement the city lifestyle were proposed.

Research Method

The aim of this research was to develop a theory between settings that are popular with open-air market and city residents who are part of the CBD. Attempts by the council to re-locate open-air markets from CBD to settings that are far removed from major movement axes have been unsuccessful.

Firstly sampling was carried out to identify motorways within CBD that were popular settings for open-air market activities. The guideline was also revised during this. Secondly, there was the grounded theory examination that was carried out to find out about two things: First was how those active in the open-air market had responded to early moves by the council to re-locate them to alternative settings. Second was to identify the level of interaction between sellers and passers-by that would be on foot (Leedy: 97). Third was to identify data on buyers.

Combined methods were used to collect the data (Leedy: 97). Archives were used to find out the response of authorities over the years to problems of open-air markets; both discussions and observation strategies were used on selected sellers and members of public; and photography was used to capture scenic photos. Combined methods helped to maximise similarities and differences among information obtained.

Data was broken down, for example the destination of each of the motorways and location of every urban open place and relationship of this to popular motorways established. Within each selected motorway, scale was attached to density of those on foot. High density was attached to motorway with lots of human traffic; and low for those where traffic was lowest. For each of the motorways, the level of activities was identified. Finally, different factors were compared and categories made (Patton: 90). The aim of this research being to develop theory, relationships are presented as propositions.

Open Space Typology

There were three types of open spaces as per use. Over time, some had been adapted to provide for different kinds of use. Following uses were identified:

Car Parking: this included those next to; Uchumi Supermarket; High Court; Railway Station and Jamia Mosque. Only the last one was private and was managed by the neighbouring mosque, the rest were council property. Jamia Mosque and High Court open places were located within the business and government sector respectively. Both the Mosque and the Council management ensured that the respective settings were only being used for car parking.

The local council had entered an agreement with curio sellers to use part of Uchumi Supermarket on Saturdays. In turn before leaving the place, curio dealers would ensure that their settings were tidied up. This place was located between the business and government sector of CBD and next to Aga Khan Walkway and Harambee Avenue.

The railway station parking located in front of the station was large and ambiguous. It had no defined use. It provided for Car Parking; a public transit point for vehicles to the south-east of Nairobi and scattered open-air market settings that dealt in diverse products.

Jevanjee Garden provided space for relaxation and is also dominated by evangelism. Features included; a monument to celebrate the garden's founder; soft landscaping and hard ground walkways. Although public, the garden is under a trustee and is controlled through two entrances. The location of this garden is within the edge of the business sector, but it is removed from the major flow of human traffic.

Kenyatta International Conference Centre open grounds was meant to be used for celebrating events. It was accentuated to accommodate both flow of activity from the centre and from outside. Entrance to the centre was controlled through two gates.

The General Post Office Gardens were semi-private. Access was controlled by design through accentuation of terraces and fencing that were not very dominant. Features here included gardens and a water fountain.

Street Typology

The CBD was laid out in gridiron form with these being used to create different levels of motorways. Through hierarchy of streets, the CBD is physically linked to its suburbs. Motorways with direct links include dual, Jogoo Road, Race Course Road, Muranga Road; Forest Hill Roads; Uhuru Highway and Haile Sellasie Avenue. Those that feed human traffic to these included Haile Sellasie Avenue; River Road; University way; Ronald Ngala Street; Aga Khan Walkway; Tom Mboya Street and Accra Road. There are also several feeder roads and alleys.

Both Jogoo and Racecourse road are dual carriageways. Jogoo Road links the CBD to high density residential suburbs that are in eastlands and Racecourse to North East and some of the northern suburbs. North-east suburbs include high density suburbs such as Mathare Valley and medium density ones, for example Pangani. During rush hours both the two roads would be filled with human traffic. Open-air market activities would be taking place within some of these settings (see photo on page 76).

Both Muranga and Forest Roads link the CBD to medium and low density residential sectors to the north and north west. The former also links up to some of the high density residential sectors. Both during morning and after work, a large number of people would be noticed to be trekking along Muranga Road.

Uhuru highway links the CBD to residential suburbs in the south east and south west. The former is a medium density sector of town, while the later links up with the medium and low density residential sectors. Both during morning as well as in the evening, there would be some crowd of people trekking along the road. After six o'clock this would slow to a trickle.

Haile Sellasie Avenue links the CBD to the southern suburbs. This is both high, as well as medium density residential. From as early as six o'clock in the morning huge crowds of people would be trekking along this avenue to the CBD.

In general it was observed that both during early hours of morning and late after five, there would be crowds of people trekking along the motorway. Also along this motorway there would be some level of open-air market activity. There were those who would continue with their movement and others who would stop for while to purchase, haggle over prices and/or purchases.

Conclusions and Recommendations

Main objective of the research was to identify possible connections between the pattern of an open-air market that is located within the CBD and some of the urban elements, for example motorways and open places. From data, open-air markets formed along those motorways with high human traffic. They would be leading to popular destinations for example residential sectors or transit points.

Second objective was to establish any relationship between open places within CBD and selected motorways, and how the setting of the latter was likely to enhance the use of open places as likely settings for open market activity. Locations of some of the open places were removed from popular motorways. As far as open-air market activities were concerned they were hardly of any value.

Firstly, this research would recommend the local council authority to adopt a specific policy on open-air markets within the CBD. This could be prescriptive. Rather than being allowed to develop in the streets, there is a high possibility that some existing or new open places could be accentuated to accommodate this. Secondly, it is necessary that the council accepts the reality of urban poverty. Some of the developments along the motorways could be direct results of this. Accepting the reality of urban poverty becomes the first step in solving the problem of open-air markets within the CBD.

Reference and Bibliography

Leedy, P. D., 1997, Practical Research, Planning and Design, Sixth Edition, Merrill, An Imprint of Prentice Hall, Upper Saddle River, New Jersey.

Nevanlinna A. K., 1996, Interpreting Nairobi, The Cultural Study of Built Forms, Suomen Historiallinen Seura, Helsinki.

Ngau, P. M., 1992, Transportation and Characteristics of Urban Travel in the City of Nairobi – a View through the Gravity Model, in: Ed. R. A. Obudho, African Urban Quartely, Vol. 7, Number One and Two, African Urban Quartely Limited, Nairobi, pp. 104-118.

Patton, M. Q., 1990, Qualitative Evaluation and Research Methods, Sage Publications, London.

Thornton White, L. W, Silberman, L. Anderson, P. R., 1948, Nairobi Master Plan for a Colonial Capital, A Report Prepared for the Municipal Council of Nairobi, His Majesty's Stationery Office, London.

Photonic Centre, Berlin-Adlershof Architects: sauerbruch hutton architects

Future Perspectives of Asian Urban Societies

Santosh Ghosh **Unmesh Kirtikar** Calcutta

1. The Rise of Megacities in Asia

The 21st Century will be an urban century and Asia will be a continent of megacities. In 1990, there were only 12 cities in the world with populations of more than 10 million each, and Asia had 7 such cities, in 2000 there were 19 such cities, and Asia had 11 of them. It is projected that by 2010 there will be 21 such cities, and Asia will have at least 12 of them. According to the United Nations amongst the top 30 cities in 2010 with Tokyo at the top with 26.4 million, there will be 18 cities from Asia, (UN 1999). The European cities will disappear from this list except Paris, and only New York and Los Angles from North America will be included.

Except in India and China, the largest cities in many Asian countries contain most of the urban population, and as per 1995 estimate, Dhaka had 33%, Tokyo 28%, Bangkok 55%. Such megacities are like dinosaurs. But unlike dinosaurs they will not be extinct. The megacities will live with poverty and pollution, health and hygiene problems and increasing deficits in services, and infrastructure. The megacities are in many Asian countries where the rate of urbanization is low, and often such cities are oases in the desert of rural poverty. The face of the Asian city has strong contrasts. Many of these countries achieved freedom from colonial rule after the Second World War and there were years of conflict and confusion. Dr. Gunnar Myrdal, the Swedish economist in the prologue to his book 'Asian Drama' painted the Asian scenario in 1968 (Myrdal, 1968) with inner conflicts, aspirations, harsh reality, desire for change etc. More than three decades have passed and the cities have changed physically very rapidly, though slowly socially, and inner conflicts still remain.

There are diverse, economic, social, demographic, spatial and environmental conditions in Asian cities. There is a long laundry list of environmental problems. The divide between the rich and poor is quite open and the conflict is expressed not only in the townscape and socio-economic strata but also in cultural identity.

The noted architect from Singapore William Lim (Lim, 1999) said that: "While Asia has changed perceptively, it is also deeply influenced by traditions and heritage as well as by its own values and lifestyles".

In a large city human beings were gradually replaced by buildings – all kinds of buildings for rich and poor. There are two faces of the city – one legal for which architects and planners work and for which building rules and land use zoning laws are applicable. But there is another city – the illegal city, built by the people out of waste materials, often on encroached land and often without sanitation at sites vulnerable to disasters, but the informal sector is connected with this. Lewis Mumford (Mumford 1961): said that the visible, tangible city was full of imperfection.

2. Globalization and the Asian City

In this conflicting situation globalization has come, though slowly in some cities. In Asia globalization is not new in cities, as the colonial rulers brought an architecture and culture of their countries.

Bernd Hamm, Director of the Centre for European Studies, speaks critically about globalization (Hamm, 1998) because of polarization of rich and poor. His prediction that "small islands of wealth will emerge in the oceans of poverty" can already be seen in megacities in Asia.

The growth rate in Asia in 2000 on average was 5.71%, which was higher than industrialized countries, and per capita GDP growth in East and South Asia was 6%, i.e. higher than other regions (Mkandnewire 2000).

Liberalism, privatization and globalization have started to grip the economy and society. Structural adjustment to market realism was the key theme, but poverty and inequality continue to grow, despite poverty alleviation programs and despite upliftment of the poor. Population living on less than $1 per day in developing and transitional economies (1987-88) has declined, but in South Asia it has increased because of increased population. Poverty is still a priority agenda. One United Nations Report (UNCHS, 2001) says that globalization has generated a spectrum of changes, but there is social exclusion and there are economic and environmental problems.

The issue, or the conflict between conservation and modernism, can be seen throughout Asian cities. There is destruction of historic buildings for real estate reasons to build modern hotels, office towers, shopping centers, banks etc. The globalization with the arrival of international financial institutions and development organizations requires a new kind of space.

Many Asian nations took up industrialization and economic development and a consumer-oriented lifestyle sometimes alien to the cultural and ethnic characters, modern, post modern and international style of architecture and western style of town planning were adopted. Schools of architecture and planning followed the western syllabus and the brain drain continued. But the rural migrants who built their own homes, tried to keep the rural lifestyle.

Fukuyama mentions Asian values and their relation to economic rise and recent decline, and says that Asian social distinctiveness has eroded, though family structure has remained stable compared to the West (Fukuyama, 1998). Chinese leader Deng Xiao Peng said, "Color of the cat is not important so long it catches mice". Many Asian cities are following Euro-American thinking.

Most of the megacities have followed a similar pattern of development. Urban development has been demonstrated both vertically and horizontally. Horizontal sprawl has filled up water bodies, damaged agricultural cover and destroyed forests, while vertical growth at the heart of cities has created congestion, pollution, strain on services, depletion of underground water and loss of historic buildings and areas. In many cities economic boom has led to ecological doom and 30 to 40 per cent of people live in slums and shanties.

The idea of 'Manhattanisation' spread with replicas of American Architecture, office towers and high-rise apartments and suburban development of housing estates, new towns and large shopping centers. Statutory town planning regulations try to control land uses in many cities but cannot check the land value. Urban development is taken up, flyovers and highways, underground sewerage, drainage, transport terminal etc. Despite narrow and winding lanes and pockets of diversity, there is gradual disappearance of Asian character and flavor. Most of the projects became uncoordinated and remain unfinished. Despite tourism promotion of ethnic culture, globalization with the cola generation has brought a consumer-oriented cityscape alien to the culture (Ghosh, 2000).

3. Architectural and Planning Trends

Till recently the architectural profession concentrated on designing new buildings and new townships until conservation of some historic buildings and areas were taken up. Often the old historic area has been neglected. Redesigning an existing city especially a megacity is difficult; Kenzo Tange, the famous Japanese architect, proposed a high level megastructure for Tokyo in the 1960s. There were other visionary schemes, mostly architecture-urban design oriented and mostly academic.

The secession movement in Vienna in the late 19th century was highly influenced by Asian culture and on the other hand Asian modernization was largely influenced by western concepts. On the occasion of centenary "Cities on the move – the Asian cities in the nineties" – between "Apotheosis and apocalypse" – was an exhibition of contemporary art and architecture of Asia opened in Vienna in November, '97. The Vienna secession says that a typical city does not exist, and there is no collective Asian identity. On the contrary there are a wide number of highly diverse urban concepts, sometimes too artistic, visionary and individualistic. Many ideas have been mentioned and some are listed: Agglomeration city, Bubble city, Fuzzy city, Garden city, Global city, Hybrid city, River city, Symbolic city, Tropical city, Water city etc. (Hanru and Obrist, 97). In Asian cities, the old city, or even old historic city, was neglected till the conservation movement came, followed by the environment movement. Conflict with modernization was evident everywhere. Historic central area became the prime real estate. A report on Dhaka, Bangladesh, states (Ali, 1996) how the transformation has taken place.

Writing about Phnom Penh, Cambodia, Christiane Blancot mentions about destructions of buildings of the colonial period by speculators greedy for large sites to build tourist hotels. While Michael Pinches writes about Manila, Philippines (Pinches, 1994): "The encounter with the pursuit of modernity in Manila, evident in the variable character of the city's architecture and spatial form, attests to a phenomenon and ideology that are largely contradictory".

There is a break-up of joint family systems, individualistic space is likely to dominate with car, computer, cell phone etc. Some Japanese architects propose hyper building – very high building with small apartments and other facilities and high-tech vertical transportation systems – in fact a typical Asian city's land use on the two sides of a road is put vertically. There are opposite ideas with eco-villages where community living with shared facilities and public transportation will be with low-rise development with greenery, fruit trees etc. This may be low-rise high-density with passive use of solar energy, use of recycled building products etc. with the concept of 'recycle, reuse, restore or rehabilitate'. The architectural vocabulary will be changed.

4. Sustainability and Vernacularism

To the millions of urban poor sustainability is survival. There are definitions and examples of green building and sustainable architecture – focusing on low energy, eco-friendly building materials, recyclable waste etc. Urban systems are not in ecosystems and the metabolism of cities produce waste. The concept of environmentally sustainable development starts with recycling of waste of all kinds.

There are however, some planning issues. The current practice of land use regulation based on town planning practice is to be replaced by ecological planning. The education and training of architects and engineers require changes. The present practice of construction of high density high-rise buildings at the center of cities and low density suburban sprawl are both expensive and unsustainable, with more energy and transport costs. Alternate solutions are to be found.

There is growing awareness for regional vernacularism in Asian cities. Suha Ozkan of the Aga Khan Award for Architecture says (Ozkan, 1985) 'Neo-vernacularism... has emerged as an approach to bringing a new life to vernacular heritage for new and contemporary functions'. Regionalism can have global relevance and Louis Khan's Assembly (new Parliament) building, Dhaka was designed to develop an all-embracing culture-linking, indeed global architectural language, intended to reach far into the 21st century (Gast 2001). Laurie Baker, a British architect who came to India more than 50 years ago to work with lepers and poor has developed architecture with the local masons and local materials. His social commitment and creativity have been admired. (Baker, 1999).

There are many architects in Asia. The urge for traditional cultural values is now everywhere in Asia, together with an urge for balance between modernity and regionalism or contemporary vernacularism.

5. Conclusion

Asian cities were united by the common colonial background of the elite (Jorgensen, 2001). With the advent of information technology these will be united again. But eco-friendly sustainable development will be a new goal. The information technology revolution in architecture and cities has yet to be assessed fully. Asian countries missed the industrial revolution, but not the information technology revolution. When the physical barriers are removed from the minds and access to information becomes easier, culture will be more developed and expanded and quality of life more enhanced (Schmitt, 1999). The next century will bring new hope for urban Asia.

The Dinosaurs have gone but megacities will remain. They will live with dragon's fire – poverty, overcrowding, unemployment, etc. Architecture and city planning will change and Asian cities are waiting to see, not replicas of European-American Architecture, but a sustainable development. The making of a megapolis into ecopolis will be the great challenge to Asian architects and planners.

The Asian city with cultural pluralism, economic diversity and physical disparity requires a new vocabulary in architecture and how it will fit in with the environment. Dr. Ismail Serageldin (Serageldin, 1997) of the World Bank compared it with the puzzle Rubik's Cube, "It takes patience, dedication and imagination to reach it but solutions are possible".

References

Ali, Prof. Meer Mobasher, 1996. "Dhaka, Bangladesh" in Santosh Ghosh, edited, "Architectural and Urban Conseravation". Centre for Built Environment, Calcutta.

Baker, Laurie, 1999. A manual of costs cuts for strong acceptable housing. Centre of Science and Technology for Rural Development, Kerala, India.

Fukuyama, Francis, 1998. "Asian Values and the Asian crisis", in Commentary, February, 1998, pages 23-27.

Gast, Klaus-Peter, 2001. Louis Kahn. The Idea of Order. Birkhäuser, Basel.

Ghosh, Santosh, 2000. Redesigning the Megacity – Conservation and Environment. Paper at International Conference on Megacities 2000, Hong Kong, February 2000, organized by the University of Hong Kong.

Hamm, Bernd, 1998. "Globalization, Urban Development and Policies for the Urban Poor", in Richard Wolff et al (eds), Possible Urban Worlds, Inura, Birkhäuser-Verlag, Basel.

Hanru, Hon and Hans Olrich Obrist, 1997. "Cities on the move", in Secession, Vereinigung Bildender Künstler Wiener Secession. Vienna.

Jorgensen, Bent O., 2001. "South Asian – an axious journey towards regionalization", in Michael Schuz, ed., "Regionalization in a globalizing world." Zed Books, London.

Lim, William, 1999. "Vision Impossible and Social Justice", Selvip Newsletter, October 1999, Cibu City, Philippines.

Mkandnewire, Thandika and V. Rodriguez, 2000. Globalization and Social Development after Copenhagen, United Nations Research Institute for Social Development, Geneva.

Mumford, Lewis, 1961. The City in History, Hercourt, Brace & World Inc. New York p. 158.

Myrdal, Dr. Gunner, 1988. Asian Drama, An enquiry into the poverty of nations, The Twentieth Century Fund, New York.

Ozkan, Suha, 1985. "Introduction: Regionalism with Mechanism", in Robert Powell, ed.

"Regionalism in Architecture". The Aga Khan Award for Architecture, Geneva.

Pinches, Michael, 1994. "Modernization and the question for the modernity, architectural form, squatter settlements and new society in Manila", in Askew and Logan, ed., Cultural Identity and Urban change in Southeast Asia, Geelong, Victoria, Deakin University Press.

Schmitt, Gerhard, 1990. Information Architecture, Birkhäuser, Basel.

Serageldin, Dr. Ismail, 1997. Solving the Rubik Cube, Cultural Heritage in the cities of the developing World, in "Urban Age", Vol. IV, No. 4, 1997.

UN, United Nations, 1999, World Urbanization Prospects, UN, New York.

UNCHS, 2001, The State of the World's Cities, UNCHS, p. 120.

Regional versus Global – Singapore Experience

Ruzica Bozovic-Stamenovic Singapore

Introduction

The city-state of Singapore represents a rare and yet very well-known example of accelerated development and radical change. Over just four decades, this city has substantially and systematically changed it's appearance and it's character. This impressive and unique experience definitely deserves comprehensive analyses and deep appreciation. However, amazing results hide even more interesting layers, like the challenging dichotomy – regional versus global. Although extensively discussed in architectural theory in the last few decades, the profession still faces the wide gap in understanding this relation. Are both terms positive? Could both coexist or the two exclude each other? Are regional and global paradigms of the future or of the past? Singapore experience might be a fruitful ground for digging up relevant answers.

Regional and Global in XX Century Singapore Architecture – Criticism

At the beginning of the XX century, Singapore architecture was a complot of Chinese, Malay and eclectic style, mixed and influenced by each other. Adjustment to equatorial climate, shared craftsmanship and same concerns for the market and trade conditions generated similarities in architecture of that time. General awareness of wider interest led urban decisions and favored common social goals over purely nationalistic ones. Remaining early twentieth century works form the core of present-day Conservation Plans.

Glimpses of the International style appeared in the thirties, mainly applied to public buildings. Kallang airport terminal building[1] was in a way associating International style with the new means of transportation. This symbiosis symbolically related the idea of progress with architecture of a-national and a-regional characteristics. The universal philosophy of timeless and rootless architecture was promoted.

The second half of the XX century introduced vibrant social events and radical changes in City-State government. Architectural development was intensive, controlled and entirely based on the adopted principles of Modern. Thus, the paradigm of Modern was born in the first quarter of the century and reincarnated in the last. The "western-born" theory paradoxically and quite unpredictably became reality in Asia, in Singapore, applied in its purest form[2]. This miraculous appearance of a model already sent to history as nearly utopian, definitely marks the end of the XX century architecture.

For the gurus of Modern the city was the apotheosis of individual freedom within collective responsibility. Denying any obstacles for the implementation of CIAM ideas[3], they rather had in mind the existence of a powerful government with mechanisms ready to pursue the model to its execution. In that sense the Singapore model has the value of a precious and pure prototype on 1:1 scale. However, the similarity to the genuine idea should not hide very subtle but substantial dissimilarities. Care for genuine heritage and tropical idioms are just superficial expressions of these distortions. Deeper and more consequential differences lie in overall negotiations of the authority of Man as the main measure in architecture. Paradoxically, regional architectural discourses and sensibility in inserting architectural identity actually embed elements of global unification. The anatomy of this phenomenon will be discussed through analyzing some statements from the Athens Charter, and the way they were reinterpreted in Singapore.

Athens Charter – Seventy Years later[4]

In its appearance, Singapore looks more genuinely Modern than many other examples elsewhere in the world. Still, a closer look at some other postulates of Modern, in particular those based on the awareness of human needs, raise serious doubts.

· Zoning

Zoning and efficient transportation are, according to the Charter, of great importance in benefiting human life. In contemporary Singapore experience mix function apartment blocks, being a condensed summary of the city itself[5] and new wireless technologies make the zoning irrelevant. The ultimate benefit of the new mix-function schemes is less time spent on transportation[6]. However, In this case "saved time" seems not to be spent as originally suggested by the Charter[7], but rather on more work. The consequence is also certain segregation of the urban professional community and its isolation from other social segments of urban life. The impersonal model becomes the archetype of a spatial and social cell for the future global community, leaving no space for regional expressions in any way.

· Historic Heritage

The Charter states that heritage should be respected on the condition that it is not obstructing the development of the city and its global interests. Application of any eclectic styles is also condemned as anachronous and false. Singapore development policy was also based on the priority of overall benefit over any historic legacy. Only after decades of radical destructions of old city blocks was Heritage Board formed, and authentic remains recognized and preserved. Although on the tracks of the best professional tradition regarding the heritage, we might argue that after establishing a global picture, this comes far too late. Although being professional and sincere, the relation to heritage in Singapore appears to be above all a tribute to the tourist industry. Some protected buildings are dedicated to new commercial functions and reconstructed with details added to emphasize the "old" or what nowadays consumers might regard as old. As a result of this misunderstanding even some new buildings appear with evident but false reminiscence on "heritage"[8]. This seems the opposite of what the Charter stated, but not entirely. The relation to heritage in Singapore experience is always in favor of the market benefiting from the overall development.

· Interests and Planning

The Charter stated that political, social and economic factors will determine the development of the city[9], that private interests will be subordinated to social ones[10], but also that architects will have human needs and Man as the ultimate measure for all decisions[11]. In Singapore, interest is formulated as a search for constant development, reflected in economic growth. Man's well-being is still the ultimate goal, but seen through the filters of numbers, interest, money. Architects have to prioritize and negotiate between wishing to be regionally conscious, and having to be globally effective. The Singapore silhouette reflects the pick of success in global terms.

Bias Dilemma: Regional versus Global

Singapore's legacy for the future is the capacity of this city to question, reposition and change the very basic and fundamental statements about city planning and development. The result should not be judged in terms of good or bad. The experience is unique and valuable as it is. Singapore is an interesting laboratory for investigating the possible destiny of cities and their inhabitants in the global world of the future.

· Particular over Global and Regional

The particularity of this city should be acknowledged and developed further without relaying to inapplicable precedent experiences in any way. Prominent local architects already successfully establish a specific symbiosis of global aesthetic quality and regional requirements.

· Regional within Global

Regional expressions are usually linked to private[12] rather than public spheres. Public functions like hotels, hospitals, students' halls, etc. are also very much related to personal, and thus suitable for "regional touch". Ambience and aesthetic qualities stimulate appreciation of space. Harmony is determined by coherence of spatial elements on the level of intensity, diversity, rhythm of visual experience, etc. These terms suggest a lot of opportunities for embedding the contextual and regional while still remaining within the global hub. Tourist industry already recognizes this fact and profits on it. In other cases, this potential is unexplored[13].

· City as a Market Place

Singapore as a city is a market place, where both spaces and rules that guide their development are a market good. This is to be accepted as such, and the insertion of regional must be explored further, keeping in mind it must have its price too[14].

· Created Needs

Manpower in Singapore is the county's main resource. Education and entrepreneurship are highly valued, and economic prosperity is the ultimate social and personal aspiration. However, the question of needs in Singapore is interesting, as those are not spontaneous, but rather created over the decades of rational and directed governmental rule. The speed of city's development gives the rhythm to all other needs, most of them related to work, consumption, compressed time and constant need for information. The need for regional identification, its level and market value are to be created and established first. Otherwise, new generations will not identify it as valid or important.

· Adapted Citizens

Social problems that existed decades ago as the consequence of Singapore's accelerated development diminished. Generations changed and the current majority of Singaporeans are adapted citizens. They successfully follow the model of preferable city life imposed on them, rather than creating one. This attitude is a logical consequence of living in developed global society whose rules do not allow for unpredictable and slow evolutionary changes. If the traditional city was an attempt to respond to human needs, the global city is creating them in response to the requirements of growth. This major shift makes the dilemma regional versus global even more irrelevant. Regional is acceptable if it has a proper place in the global picture of unrestricted development.

· Mimicry

Although global interests by definition overlap the genuine regional aspects of space, this is not the case in Singapore. Both terms stand for a tool of development, being in that sense of same importance, similar value and applied with same devotion. Regional and global in Singapore are mimicry of what they usually stand for, as they are actually not opposing or excluding each other. The fact is that Singapore is a valuable collection of architectural artifacts designed by world-famous contemporary architects. Local architects also contributed many distinctive buildings of high value. Some of these masterpieces have regional characteristics, but many express their author's style rather than the context. In both cases this difference does not mean taking sides in the dilemma regional versus global. Diversity is welcome as it contributes to the unique global value of this architectural collection.

Conclusion – The Future?

· Managed City instead of Autonomous City

Regional versus global is not just a problem of space or form. It might not be a problem of the future at all. In global cities overwhelmed with congested time, speed of communication, changed social patterns, and ultimately the reduction of individuality, the aspect of regional might become just an optional preference. The city has been managed through history. This century promotes the globally managed city that is managing the lives of its citizens. Paraphrasing the famous "form follows function" we might say, "form follows the purpose". The regional approach is just one aspect of doing it. In the typical developed city of this century regional differences will appear only on the tracks of global needs. The Singapore experience is a testimony to the fact that, before being solved, the problem of regional versus global actually diminished. It was absorbed within the process of managing a traditional city pattern with respect to interests of the global future.

However, 'regional versus global' remains ultimately the question of individual freedom, and choice to think, to act and progress, cherishing the regional heritage within the hub of global development. Making it possible is the real challenge for the profession.

---

1 Completed 1937. Architects: Public Works Department (old), Architects 61 (refurbished).

2 That was not the case in many developed countries where the principles of Modern were changed, criticized, transformed to new "isms", etc.

3 Chandigar is the proof for that.

4 "La Charte d'Athenes" – Athens Charter is the beacon of Modern Urbanism and a key document containing major statements of this architectural movement that shaped the development of the cities in the last century. It is a collection of documents from the CIAM Congress in Athens held in 1933. The charter was first published in 1942 in France, but the turbulent wartime was not an advantage for the popularization of its doctrine. The second edition was published sixteen years later with the introduction of one of the giants of the Modern movement, Le Corbusier.

5 With apartments, offices, recreation spaces, green areas, garage blocks, etc.

6 This model was actually promoted decades ago with Le Corbusier's famous Marseilles Block.

7 For leisure or cultural needs.

8 A lot of Cornish granite and Greek columns being applied.

9 Statement number 92.

10 Final statement, number 95.

11 Statement number 93.

12 Many Singaporean homes still express strong national distinction.

13 Healthcare spaces in particular provide a lot of opportunities for implementing hints of the regional, in order to encourage healing spaces.

14 For example, Changi airport welcomes all the visitors to the city, but it is also a popular place for Singapore's youth to "hang out". Both provide opportunities to promote cultural and traditional values, and to be profitable at the same time, once that Changi is not perceived just as another airport.

The New Economy and the City Image

Anne Stenros Helsinki

The Environments of the New Economy

We live in a rapidly changing technology environment – amidst the third industrial revolution whose speed and ultimate configuration are hard to predict. In this process of ever-faster change, the built environment has had to adapt to ongoing growth and change that are quick and often uncontrolled. The old urban fabric is being renewed at a growing pace. Also the symbols of cities are being reformed and revised. The motors of the economy – the biocities and the technology parks and centres – have taken the place of town halls and cultural centres in building and development in recent years. But above all the new economy means global competition, involving regions and areas in relation to desired companies, firms, labour and residents. This is all reflected in the competitiveness of areas. Clusters and concentrations of skill and achievement will emerge.

According to the report issued by UNICE – the Union of Industrial and Employers' Confederations of Europe – the competitive ability of the new economy particularly depends on four factors. 1) A strong entrepreneurial spirit, which is the key to growth; this entails, among other factors, how private enterprise and its opportunities are supported socially and regionally. 2) The second factor is a competitive environment, i.e. with labour market and the market for goods and services sufficiently flexible for change. 3) The third factor is a world-standard infrastructure of skills, meaning, among other things, that researchers have strong incentives to seek new knowledge and to give it commercial applications. 4) The fourth factor consists of societies able to support change in better ways, i.e. the ability of individuals to adapt to technological development, which means that states and governments should encourage private individuals to try out new technologies, products and services.

The desired state of society is also reflected in planning, decisions and realization concerning the physical, built environment. In view of this environment, the competitive factors of the new economy could also mean the following. Entrepreneurship can also be reinforced by giving an area an image as the concentration of a certain sector of industry. The result will be information concentrations of specific fields, which will attract new firms and companies into the area.

A competitive environment is also a "home base" for companies, in broader terms the basis of good operations as well as quality of life. Competitive companies must also be able to offer their employees a high-class living environment and services. A skills infrastructure of world class requires, among other things, a functioning relationship among universities, institutes of learning and local firms and companies; in the future interaction between research and practical operations will gain even greater emphasis. A society supporting that ability of the individual to change in turn requires the individual to have an opportunity to have local influence on the environment and its development.

How can the physical environment respond to these challenges of the new economy? Production is no longer dependent on local environmental conditions, such as raw materials, water or physical communications; the localization of firms and companies is now dictated by immaterial local properties such as general operating circumstance, the level of innovation, the image of the area etc. The attractiveness of an area in the eyes of industry and local residents is an increasingly abstract concept that can be explicitly controlled to a certain extent.

At the level of the individual, change in society means wider and better availability of information, a greater range of available goods and services, and lower prices. It also means work of a more independent nature, but also greater responsibility and authority to make decisions. Moreover, the boundaries between work and leisure, home and the workplace, and public and private will be blurred, leading to changes in lifestyles. On the one hand, people will become increasingly dependent on their physical dwelling environment if they are engaged in tele-work. But on the other hand, it will be easier than before to choose jobs, and people will change homes and places of employment more often. In the future the attractiveness of residential areas and public spaces will become a regional competitive factor of growing importance. Local advantages and strengths must be the starting points for competitiveness. Decision-makers must know the history, narratives and stories of an area, and they must reinforce their visibility. It is also necessary to attend to and take care of diverse, modern public services, which also means the public spaces and areas of cities. Residential areas must quickly respond to different, changing lifestyles and ensure both privacy and communality to a sufficient degree. The individual's opportunities to develop himself and to influence his environment are to be given support via the environment: there must be sufficient and diverse cultural and pastime facilities, and there has to be easy access to sports and exercise facilities. There must also be support for the participation and influence of residents in local decision making.

The image of the competitive environment of the new economy is strong and distinctive; it is also a positive one in relation to the individual, entrepreneurs and the business community. All these aspects can be influenced through the physical environment and the regional image based on it. Previously, there was talk of the "image of a city", the physical townscape in the minds of its inhabitants. But today, experts speak of the concept of the "city image", containing not only the physical environment but also the properties and aspects of function, image and association. The city image steers regional desirability and thereby its welfare and further development.

The Environmental Responsibility of the City

There is growing current discussion and debate on the social and ethic values of companies, and especially their responsibility towards society in terms of economy, environment and people. The corporate citizenship entails economic, environmental and social responsibility. Traditionally, environmental responsibility has mainly encompassed the dimension of sustainable development. In the future the importance of responsibility for the environment will, however, be seen in broader and more diverse terms than previously. Alongside sustainable development and environmental conservation values of equal importance are welfare of people and their well being in relation to the environment, and the development of the regional image in a positive and affirmative vein. The "friendly community" and the "walkable community" are keywords in developing the environment. One of the main environmental criteria of the new economy after the "event cities" concept will be the "ecocity" properties of the area. Security, cleanliness, scale, sustainable development and even an organic society are objectives with which the environment is evaluated and marketed.

In discussing the values of a city it is necessary to take into account an interpretation of environmental responsibility that is broader than previous ones. By making the whole built and natural environment a qualitative competitive factor and an important functional and societal value, it is possible to creative in a genuine sense a city image that is attractive and easy to market even at the international level. In speaking of the environmental responsibility of the city, its main criteria are the quantity and quality of public space, the comforts of residential areas and the standard of their physical communications, the range of business and services, the state of the natural environment and access to recreational areas, and the spirit of the area and region itself – the genius loci.

There is never a good image without due grounds for it. Even good marketing will not save a poor product, while a good product will often sell even without major marketing efforts. The content of the city image is largely composed on concrete factors: the purity of the environment, security, comforts, the range and level of services, special local features and history. On the other hand, the image also entails factors that cannot be predicted, such as personal feelings, desires and events, and values.

In practice, environmental responsibility is also realized through the genuine interaction of residents with their environment. Active neighbourhood committees and civic organizations can play a key role in creating a network of opportunities to influence things at the grass-roots level. Care for their own immediate surroundings will permit residents to make a commitment to the environment and also to take responsibility for it.

The Urban Genius Loci

Throughout the centuries our image of the city has been marked by the history of cities: important historical figures, locations and related events of the past. These aspects are underscored with monuments, squares, parks and buildings. The tales and narratives of history have created the content and richness of the city, and the attendant framework has been in keeping with these tales. The stratification of the urban fabric has also spelt continuity from one generation to another.

There can be no real city without history. It is from history that the stories and tales of the city arise, the content that is read and interpreted by its inhabitants. History creates stratified narratives in which urban identity is created to find concrete expression in the physical, built environment. Narratives evoke emotions and experiences together with the environment. Nor can there be a city without a centre. It is from its centre that a city begins. The centre is its deepest and most significant narrative, and often its most stratified part in terms of history. The centre is the heart of the city and the basis of its identity. Despite regional scope or diffusion the city image is largely articulated through the nature of the centre.

In "The Seduction of Place" Joseph Rykwert reminds his readers that the city is steered by other forces than rational-economic ones alone. Conceptions and ideas, emotions and desires have steered and will steer the objectives of those who build cities. Thus, the city is not developed according to quasi-laws of nature as taught by economists, but, as Rykwert underlines it is always an artefact of expressing will and a human construction, in which conscious and unconscious factors play a role. Rykwert also points out that the modern city is contradictory, containing many cultures and its image can thus not be explicit and coherent. Its diversity should rather be seen as a positive factor and an asset. The "urban image" of the contemporary city is inevitably multicultural and multi-faceted, and continuously changing. The changing of images can, however, be guided and controlled through the urban fabric.

At the end of his book Rykwert summarizes that instead of the process of making space that has continued for long we should focuses on creating places in cities. Places are the fixed points of inhabitants and the true materials of the city image. They are the nodes of history and are full of narratives and tales.

The genius loci cannot be constructed but it can be supported. By underlining the stratified nature of history and by increasing the natural interaction of residents with their environment conditions can be created for the existence of a genius loci. The home, locality and the importance of place have not disappeared in our global world. On the contrary, they often provide identity and strength in a changing world.

Urban Spaces in Latin America – Recent Projects

Silvia Arango Bogotá

In many Latin American cities, during the last two decades, an interesting and meaningful phenomenon is taking place: the design of public spaces that affects important portions of the existing city. These urban designs, of different scale and character, have some things in common: they imply great inversions, have been pushed forward by local authorities and seek for a big social impact.

Why is that social space within the city, long abandoned, has become a main political concern ?

There are several reasons that could explain such a change. On the one hand, the increasing globalization of the economy has impelled the cities to compete in order to attract tourists and highly qualified investors who demand not only efficient services but also a nice and enhancing environment. On the other hand, there are reasons that come from the historic characteristics of the urban development of the Latin American cities; in the second half of the 20th century they suffered a big decay in quality of life terms, due to the vertiginous urban population growth, but also because the modern urban models applied, priviledging automobile circulation. Social pressure towards the improvement of open spaces to develop an intense social interaction, which is culturally meaningful in Latin America, could be an additional factor to explain the interest that the local authorities manifest in public urban space.

In this brief presentation, I won't refer to those projects that are relatively scenographic and superficial, but concentrate on some examples of recent interventions on urban spaces willing to recover ancestral memories and uses, and thus, more capable to interprete the cultural identity of the Latin American city nowadays.

Punctual Urban Spaces

Plazas: The re-design of highly symbolic plazas, like the Plaza de Armas of Santiago (Chile) or the Plaza Mayor of Quito (Ecuador), seeks to cover both civic life on the scale of the metropolis (like processions or political meetings) and daily common activities.

Neighbourghood Parks: In several cities, spaces have been designed on a domestic scale to rescue the neighbourhood's life and link up their communities. Such spaces, commonly built nearby schools, markets or playing courts, can be the trigger for a sense of belonging, specially when made in precarious contexts or deprived areas, like the favelas of Rio de Janeiro.

Promenades: From the colonial times, the use of the afternoon "walk around" still has a significant place in social life, particularly in cities with a tropical climate. The Princess Paseo in San Juan (Puerto Rico), the malecón (sea-side walk) of Guayaquil (Ecuador) or the Chabuca Granda Paseo in Lima (Peru) are successful recent examples of the recovering of this tradition.

Urban Interventions at a Metropolitan Scale

Transport systems: The use of bycicles as an alternative transport system, has caught the attention of people in big cities. In this sense, the experience of the 7 million people city of Bogotá (Colombia), that is building an extensive network of cycle paths, is an interesting experiment to follow.

Hydric Systems: One of the biggest problems faced by Latin American cities is the different types of pollution. I would like to present just two examples of the efforts that are being made concerning de-contamination of hydric systems. Both in Bogotá and Mexico, the recuperation of the regional hydric system includes ecological long term considerations as well as a geographic memory of the original landscape and of a peculiar way of relating with nature.

Before the Spanish conquest, the high land savanna of Bogotá was a lake that urbanization slowly drained; there are still some marshes, streams and brooks, with its own flora and fauna, as remains of that past. This surviving water presence within the actual urban tissue began to be recuperated, as well as the main, greater river that borders the city. Similarly, Tenochtitlan, the Aztec capital was built in the middle of a lake that disappeared under the continuous growth of modern Mexico city. From the lakes of the valley of Mexico, only Xochimilco remains. An ambitious project is regaining this in a landscaping and architectonic sense and also in social development. The enormous park includes the re-construction of callampas – floating cultivated islands of indigenous origin – to provide means of life to the surrounding population.

If during the 20th century Latin America suffered a hasty and uncontrolled process of urbanization that only allowed, and with great difficulties, to face quantitative urgencies, now, at the begining of the 21st century there are hints of a hope that we will be able to face qualitative challenges. There are reasons to think that maybe we won't be richer, but it might be that the 340 millions of people that inhabit our cities will have a better urban life in the days to come.

weijen wang
hong kong
learning from antipodes
vienna
peter lorenz
www.peterlorenz.at
UIA Berlin 2002
XXI World Congress of Architecture
July 25th 2002
© Luftbildstelle MA 18 Wien

City seeks Dialogue – From the Stadtforum (City Forum) to Cooperative Development Planning

Regula Lüscher Gmür Zurich

In the mid-90s, on the initiative of the Mayor Josef Estermann, preparations began for the Zurich Stadtforum. At this point in time, the opinions of administration and real estate owners diverged with regard to the implementation of building proposals. The legal route proved to be less than promising. The "contrast between the conflicting interests" had "become caught up in issues of legal proceedings to such an extent" that the "material problems faded into the background to the point of losing their focus altogether," explained the preparatory paper on the Stadtforum from 1996/97. The Stadtforum formed the prelude to a new "Culture of Communication" between the City Administration, the business community and the public in planning proposals. Using "open and public discussion", an attempt was made to achieve a "better understanding between those individuals involved or affected from all sections of the community". Added to this, interested representatives from the business community and the general public, from the authorities, councils, the field of science and private organisations were invited to discussions.

The result of the Stadtforum was the basic conclusion that the image of Zurich West would have to be enhanced from the point of view of town planning. The aim of this enhancement is to "bring benefits to all in general". The participants in the forum established the following guiding principles for raising the image of the city through urban planning:

- Zurich is to be made more competitive and attractive in the European and global contest between cities.
- The marketing of locations is to be improved for real estate owners.
- A climate should exist for investors that encourages them to realise projects.
- For the general population, the environment, and thus the quality of their homes and lives should improve.
- Attractive districts should contribute to retaining or winning back higher earners to the city.

Cooperative Planning Zurich West

In 1998, Zurich West took up cooperative planning. Under the overall control of the Office of Urban Development, representatives of the city, the canton, the districts and the real estate owners worked out a development concept. A fundamental part of the planning process was that people got to know one another and built up trust. Using concrete plans, city proposals could be discussed and common goals and wishes could be developed. OMA (Rotterdam), Ernst Basler+Partner, Max Dudler and Morger & Degelo submitted various approaches, which took into account links with the city as a whole, the expansion and the meaning of increased infrastructure and the planning itself. On the basis of these contributions, the city and real estate owners were able to put their own ideas in concrete terms in the following phases. A decision was taken not to follow any of the approaches, but to work out an individual development concept.

In February 1999, landowners and the city unanimously adopted a letter of intent. In this, among others, criteria for sustainable development and the development concept were defined. The sustainability should take into account three areas:

- Economic sustainability (investment should have a long-term effect, guarantee yields compatible with market conditions, improve the financial support of infrastructure through taxes and contribute to international competitivity)
- Social sustainability (the development of Zurich West is to be transparent for those involved and include the affected persons, create high quality with regard to living space, health and safety and promote the integration of different sections of the population)
- Environmental sustainability (statutory levels for emissions and the effects of pollution and noise etc. should be bettered if possible, non-renewable resources should be used sparingly, the land protected in quantitative and qualitative terms and the rich flora and fauna encouraged).

The letter of intent saw sustainable effects as a chance, and emphasised that the effects of the development on the economy, society and the natural environment must be comprehensively taken into account.

General aims of the development concept are

- To turn Zurich West into an attractive area of the city
- To target the planning to stages and flexibility
- To emphasise the cocktail of benefits (living, culture, industry, services) and setting trends with regard to town planning for both buildings and open spaces.

Quantitative standards of values for ground usage establish the requirements for exploitation, living areas and free spaces.

Under the 'quality demands', among others, the emphasis will be placed on preserving the specific character of the area: the high density of buildings, the need for a significant residential area, the establishment of culture, services, research, trade and industry, the creation of new green areas and a marked expansion of public transport.

The process of bringing people together and the creation of trust between the city and real estate owners was the basis for a successful urban development strategy. After this first step, the administration, together with Pierre Feddersen, used a kind of workshop concept to work out where open spaces, such as squares, paths and green areas, should be sited. The district can be reached on foot and can be experienced, and that is the key requirement for linking and networking the private buildings. A comprehensive plan is needed for this network of streets and paths.

Added to this, the industrial areas were looked at and carefully analysed. The aim was to identify the spatial qualities such as large halls, which are separated by small lanes. The right-angled arrangement is also typical, but it is broken up by railway lines, resulting in diagonally cut-off areas. A further example is the second urban level, i.e. the bridges from which you can get a view into the area from above.

On the one hand, the area next to the river Limmat is good for relaxation, whilst the tracks give the area a pleasant feeling of space. At the same time, important transport arteries and radial axes cut through the area.

The planning in this area of the city is currently at completely different stages. For example, in the Sulzer-Escher Wyss area, planning had begun, but at the same time the company was still in production. Parallel to the planning, existing buildings have been put to a different use or refurbished. So, less developed areas are only to be found under the railway viaduct.

To plan a city area that is already built up, and which will change in a process over many years through individual demolitions and new buildings, is a contradiction in itself. A development process which is characterised by the juxtaposition of forward thinking, planning, projects, building and breaking up requires different ideas of a city than the re-building of a district from soot and ashes, a tabula rasa process.

The position of squares and parks is secured in that we are setting them down in a model. The design quality of these open spaces guarantees competitive procedures. Tomorrow there will be good architecture, in that established architects will build together with construction professionals. That is our wish!

The new should create a link to what exists, in order to give what's new a sense of belonging, uniqueness and distinctiveness, in order to give the oft-quoted authenticity. For, at the beginning of the development of Zurich West, there were niche uses, alternative culture which could occupy spaces thanks to the low rents. These people made Zurich West well known, the cultural scene began to establish itself. And to a large extent that was thanks to the atmosphere in the old factory halls.

Capitalist Realism

Bart Goldhoorn Moscow

My contribution on the theme of urban design and democracy may be a bit untypical – I will not speak about their presence in Russia, but about their absence – urban design and democracy both don't exist in Russian urban planning. Having said this, you might want to have an explanation, and you are quite right to demand one. Although few people abroad are aware of it, democracy in Russia, the emergence of an embryonic 'civil society' was closely connected with the emergence of environmental concerns among the population, that is, concerns about the quality of their surroundings – and that is the subject of urbanism.

Perestroika

Russia has a specific position among the former Communist countries in Eastern Europe. In contrast with these countries, it had no occupying force that could be made responsible for all wrongdoings of the Communist era. It is important to realize this when we look at the processes taking place during the perestroika era. In countries other the Russia, the aim of 'dissident' movements was clear – to get rid of the occupant and its marionettes. Consequently, the disappearance of Communism and the Russians could lead to a complete change of the structure of power.

In Russia the position of dissidents was not so clear, although in the Western press they were often presented as people who were opposed by the regime. Cynically enough this was also the text used by the Communist regime as an argument to put these people behind bars, because of anti-Soviet activities. Both of them thereby ignored the reason why these people were accused in the first place and became dissidents – the cause for which he of she was fighting, that was not, as in Eastern Europe, to fight a occupying force, but to protest against specific problems in society.

Questions related to urbanism are connected to the daily lives of people, and less to big political structures. Moreover, they are local – even more reason for the foreign press to ignore them. In the perestroika era the protests focused on two aspects: ecology and the destruction of the historical city – both conservationalist movements, meaning to preserve natural and cultural heritage. Each Russian city had its own case where protesters with banners formed picketing lines or chained themselves to the objects of protection – trees or buildings – being threatened by the construction of yet another road or concrete apartment building.

Before the changes in Eastern Europe, most democratic revolutions in which totalitarian regimes were replaced by democracies had been coupled with reforms of the economy according to social-democratic principles: a more even distribution of wealth, introduction of norms and standards to ensure the quality of life for the whole population and not only for a few, social security, housing for the poor, etc., in short – emancipation of the repressed. Infrastructure was built up on the basis of the forces that had led to the dismissal of the regime, thus forming the basis for a 'civil society'. In the capitalist revolution in Russia the situation was different. The socialist state actually already had an infrastructure that was meant to guarantee minimal living standards: norms and regulations, labor unions, agencies for ecology and conservation. In the perestroika era, these agencies were not replaced but on the contrary, enforced. They gained power as a result of the popular protest that suddenly became possible and was stimulated by Gorbachov's policies. Apart from that, new institutions were formed to canalize the protests and regulate the participation of the people in the planning process. In Moscow for instance a 'social council' was set up, consisting of representatives of the 'intelligentsia' – writers, scientists, filmmakers, artists, etc. This council has to approve every new project, next to the existing 'urban council', consisting of architects and urban planners.

All in all the introduction of democracy led to an increase in norms and regulations – Soviet standards and norms were superimposed with the new demands of preservation. This led to a paralysation of the planning process. In order to understand the scale of the problems one should understand the nature of the Soviet norms. They were not designed as minimal requirements, but had the character of prescriptions – they were meant to describe an ideal situation. A good example is the use of the standard floor plans for housing that were developed to be applied all over the Soviet Union. Using the logic of industrial building production, one would assume that particular floor plans were linked to specific construction methods. However, in reality identical floor plans were used for housing regardless of the construction type. Instead of defining the borders of an infinite number of possibilities the norms describes a limited number of solutions that exclude others.

The Capitalist Revolution

In the beginning of the 1990s, democratic reforms were superimposed with the capitalist revolution. For the planning process this meant that private capital was introduced as a means of achieving urbanist goals. At the same time public finances were in a dire state. The salaries for people working in the public sector (that is practically everybody) were almost completely wiped out. This was also the case for the above mentioned agencies and regulating bodies and councils. It is not difficult to guess what happens in such a situation, especially when you are surrounded with the preachers of capitalist ideology. In order to survive, the only possibility is to 'privatize' one's activities – to ask payment for services. And since we speak about monopolies – without their signature and stamp you can't get a building permit – success is guaranteed, especially when a lot of money can be made in the construction business. It is too easy to call this corruption. If according to regulations you have to get an historical research on the site you want to build, and this research takes one month of work by a well-educated historian, is 2000 dollar too much to ask for? (By the way, why should a Russian specialist have to earn less then a foreign one – let's make it 4000 dollar.) And if you want to speed up the process of approval, you can hire your fireman as a consultant – he will advise you how to optimise your project and see to it – against a high fee of course – that your project won't meet any resistance.

This system functions on the basis of the fact that it has virtually become impossible to realize a project in Moscow without breaking the law. This doesn't mean that nothing is being built – in fact Moscow is experiencing a construction boom. That's why nobody is really interested in changing this situation, for instance by replacing the jungle of regulations, norms, committees and councils by a transparent system of norms and zoning laws. And it will deprive them of their source of income, the investors. But in the system there is always a way to get around the rules, the municipality, because it is an investor itself.

Post-modern Architecture

So what does this new architecture of democracy and free market look like? In order to answer this question we have to revert to the 1980's – to the emergence of post-modernism in Russia. At that time architecture had its own 'dissident' movement – the so-called paper architecture. This movement became famous as a result of a series of competitions organized by The Japan Architect, that were almost all completely unexpectedly won by young Russian architects surprised the world by their beautifully drawn conceptual projects. The projects all expressed a nostalgia for the historic city with its authenticity and ruin-like qualities opposed to the brutal rationality of the Soviet city. It was the most artistic translation of the protests against the technocratic character of Soviet planning with its demolition of the historic urban fabric. When the regulating bodies were enforced in the perestroika era this nostalgia became a main factor in the regulation of city planning. This does not mean that the 'paper architects' became the ones to build the city. The big Soviet design institutes were very fast in adapting to the new situation and catered to the need for historical styles. Apart from this, the newly created councils saw no problem in correcting projects from other architects that didn't answer to this. Investors have even more reason to pressure the architects to incorporate these preferences in their projects: not only can they get their projects easier approved, they also cater to the love of the new rich for 'representative' architecture.

So one could call the new Moscow architecture post-modern. In an international perspective this may look out-dated, but it is explainable if we look at the Soviet city of the last 30 years, consisting of a compilation of concrete boxes. The fact that Russian architecture doesn't answer to the international architectural fashion is not the biggest problem. Occasionally it produces quite interesting buildings, especially in areas that are less controlled by the authorities. A bigger problem is that the lessons of post-modernism have not been incorporated on the level of urban design.

Here I want to step out from the Russian situation and take a look at post-modernism in Europe. Some people might call me old-fashioned – post-modernism is dead, and now we have neo-modernism. I would call them short-sighted. Let me explain.

In fact the emergence of post-modernism in the West in the 1970s had the same origins as that in Russia in the 1980s: the change of a supply-oriented economy into a demand-oriented economy and the breaking up of traditional power structures. It would go too far to give a full overview of the links between these processes and the development of post-modernism but I want to mention two aspects here: the emergence of the historic city center as a center for communication and trade, and the emergence of a market for architecture.

In my opinion the change in the attitude towards the historic city in the 1970s is one of the biggest cultural shifts in the 20th century. If at the beginning of the 1970s the centers of European cities were empty and historic buildings falling apart from neglect, now they are the most beloved places to live in with the highest real estate value. It would be ridiculous to say that the popularity of historic architecture has disappeared with the emergence of neo-modernism. What has happened is that post-modernist architecture has disappeared from the architectural scene, that is increasingly getting the character of a market-driven fashion industry. Nevertheless, almost half of the furniture fairs in Milan and Cologne are occupied by producers making 'classical' furniture. The consumer hasn't forgotten, but the consumer of architecture depends in densely populated Europe strongly on supply that is largely controlled by the architects who have successfully linked up with social democratic governments.

(By the way, one can only be happy with this, but please don't call this democratic or a free-market.)

If we look at urbanism, we see that in contrast with architecture, post-modernism generated a new form of urban design, where the various interests of market and society are regulated before actual investors with their architects get involved. Nobody in Europe is proposing a tabula rasa urban planning, where huge parts of the city are planned and built at one time. Urban plans are based on traditional urban typologies. Architecture has become the generator of pluriformity, the individual handwriting of the architect within the strict borders of the urban design creates the simulation of the variety of the historic city.

Conservative Urban Design, Progressive Architecture

Capitalist Realism

If we look at Russia from this perspective, we see the opposite: post-modernist historicist architecture, but the absence of post-modernism on the urban planning level. Style has the meaning of an excuse for urban schemes that are based on the maximization of profits of all parties involved. This phenomenon can be called Capitalist Realism – the counterpart of Stalin's Socialist Realism. Realism has two meanings here: a) Realism opposed to abstraction, that is – post-modernism vs modernism. b) Realism as opposed to idealism – Capitalist realism reflects the reality of the market, where the population demands historical styles, investors maximal profits, and civil servants a normal payment for their services.

To be called a democrat has become an insult in Russia. To be called a capitalist is better – it's better to be honest in wanting to make money for yourself than to pretend you're working for the common good while in fact pursuing the same goal.

Federal President's office, Berlin Architects: Gruber + Kleine-Kraneburg, Architekten

The Community Development Corporation as a Nucleus for a Civil Society

Bernd Zimmermann New York

New York City's Dual Structure of Participation

Citizen participation in NYC's planning process has a history of over 30 years. Today it is a formal part of its Charter. Community affairs of 59 community districts, generally drawn around areas of between 100- to 200,000 residents, are each governed by a community board. These boards may have up to 50 members appointed by Borough Presidents, including nominations by local councilpersons. As honorary members, who ideally represent a cross-section of community interests, they vote in an advisory capacity on land use matters and monitor the delivery of city services.

Board offices, "little city halls", have small paid staffs headed by District Managers. For area residents they are supposedly the first line of contact with government.

Despite their advisory nature, board influence on land use matters, which are subject to discretionary approvals, has gained substantially over time. The so called Uniform Land Use Review Procedure (ULURP) governs this process. It involves first them, then the Borough Presidents, followed by the Planning Commission and, in certain instances, the City Council.

Community boards are not the only way citizens participate in the city's planning process. While boards may work well in wealthy neighborhoods where professional expertise may rest in their membership, this is typically not the case in poorer communities. There, generally speaking, self-help groups, organized under the non-for-profit corporations law, have sprung up in great numbers, typically filling in where government left off. In the past they sought to by-pass community boards, which they deemed as too political, contrasting with their orientation on delivering tangible results concerning neighborhood improvement.

Today, after remarkable success in rebuilding devastated communities and the advent of a new generation of self-help groups, the relationship between both is improving as the nature of the issues is changing.

Planning Democracy by the Grass Roots

The surge of organized community self-help in inner-city communities, such as in The Bronx during the 1970s and 80s, was a response to a traumatic housing and neighborhood crisis which resulted from government's failure to effectively address profound changes in America's urban landscape. There, technological innovations, a massive corporate and middle class exodus, socio-economic and racial segregation and other factors created an urban nightmare so extreme, that the role of the city in the national economy was in serious jeopardy. Official planning capacity, limited by its architectural and engineering orientation, gave up in view of the complexity of the issues. Unable to comprehend the causes and effects of sprawl and the interplay of the collapse of the human and social infrastructure with that of the physical environment, planning rather retracted to attend to Manhattan's development agenda. By concentrating on corporate retention, governmental planning made itself an instrument of a fierce, sprawl induced, economic warfare which had erupted in the region.

This housing and neighborhood crisis combined with the breakdown of the civil society gave rise to a spontaneous and innovative planning culture of the grass roots as government abandoned neighborhoods wholesale. Local leadership sprang up against a back-drop of despair, determined to find solutions based on people and places directly affected.

It was a layman's movement in need of help which came from American philanthropy. Intermediary organizations channeled the emerging local initiatives into structures by developing their professional capacity in organizing, negotiation, development and business skills. They also provided early grant financing for housing development, helping groups to establish a sustained record of accomplishments. Thus the grass roots transited from loosely organized groups to well-run not-for-profit community development corporations (CDCs).

CDCs became synonymous with community empowerment. They represented a new "industry", which made community development a private sector enterprise.

In contrast to the focus on the "physical city" of governmental agencies, the CDCs re-invented community development as a holistic undertaking, whereby both the rebuilding of the physical city as well as the human and social infrastructure go together. Guided by principles such as participation, pragmatism, flexibility, human scale, and tangible results, the CDCs offered something which governmental bureaucracies with their rigid programs could not: responsiveness to local conditions and trust of the client population of which government had none.

Their successes gained them respect and acceptance by government which needed credible partners, particularly in the field of affordable housing. The total collapse of the civic culture, unemployment, drug addiction, and social pathologies of all shades, however, soon made it obvious that the additional marshalling of a full complement of social services was required to stabilize neighborhoods and protect capital investment in them. Housing had thus become largely a social issue.

Job training and workforce development, day- and healthcare were the main foci of the social programs developed by the CDCs. Economic development was added later. "Neighborhood" as a concept was reinvented and painstakingly rebuilt as reality from the ground up.

For this to unfold an unusual act of courage was required by the political establishment. i.e. willingness to acknowledge its failure and to risk the loss of control by delegating responsibility to new players in the field of urban development.

That was not all. The following external circumstances helped create a favorable environment for the movement: 1. the city's dangerously growing homeless crisis resulted in a massive tax base financed production program for affordable housing, 2. the down-sizing of Washington's bureaucracy by President Reagan which included the revamping of unproductive funding streams in favor of leveraging private risk, created new opportunities for non-for-profits, 3. both the "Community Reinvestment Act" of the 1970s and the 1987 "Low Income Housing Tax Credit Program" reanimated the financial markets for affordable housing, 4. the retail sector rediscovered the profit potential of the buying power of rebuilt neighborhoods and sought neighborhood partners, and, 5. an unending stream of new immigrants.

In this environment community development in the full sense of social and physical rebuilding was privatized via the non-for-profit sector. The investments of the CDCs in the communities from which they sprang, where they lived, and which they were part of, were clearly made in self-interest. They became an important mechanism whereby "public benefit products and services" would be delivered to the socially marginal.

Changing the Issues of the 90s

Planning is not the strongest suit of NYC's government. In fact, the last city-wide comprehensive planning effort dates back the late 1960s due to Federal mandates. Every mayor after John Lindsay downplayed its relevance, making the city's bureaucracy a collection of uncoordinated agencies, each with their own mission and agenda, often working at cross-purposes. The 1989 Charter amendment weakened the role of the Planning Department further by taking away its budgetary power, its only coordinating instrument. That and budget cuts relegated it mostly to guardian of the zoning law.

Yet section 197-a of the Charter encourages planning on all levels, including the borough presidents and community boards. While the Planning Commission would review and approve those plans, their lack of legally binding force remains much decried, however. Nevertheless, section 197-a should become key to the emergence of a new drive for planning by community based organizations, filling the planning void of government.

The crisis of the 70s, which caused community empowerment through the CDCs, is no longer valid in the 1990s. For the new generation, which grew up in the culture of the 70s, the issues shifted to social and environmental justice. With the consequences of suburban sprawl, mostly top-down, compartmentalized government, and often devious agency cultures, they question the racial bias of public policies and plans, and their discriminatory environmental impacts on inner-city communities of color.

Political expediency had prompted government to use industrial zones in poor communities to off-load burdensome land uses such as water treatment plants, solid waste facilities and other noxious uses which seemed to be "inappropriate" to wealthier ones. This was clearly a violation of a Charter principle: "the fair and equitable distribution of beneficial and burdensome uses" and thus an issue of social justice relative to public health, barriers to equal opportunity, and quality of life.

On specific facility siting decisions which constituted egregious breaches of the fair share principle, the groups took government to court and often prevailed. But by invoking the privileges under section 197-a they engage in the development of pro-active neighborhood plans including plans for sustainable growth in their industrial zones. This is fostering lines of cooperation with their respective community boards.

Seeking to Strengthen Planning

City-wide coalitions, supported by planning schools and intermediary organizations, such as the Municipal Art Society are pressing for a new planning culture in NYC today. As members of the Civic Alliance they are also carving out a role in the planning process for "ground zero" of the WTC and Lower Manhattan as part of their effort to assure socially equitable development.

During the 2001 mayoral campaign they made participatory and community based planning a forceful issue and continue the call for a stronger role of the Planning Department in government affairs. They seek a city-wide plan, which is informed by community based 197-a plans. It is a call for a profoundly democratic planning process in NYC.

Seattle, Wash. and Rochester, NY are looked upon as the most sophisticated models. The mayors of both cities bought into "bottom up" planning processes and made their governmental structures responsive to new needs. While cumbersome, this gives rise to expectations that cities can overcome segregation and inequities so that nobody wins at the expense of others.

Beyond Local Plans

"Top-down" versus "bottom-up" planning is still hotly contested. Particularly in the era of globalization where cities need to be strategically well positioned, one would think that top-down visionary planning and intervention by government is the way to go. But there really is no reason why a city's strategic framework cannot be developed in a participatory planning process as well.

If the city's post 9.11. planning effort for Lower Manhattan is an indication, an open process which balances between expedience and prudence is at the verge of unfolding. The vision for a "21st Century City" is of city-wide and regional importance, forcing a broad based process as so many stakeholders and interests have to be satisfied. It is generally acknowledged that inclusion brings out a wealth of creative ideas, an experience widely shared by CDCs.

Some feathers may get ruffled, mostly of those who have won in the past. That may not be a bad thing. Although such a process will be expensive in time and money, the price for plans with built-in social conflicts and political volatility may no longer be worth paying in today's world. While social inequities may rise worldwide through globalization, a democratic planning process may deliver at least a partial answer on the local level and be testimony to accountable development.

New Urbanism in the USA and Urban Design Reform in Europe: Common Ground, Differences and Contradictions

Harald Bodenschatz Berlin

The large post-industrial towns and cities in the wealthy countries of Western and Central Europe are not in top shape. Most of the inhabitants – or at least those who earn enough money to have the choice – have moved away from the compact, dense cities. Suburbanisation has long since become established, and housing in the surroundings of the large towns and cities have been followed by shopping centres and later by jobs in the tertiary sector and leisure centres, all on very separate sites.

For many, this development is regarded as a form of individual liberation. As a liberation from the lack of space and the dangers of the city, from the difficult mix of social groups and from pollution and crime. However, this development also poses a threat from two points of view: in spatial terms it encourages over-development of the landscape, and in social terms it promotes the isolation and the drifting apart of different social groups. Large-scale social segregation and isolated living are placing a burden on the big cities. The term "city" itself has been hollowed out: it conjures up idealistic images that no long have very much to do with reality. Cities are disintegrating into functional islands, and urban design is turning into isolated development. We are no longer dealing with cities, but instead with city regions, which are not yet associated with any fixed images.

In the European debate, this process is often referred to as the "Americanisation" of the city. This is because, from the European point of view, U.S. cities have broken down and disintegrated into suburbia, are divided up by motorways and are marked by ethnic and social collapse and by architectural chaos. U.S. cities have become a negative role model for European ones. European cities, people say, have to be protected from becoming like American ones.

However, over the past few years, Europeans have become a little confused about what they see on the other side of the Atlantic, because an autonomous form of criticism of urban development in U.S. cities has arisen, primarily criticism of suburban sprawl. This criticism is based on numerous networks under various labels: Smart Growth, Livability, New Urbanism.

Of these networks, New Urbanism is particularly successful, but also particularly controversial. For the Europeans, New Urbanism is a provocation. The first reaction to the movement here was to say "how awful, how terrible, typically American!" The main point of criticism is that New Urbanism often uses a traditionalist architectural language, which is seen as being backward-looking and a falsification of history. This negative codification creates an almost impenetrable shell that prevents an objective view of New Urbanism.

During the unprecedented building boom of the 1990s, things started to happen again in U.S. cities. The conditions in many inner-city areas improved, but the main development was that the cities continued to encroach on their surroundings, and suburbs went into mass production. The new aspect here, however, is that this latest episode of suburbanisation is no longer simply celebrated as progress, as the realisation of the American dream. A fundamental change has begun to take place in public opinion in this context. You often hear people say that the development of suburban sprawl has reached a stage that is harming U.S. society. We only need to recall the film American Beauty or the countless studies conducted in the social sciences showing that the "ideal world" of suburbia is an illusion.

In this context, the call for alternatives to the traditional suburbs makes sense. There are two major alternatives in practical urban design: firstly, gated communities, closed areas of housing with surveillance that are no longer open to the public, and secondly, the distinctive products of New Urbanism. In contrast to gated communities, New Urbanism appears to many to be the urbanistically correct response to the misdevelopment of U.S. cities. New Urbanism is not an architectural movement, but an urban design movement. Its programmatic aim is to achieve mixed-use communities with a social mix, greater density of development and architectural variety within a regulatory framework of urban design. New Urbanism calls for the creation of walkable neighbourhoods and promotes local public transport and the reduction of car traffic. It calls for open cities with maximum links to their surroundings, the opposite of gated communities. The basic assumption is that a form of urban design guided by the principles of historical cities will promote social integration.

However, a number of compromises are made when these principles are put into practice. Most New Urbanism projects are built in suburban areas by private developers, and projects such as these often differ from conventional suburbs only in degree. This is true even of what is perhaps the most important development principle of New Urbanism: the rejection of self-chosen ghettos and the advocacy of maximal links between a new estate and its surroundings. Even the goal of a social mix is usually only achieved to a rudimentary degree. However, it should be emphasised that this problem is being discussed within the New Urbanism movement, and this debate is becoming increasingly intense, as shown by the congress in Portland in the year 2000, for example. The search for strategies to counter social division in cities was a central topic at the congress.

The most ambitious goal of New Urbanism is found on a different spatial plane, however: at the regional level. At least part of the New Urbanism movement believes that it is only at the regional level that social and spatial goals can be achieved on a sustainable basis. The aim is to establish interlinked, socially well balanced regional cities with flexible limits to expansion. One of the advocates of this view is Peter Calthorpe, who, together with William

Fulton, published a manifesto on this topic in 2001 in their book entitled "The Regional City. Planning for the End of Sprawl". As can be seen here, New Urbanism deliberately avoids polarising the debate by not making a sharp distinction between compact cities and suburbia. The question of where the focus should lie is obviously the subject of debate within the movement, but the issue is always about the focus, not a question of either/or. New Urbanism has set itself the task of upgrading the entire urban region – downtown and suburbia, not downtown or suburbia.

New Urbanism is primarily a U.S. movement. Only few foreigners attended the most recent congresses in Portland and New York, and most of these came from Canada or from as far away as Australia, while hardly any came from Europe. The exclusive reference to the USA demonstrates the self-confidence of New Urbanism, but it is also a potential weakness in strategic terms. Debate across the Atlantic is hampered by two factors: by the introverted nature of the U.S. debate and by the sceptical attitude in Europe.

While New Urbanism is essentially a response to suburban sprawl, the European debate on urban design reform focuses on re-organising compact cities, particularly on reorganising the city centres and converting urban wasteland, on new uses for industrial estates, military land, and land belonging to railways, ports or airports. From the European point of view, the main issue is to stabilise historical compact cities against the competition from suburbia, and the public sector plays an important role in steering this development. One of the advocates of this view is Richard Rogers, who published his manifesto on urban design, "Cities for a small planet", in 1998.

European cities are regarded as having structural features such as a relatively high density of development, an interlinked system of public spaces, a social, functional and architectural mix and a spatial hierarchy with the city centre at the top. The specific form that these structural features take differs from region to region, it is claimed, reflecting the cultural variety of European cities. In this sense, a specific European city is the material interpretation of its own particular history, which needs to be maintained or reproduced. In social terms, this usually means that the middle classes are reclaiming compact cities.

But what does this type of focus in urban design mean for suburbia? Here, a major deficit of the urban design debate in Europe needs to be pointed out. In Germany, in particular, knowledge about suburbia is still very limited. In contrast to the situation in the USA, there are still too few studies available in Germany, in particular, about the origin and discrepancies of suburban development. Attention has repeatedly been drawn to this deficit by one of the protagonists of the urban design debates in Germany, Thomas Sieverts, whose manifesto Zwischenstadt zwischen Ort und Welt, Raum und Zeit, Stadt und Land ("In-between City between Place and World, Space and Time, City and Country") appeared in 1997.

We need to debate suburbia on two levels, and these should not always be confused or played off against one another: the overall urban level and the level of the particular suburban fragment. The individual fragments of suburbia can each be regarded as social and urban development islands in themselves. The only question is how they can be optimised, both internally and in comparison with other islands. However, they can and must also be looked at on a large scale at the level of the city region – as islands that potentially take away resources from the mainland and that play a major role in its structural change. Here, the politically awkward question arises as to the extent to which the price of suburban living can be made to reflect the actual costs, or whether, in a society otherwise geared towards public savings, we should exempt suburbanisation – of all things – from these constraints. Christiane Thalgott has drawn attention to this aspect on several occasions. The principle of sustainability can ultimately only take effect at the overall urban level.

On the whole, the debate about urban design reform in Europe is currently marked by considerable programmatic fragmentation. In fact, it is not correct to speak of one debate, but of several debates. The protagonists of these debates are isolated prophets who make little reference to one another. There is no relatively uniform, organised movement to renew European cities. Thus the advocates of sustainable cities, social cities, traditionalist cities or in-between cities dissociate themselves very definitely from one another.

In summary, the U.S. debate and the European urban debates start out from different premises, because cities on either side of the Atlantic are in a very different state. In addition, the main topics of the two debates are projects in different areas: in suburban areas in the USA, and in compact cities in Europe. Even the preferred form of architectural and urban design is different. There are differences in the groups of actors: in the USA there is a tendency towards private urban development, whereas in Europe it is still largely controlled by the public sector. Finally, the programmatic regional targets are also different, namely protecting compact cities from suburbia in Europe, and upgrading both suburbia and compact cities in a "regional city" concept in the USA. The forms taken by the debates differ too: while New Urbanism is an organised reform movement with various schools comprising not only experts but also politicians, developers, and social and environmental activists, the European debate is poorly organised and fragmented, dominated by isolated experts and government institutions.

In contrast, the similarities between the principles of urban design advocated in the U.S. and European debates are striking: both are based on general, complex criticism of the development of cities. Criticism of gated communities, social exclusion, over-development, car-dependent development and modern urban design. Criticism of the social segregation that occurs in a suburbanised city region, the separation between living and the city, and the social disintegration of cities in general. Both debates are guided by the social reintegration of living in the city, by social and environmental sustainability, and both advocate new projects of urban design that should fulfil at least the minimum criteria for a city: social and functional mix, and walkable public spaces.

A joint message here might be the call for urban design instead of isolated housing construction. That means not only a functional mixture, but also a social one, a careful, situation-dependent mix in terms of income groups, ethnic groups, lifestyles and age groups. Or at least an attempt at achieving this kind of mix. It also means networking at the level of the city region in spatial terms and, most importantly, in socio-economic terms – as a balance of values based on solidarity within a city region. Just as the pre-industrial cities had to be adapted to the industrial cities in the nineteenth century, the compact cities of our modern industrial society have to be adapted to the conditions of the post-industrial city region. Now, as then, there needs to be a lively debate as to the best way of achieving this.

The projects that are going to be presented now are examples of a reorientation of urban design on both sides of the Atlantic. Harald Kegler from Dessau played a major role in preparing this workshop. Since the Second World War, Munich has been a successful model of the preservation and renewal of historical cities. While for a long time Munich was a special case in Germany, Elblag in Poland is a current example of a Polish tradition, unique in Europe, of reconstructing historical centres that were destroyed during the war. The two projects in North America are not set in suburbia: the Aqua Project in Florida is a conversion project for Allison Island between Miami and Miami Beach, and finally, Vancouver, Canada, is an ambitious model for revitalising a city centre.

All these examples are primarily geared towards the middle classes, and they usually take place in prosperous city regions. This is true at least of Miami, Vancouver, Munich and Freiburg. And on a more modest scale of Elblag too. Does that mean that we are running away here and now from the "real problems"? Not at all in my opinion. The "real problems" are not only, and not primarily, other people's problems, problems in poor regions. We, the rich regions that determine the pace of globalisation, are creating the "real problems" ourselves. In the rich regions, solutions need to be sought for the socially and environmentally sustainable integration of living in a city region characterised by solidarity – in a debate involving all the actors, including private investors, who are unpopular in Europe, and the middle classes, who have choices about how they live. Unless the high-income sections of the population, in particular, are actively involved, an urban design reform project will be impossible.

Different starting points, different preferences regarding design, similar principles – this is actually a very good basis for cultural exchange. However, cultural exchange across the Atlantic only works to a very limited extent. There are few links between the two debates, and this exchange is hampered by prejudice and a lack of knowledge. However, both sides can benefit from the exchange of ideas, while they cannot benefit from setting up ideological gated communities of urban design. The first step towards the exchange of experience and ideas today should soon be followed by a second one: giving specific shape to a transatlantic urbanistic dialogue.

Audi
Resources are limited. But our ideas aren't.
And this is why we are always setting new goals for ourselves in our quest to live up to the highest environmental standards. To learn more about environmental protection and order our "Interim Report" and "Environmental Statements" pamphlets, please get in touch with us: +49 841 8934900.
Vorsprung durch Technik www.audi.com

THE BUILT AND THE NATURAL
DIALOGUE BETWEEN DISCIPLINES

Introduction and Abstracts

NATUR UND GEBAUTE UMWELT
DIALOG DER DISZIPLINEN

Einführung und Abstracts

Kisho Kurokawa Tokyo

Ken Yeang Ampang

Luc Lévesque Montréal

Friedrich Kuhlmann Berlin

Harald Kegler Lutherstadt Wittenberg

Leigh Woolley Tasmania

Sonia Ricon Baldessarini Berlin

Don Grant Sydney

Jörg Schlaich Stuttgart

Newton Harrison San Diego

Introduction
Einführung

As traditionally contradictory phenomena, the built and the natural environment remind us that nature and landscapes of primary nature find themselves in conflict with a man-made environment that we are accustomed to treating as secondary nature – a paradox. Nature is now calling us to account for this attitude.

Environmental development and cyclical economies are integral factors in a form of social development that is rooted in the horizon of the earth and does not collide destructively with the natural conditions of existence. The "environment" is the product of historical processes, by means of which natural habitats and landscapes are reinterpreted and subordinated to man-made structures ("secondary nature") until they have been adapted to the levelled-out and unnatural vital needs of societies. The conditions of life and civilisation, modified above all by technology, change in retrospect into crisis factors arising from social change.

Context

The term resource architecture encourages us to regard architecture, and good architecture in particular, as a scarce commodity, one that should be treated with care and managed on a sustainable basis in harmony and exchange with natural resources, the landscape and towns, with their cultural, social and urban resources. The durability of architecture and the prolongation of its lifecycles provide the foundation for new quality standards.

Innovative Architecture within the Horizon of the Earth

The changes in the global climate, the pollution of the air, the water and the soil, overpopulation and the scarcity of natural resources are concomitant features of architectural and urban development, which are responsible for up to 60 per cent of the earth's energy consumption. The effects of urbanisation are partly "anthropogenic problems, from which the majority of human beings are suffering both materially and culturally and which are impoverishing and ruining the natural environment we share" (Hartmut Böhme). The traditional separation into zones regarded as existing separately from one another – natural landscapes, rural zones, suburban outskirts, densely inhabited centres and metropolitan hubs in various regions of the world – is no longer tenable. The urbanisation of continents is creating generalised conditions affecting every aspect of life. "We can assume that our natural environment, as it exists and has developed during the history of our planet, as well as all that has developed in such diversity in the many cultures of the world, will change dramatically." (Hartmut Böhme).

Gebaute und natürliche Umwelt erinnern als traditionell kontradiktorische Positionen daran, dass Natur und Landschaft der Ersten Natur in Widerstreit liegen mit einer Umwelt, die zivilisationsgemacht ist und die wir durch Gewöhnung der Wahrnehmung als Zweite Natur begreifen – eine Paradoxie. Die Natur präsentiert uns dafür die Rechnung.

Umweltentwicklung und Kreislaufwirtschaft sind integrale Faktoren einer gesellschaftlichen Entwicklung, die sich in den Horizont der Erde einlagert und nicht über die Maßen gegen die natürlichen Entwicklungsbedingungen verstößt. „Umwelt" ist Produkt historischer Prozesse, durch welche natürliche Lebensräume und Landschaften so umgedeutet und mit zivilisatorischen Strukturen („Zweite Natur") überlagert werden, bis sie an die nivellierten, naturfernen Lebensbedürfnisse von Gesellschaften angepasst sind. Die vor allem durch Technik veränderten Lebens- und Zivilisationsbedingungen wandeln sich rückwirkend zu Krisenfaktoren des gesellschaftlichen Wandels.

Kontext

Der Begriff „Ressource Architektur" leitet die Perspektive, Architektur, gute Architektur insbesondere, als knappes Gut zu begreifen – als Gut, mit dem in Einklang und im Austausch mit natürlichen Ressourcen, mit Landschaft und Stadt, mit kulturellen, sozialen und urbanen Ressourcen sorgfältig und nachhaltig zu wirtschaften ist. Dauerhaftigkeit der Architektur und Verlängerung ihrer Lebenszyklen begründet neue Qualitätsmaßstäbe.

Innovative Architektur im Horizont der Erde

Veränderungen des Weltklimas, Verschmutzung von Luft, Wasser und Boden, Übervölkerung und Knappheit der natürlichen Ressourcen begleiten die Architektur- und Stadtentwicklung, die bis zu sechzig Prozent des globalen Energieverbrauchs bindet. Auswirkungen der Verstädterung sind Teil „anthropogener Probleme, unter denen die Mehrheit der Menschen materiell und kulturell leidet und unter denen die natürliche Mitwelt verarmt und ruiniert wird" (Hartmut Böhme). Die traditionelle Trennung in unabhängig voneinander gedachte natürliche Landschaften, ländliche Zonen, suburbane Peripherien, verdichtete Zentren und metropolitane Kerne in den Regionen der Welt trägt nicht länger. Die Verstädterung auf den Kontinenten schafft übergreifende Bedingungen, die alle Lebensbereiche umfassend beeinflussen. „Es ist davon auszugehen, dass der Bestand der natürlichen Umwelt, wie sie erdgeschichtlich gebildet ist, und der Bestand dessen, was in den vielen Weltkulturen als menschenwürdiges Leben ausdifferenziert wurde, sich dramatisch ändern werden." (Böhme)

In Europe, such terms as resource architecture and the endangerment of resources go to the very heart of the discipline (of nature and landscape protection as well as of landscape planning, architecture, town planning and urban development), not so much because of a lack of knowledge as because of a failure to act. Resource architecture not only defines architecture as a material good, but as a highly complex cultural resource, which needs to be developed and cultivated as a sustainable basis for life – like self-determined culture, natural landscapes, clean water, clean air, and healthy food.

In Europa treffen die Begriffe Ressource Architektur wie Ressourcengefährdung den Kern der Disziplin (des Natur- und Landschaftsschutzes wie der Landschaftsplanung, der Architektur wie der Urbanistik) – weniger im Sinne eines Wissensdefizits als im Sinne eines Handlungsdefizits. Ressource Architektur definiert nicht nur das materielle Gut Architektur, sondern Architektur als vielschichtige kulturelle Ressource, die als nachhaltige Lebensgrundlage zu entwickeln und zu pflegen ist – wie selbstbestimmte Kultur, wie naturhafte Landschaft, wie sauberes Wasser, klare Luft und gesunde Nahrung.

Responses

Architecture that is aware of environmental conditions is architecture that develops and uses energy-conserving technologies and places its hopes in the use of renewable energies. The development of architectural forms that conserve their value and are open to change goes hand in hand with acceptable urban densities and concentrations and the renaturalisation of the landscape wherever it can be liberated from the burdens of urbanisation. The built environment can represent the most important part of a nation's wealth. As a forward-looking manner of securing our built heritage, architecture is developing a mode of thinking in extended life cycles. Cultural values are preserved under the condition that architecture remains capable of change.

Erwiderung

Architektur, die sich der Umweltbedingungen bewusst ist, entwickelt und nutzt energiebewahrende Technologien und setzt auf die Nutzung erneuerbarer Energien. Die Entwicklung werterhaltender und wandelbarer Architekturformen geht einher mit einer verträglichen Verdichtung der Städte und einer Renaturierung der Landschaft, wo immer sie von den Lasten der Verstädterung befreit werden kann. Die gebaute Umwelt kann den bedeutendsten Teil des nationalen Wohlstandes repräsentieren. Als vorausschauende Sicherung des gebauten Erbes entwickelt Architektur ein Denken in verlängerten Lebenszyklen. Die Erhaltung der kulturellen Werte steht unter dem Vorbehalt der Wandelbarkeit der Architektur.

Proposition

A redefinition of the resource-based concept of the landscape as the repository of aesthetic, social and historical potential is proposed not least because this concept preserves historical patterns of cultural experiences of life under natural conditions that must be continued. The assimilation of elements of nature and the landscape into towns and cities enables architecture and the city to be seen as integral parts of natural cycles (cyclical economies). Architecture must exert a far greater influence than ever before on the conception and planning of urban structures, buildings, the use of materials and system components and thus on energy consumption, too. The goal of architectural development is to achieve the maximum degree of flexibility without any loss of identity.

Thesis

Eine Neubestimmung des Ressourcenbegriffs der Landschaft als eines ästhetischen, sozialen und historischen Potenzials wird angeregt – nicht zuletzt, weil in ihr historische Muster von kulturellen Erfahrungen des Lebens in natürlichen Zusammenhängen bewahrt werden, an die anzuknüpfen wesentlich ist. Die Integration von Elementen der Natur und der Landschaft in die Stadt schafft Voraussetzungen dafür, Architektur und Stadt als integrale Teile natürlicher Kreisläufe (Kreislaufwirtschaft) zu verstehen. Architektur hat erheblich mehr als bisher entschiedenen Einfluss auf die Konzeption und die Disposition von Stadtstrukturen, Gebäuden, auf die Verwendung der Materialien und Systemkomponenten und damit auch auf den Energieverbrauch zu nehmen. Ziel der Entwicklung der Architektur ist, ein Maximum an Flexibilität ohne Verlust an Identität zu erreichen.

Reinhart Wustlich Hennef

Eco-City: The Philosophy of Symbiosis From the Age of Machine to the Age of Life

Kisho Kurokawa Tokyo

Road Network

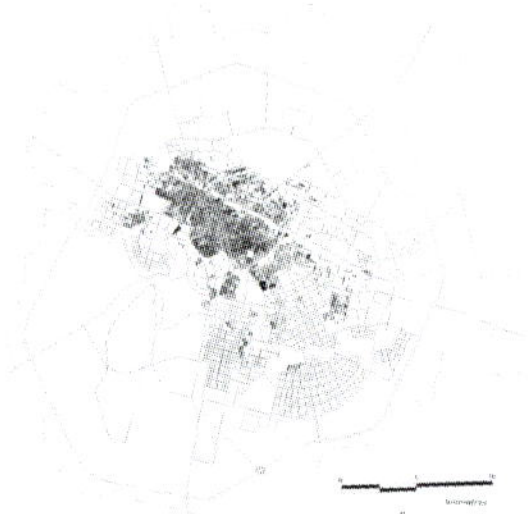

Existing Building

Existing Wet-land, Rivers and Lakes

Green Network

It was in 1959 just after the collapse of CIAM movement, I declared that the world will make a shift from the Age of Machine Principle to the Age of Life Principle. Then I began metabolism movement in 1960. Since 1960, my architecture and city planning have been designed with the keywords 'metabolism,' 'recycling,' 'ecology' and 'symbiosis' in mind. These are all the most important keywords of the life principle. At the beginning of the 21st century, my prediction that the world will make a shift from the age of machine principle to the age of life principle, has gradually proved true. If the machine represents an age of homogenization, economy and globalism, life represents an age of pluralism, or diversity, culture and economy.

And so following the architecture of the Age of Life is an expression of this age of diversity. An intercultural age, the value system and lifestyle of the West and many other cultures of the world, exists in symbiosis, is dawning on us. The doctrine of humanism that has made man the rational being second only to God, and given him control over other living things and the natural world, is facing a crisis. We have gradually come to realize that the survival of humanity, even from an ecological perspective, depends solely on the symbiosis of the many life forms or ecosystems on our planet. Eco-architecture and Eco-city are perhaps the goals of the Age of Life. We can no longer believe that the machine, scientific technology, and the human intellect are almighty.

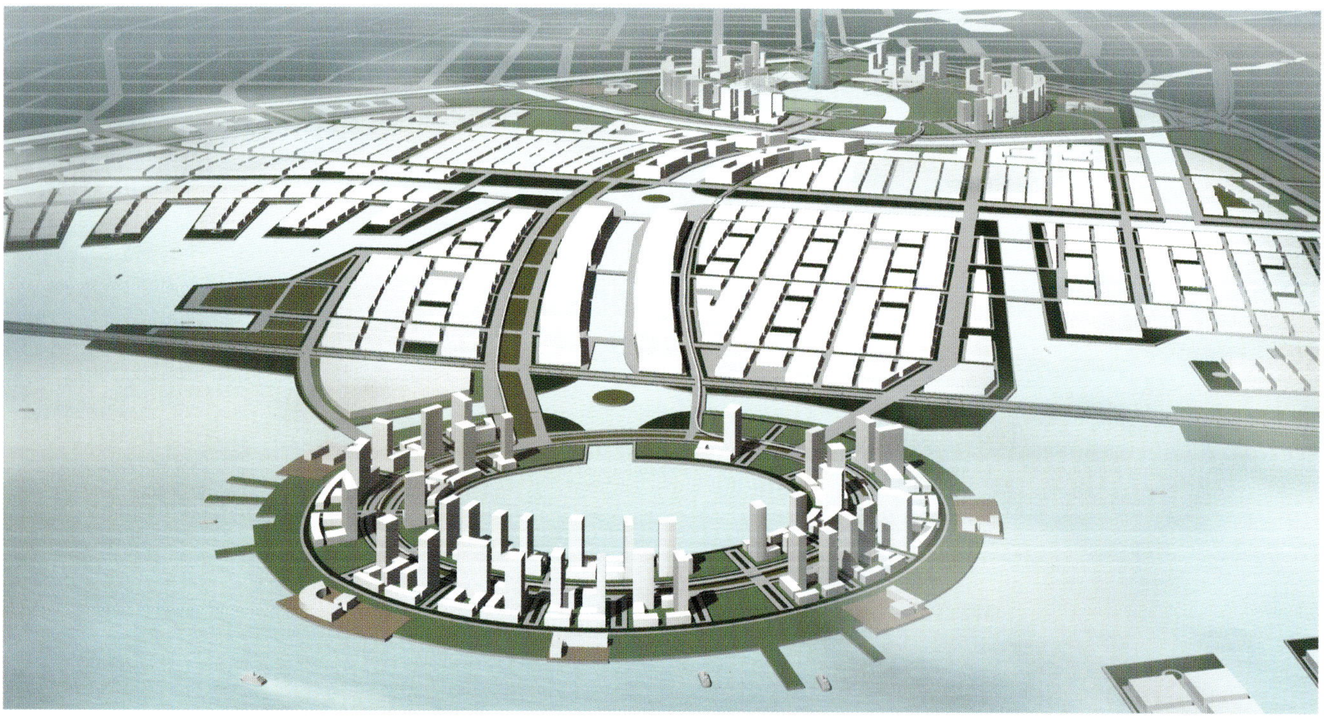

Bird's-eye view, north

Master Plan 2010

Master Plan 2030

My Plan for the new capital of Kazakhstan won the first prize in an international competition and was formally approved by the Parliament of Kazakhstan in June 2001. This plan was created to achieve a number of themes including symbiosis of history and the present day, symbiosis of the city and nature, symbiosis with other organisms, or life systems, recycling of water (rain and groundwater), recycling of waste material, and sustainable growth. The new capital city of Astana will be a City of Symbiosis with History, by preserving old streets and architecture in the existing neighborhood of Akmola along with the new city on the left bank of the Ishim River.

It will also be a City of Symbiosis with Nature thanks to the River City Plan that changes the role of the Ishim River from the edge of the city as waste water channel to its center for living with nature, and the plan for an artificial forest that will protect the city from the winter winds. It will also be a completely new City of Symbiosis where the forest, Ishim River, and the urban parks are incorporated in a network by Eco-Corridors creating spaces that serve as both public parks for the citizens and as habitats for other organisms, insects, butterflies, and birds. Discharged sewage water will be purified and recycled underground for use in irrigation of farmland.

The plan for a new city of 1500 ha with a population of 1.5 million to the east of existing Zhengzhou City, which is my most recent plan, won the first prize at an international competition in December 2001. The plan preserves all the 34 existing rivers and restores their green vegetation to create riverside parks that are incorporated in a network of Eco-Corridors. The plan for these Eco-Corridors that link isolated ecosystems is an important method of preserving biological diversity and is an important theme in the Symbiosis City concept. It uses an existing fish breeding pond to form an artificial lake with a surface area of 700 ha. I gave this lake the name Dragon Lake because of its shape, and there is a legend about a dragon in this region.

The City of Zhengzhou will be reborn as a City of Symbiosis with Nature, where lakes and canals that are artificially created natural features are emphasized. The core of the new city is planned as a ring city without a center, and the new city is planned based on the cluster method so that it will grow like cells.

This Symbiosis City or Eco city will be in a symbiotic relationship with the natural environment of its rivers and lakes and with the 3,500 year-old Central Chinese cultural traditions of the old city. It will begin in 2002 and be completed in 2015.

The Ecological Design of Large Buildings and Large Sites

Ken Yeang Ampang

The paper contains a set of propositions for the design of the skyscraper and other intensive building types as ecologically-responsive buildings.

Keywords: Ecological Design, Bioclimatic, Building Configuration, Natural Ventilation, Passive Low-Energy, Sunpath, Roofscape, Vertical Landscaping, Materials, Waste, Ecosystems.

Abstract

Already well-argued elsewhere is the case for architects and engineers to design our man-made environment to be sustainable or 'green' (or ecologically-responsive). Design objectives if simply stated here, are to build with minimal impact on the natural environment, to integrate the built-environment with the ecological systems (ecosystems) of the locality and if possible, to positively contribute to the ecological and energy productivity of the location. For many designers today, these objectives are regarded as given pre-requisites for all their design endeavours.

The proposition here, of the skyscraper as an ecologically-responsive building, might well be regarded by some as a conundrum. After all, the skyscraper is one of the city's most intensive building-types (besides say, the shopping mall or the convention centers). Such buildings, in comparison to those other considerably smaller experimental low-energy ecological buildings built elsewhere, are by all means and purposes not low-energy nor self-evidently ecological building types by virtue of their enormous size and high consumption of energy and materials.

The point then is that regardless of whether we are for or against intensive urban buildings such as the skyscraper, in reality this is a building type that will not simply go away. Current trends in fact, indicate that the skyscraper as a building type shall continue to be built into the next millennium and to be in existence in most of the world's cities. This condition shall likely remain until we invent another economical alternative, or until radical changes are effected in planning administration or to the current high trend of rural-to-urban migration into our cities, especially in Asia.

The pressing key issue then is how the designer today can design these massive building types in an ecologically-responsive way? This issue has to be urgently addressed immediately. For otherwise, the alternative would be a situation where our rapidly developing cities continue to be covered with multitudes of high-energy consuming, polluting-waste-producing intensive buildings. Our efforts should be urgently directed to designing ecologically-responsive skyscrapers and other intensive building types to reduce the aggregate negative impacts on the biosphere.

In this paper, the key factors for the green design of the skyscraper (and other intensive building types) are explained in the form of a simple General Systems model of the crucial interactions that any built system will have on the natural environment. As a model for design, it provides a check of the items to be considered if design is to be regarded as ecological. Illustrated are a number of built and designed precedents, using a case-study to explain the set of design principles and ideas for the ecological design of tall building.

Specifically, the paper discusses the ecological benefits of the following: passive low-energy strategies and related bioclimatic design principles as a sub-set of ecological design (such as principles on the location of the service-cores, natural ventilation of spaces, orientation of builtform, building configuration options, sunshading, wind-scoops, atriums, skycourts, etc.), the use of transitional spaces, façade design, roofscape design, building M&E Systems design, wind and natural ventilation, natural light and lighting systems, vertical landscaping, energy embodiment in buildings (primary and delivered), materials strategy (reduce, reuse, recycling, reintegration, etc.), urban design implications, life-cycle impacts, etc.

Deutsche Bank Unter den Linden, Berlin Architects: Benedict Tonon; Novotny, Mähner und Partner

Interstitial Landscapes as Resources
A Few Thoughts about a Tactical Approach to Urban Intervention

Luc Lévesque Montréal

Reducing the landscape to an idealised modality of nature or built heritage is to forget that its history is inherently linked to the progressive taming of the most inhospitable places. Forest, mountains, seas and deserts haven't always been considered as landscapes. Aren't the real territories to explore today the ones that we don't see simply because of our immersion in them – these extensive lands of the urban? At a time when we don't exactly know how to deal with the various phenomena that transformed the urban substance, the concept of landscape, as a cultural construct, liberates a zone of indecisiveness that allows us to think about the potentials generated by a multi-layered reality[1].

It is from this perspective that the "interstitial landscape" draws our interest. In urban sociology the expression "interstice" is currently used to describe places of "otherness" and informal practices[2]. The interstice can also be defined as a space without precise use, located for an indeterminate period of time between functionally determined, built configurations. The interstice speaks about porosity. The pore is both cavity and passage, a place propitious to the development of processes that escape control and contaminate normative representation. In a world more and more mediated and virtualised, the interstitial condition notably offers the possibility to learn from the experience of a new type of wilderness – the raw reality of an impure urban wilderness. In following this thread, the interstitial landscape constitutes a resource for experimentation and is, in itself, an experiment[3]. This observation calls for approaches to urban intervention that focus less on imposing order than on inflecting the existing dynamics with a tact similar to the one of acupuncture.

The tactic[4] of furniture punctuation constitutes an example of intervention which follows this path. What is specifically interesting about furniture is less the sophistication of its design than its capacity, through strategic positioning, to generate a field of interrelations and situations that may enrich urbanity[5]. Furniture as a minimal and potentially nomadic dwelling unit (as the French term "mobilier" expresses it more clearly by directly referring to the idea of mobility) seems to us indeed an interesting tool for taming the interstitial without denying its peculiarities or, as Paul Virilio puts it, to incite urbanites "to inhabit the inhabitual"[6]. Viewed from that angle, a simple generic picnic table can be the vector of unsuspected urban experiences. Thus, a moving network of picnic tables infiltrating the city's interstices could be an example

of an alternative way to deal with the question of urban spaces in a context of incertitude and transformation.

The Hypothèses d'amarrages (Mooring hypothesis) project, initiated by SYN- and ongoing since May 2001 in Montreal, concretely explores this conceptual framework. Starting from the fact that many spatial residues are produced and abandoned by contemporary urbanisation, that a great number of them are not going to be developed in the near future, and also that a lot of these sites have spatial and landscape qualities propitious to temporary occupation, the first phase of the project has proposed using picnic tables to squat an array of selected interstitial sites of the Montreal metropolitan area. The intention is to exploit the potential of forgotten, banalised or underused spaces to offer to the urbanites new possibilities of interaction with the urban landscape. These are, for example, uncommon views toward the city, other kind of temporal and sensitive experiences, opportunities to tame fragments of the urban wilderness and chances to reinvent, more freely, urban attitudes outside preconceived representations and organised entertainment. Finally, the project Hypothèses d'amarrages uses an operational field inherent to the concept of landscape but often forgotten: the fact that an intervention can or could, in many cases, limit itself to almost nothing, that the existing conditions conceal potentials that can be made visible or accessible by discreet, almost immaterial, gestures.

1 see: Luc Lévesque, "Hyperpaysages: à l'affut de territoires réticulaires et mentaux", in CV photo, no. 54, Montréal, 2001, pp. 5-6.

2 see: Frederick M. Trasher, The Gang. A study of 1313 gangs in Chicago, Chicago and London, The University of Chicago Press (1927), abridged edition 1963, p. 20; Jean Remy et Liane Voyé, Ville, ordre et violence, Paris, PUF, 1981, p. 71; Nicolas Bourriaud, Esthétique relationnelle, Paris, Les presses du réel, 1998, pp. 14-16.

3 see: Mirko Zardini, "The prevalence of landscape", in Nuevos paisajes/New landscapes, Barcelone, Museu d'art Contemporani, ACTAR, juillet 1997, pp. 203-209; Ignasi Sola-Morales, "Terrain vague", in Anyplace (Montréal), Anyone Corporation (New York), Cambridge, London, MIT Press, 1995, p. 118-123, and Architectures in Cities. Present and Futures, Barcelone, UIA Barcelona, 1996, pp. 21-23/266-309.

4 about the notion of tactic, see: Michel de Certeau, L'invention du quotidien (Arts de faire I), Paris, Union Générale d'Éditions, 1980, I, chap. 3, pp. 19-23/82-94; Sanford Kwinter, Architectures of Time, MIT Press, 2001, p. 122-123.

5 see: Luc Lévesque, "Montréal, L'informe urbanité des terrains vagues. Pour une gestion créatrice du mobilier urbain", Les Annales de la recherche urbaine, no. 85, Paris, pp. 47-57.

6 Paul Virilio, L'insécurité du territoire, Paris, Stock, 1976, pp.199-208.

Fastland – Urban Landscape in Condition of its Disposal

Friedrich Kuhlmann Berlin

Quotation about a woodcut by the painter Wolfgang Mattheuer from 1973 entitled 'Hinter den sieben Bergen' (beyond the seven hills): "We are constantly surrounded by technology, it has become part of our life. I don't think much of romantic, idyllic holiday landscapes, I prefer 'modern idylls'. I seek and find many ideas for my work in the immediate and at times powerful clash between nature and technology and in our response to it. In my opinion, this tension is crucial for our time, it reflects the complexity of the world we live in." We have passed this point a long time ago.

Fastland One (Fast Land)

Brian Wilson wrote Pet Sounds in 1965. As a pop album, it represents an eccentric perception, in which apparent integrity is unmasked, destroyed and built up anew. Pet Sounds conveys a sense of fragmentation instead of synthesis. The integrity of the promised image is undermined, because the sounds are in opposition to each other, so that instead of being worn down, perception is deepened every time one hears the sounds again. The familiar image of the sun, girls and cars was dropped in favour of a diffuse mood which established its own frame of reference.

In this context, landscape is seen as a pictorial construction as opposed to landscape as a structure or a complex system – structures as well as systems generate images. Every action generates images – it refers to the given whole or to a specific individual atmosphere. Landscape is created through conflicting actions. It is a permanent process enabled through known or unknown overlapping intentions. Landscapes are subject to continuous interventions of the urban and of the halflife of planning paradigms involved in the process.

Urban landscape evolves from the so-called cultural landscape and from the city. It does not have a genius loci, ideas come only from external terms of reference. In the meantime, the urban exists only as an idea which could be anywhere, but it is no longer tied to a familiar image. Urban landscapes are places of experimentation with a permanent fuel feed. They follow interventions which are subject to disruptions and fashions. These facts are of no consequence for future action – one seeks the outside. The gap between everyday life and images from the past makes the new difficult to read. Paul Valery wrote in 1894 in his Leonardo essay about concretised habits of language that have set up barriers around our perception: the terms referred to are "landscape" and "nature". Beautiful landscapes and the concept of nature have attained such control over our behaviour because we prefer looking at things through the spectacles of a lexicon to seeing them with our own eyes; by doing so we block off our view of the real[1].

Fastland Two (Scandinavian for Mainland)

As concepts, urban landscapes, intermediate landscapes or universal landscapes evoke different im-ages forming a composite whole. They describe a kind of intermediate stage to be "in time". From this vantage point it is much easier to understand regional signs and to deal with them, because they are not historically connoted. Universality does not replace the search for places, it occupies places be-yond one's home environment. The Swiss geographer Andreas Huber gives a good example of places within universality. In his book entitled "Heimat in der Postmoderne" (home environment in the post-modern era), he describes how Swiss retirees construct their own home environment in the Spanish Costa Blanca. He says that home environment comes into being when one is in his "own time"[2].

Ferdinand Fellmann says that our behaviour is controlled not by concepts, but by images which we form of situations in which we find ourselves. Thus, images control behaviour as paradigms[3]. He goes on to say that images are the sum of perceived elements in a landscape. They combine various elements to form aesthetic units. If landscape is created by action, then images are created by shaping circumstances. It is the act of shaping that follows various actions in space. Thus, an image is a consequence of the substance of these actions. The image is formed from conditions which create landscape images not defined in space. A landscape is a state and a wish.

A cultural landscape is perceived as a wish. It provides a retreat from false culture while asserting permanence, substance and sense. The planned image conceals fragmentation and blends the image into a utopia by tapping into something that appears to have been there all along. A landscape is thus divested of its historically negotiated compensatory function, because an outside is no longer there. Compensation becomes an aesthetic mask, landscape planning always relating to compensation.

Planning is a form of technical action in which landscape is rendered effective with one's own means that are available. This includes conserving images typifying cultural spaces. Landscape planning systematises the use of traditional landscapes and their associated uses without those images that would necessarily result from new intentions. Simulations of a new landscape are created which lack an economic basis and live from subsidies. Planning creates weak images and constructs "anti-worlds". It can only minimise effected interventions and must capitulate as a creator of future development.

By charging collective longing, cultural landscapes become something whose intentions have passed. But reality is different: Landscape is not only something that is already there and that consequently is only protected and modified. Since it represents a way of thinking and is therefore not constant, it needs to be imagined and designed. In this connection it would be too simple to maintain that planning is conservative. The problem of planning consists in the mode of its simulation and authenticity. It creates redundancy by using serial images that claim a perfect past.

Slowland

Landscapes can be described as the authentic result of actions on the one hand, and as the result of simulation on the other. Simulation is the result of an intended creation of authenticity. In connection with the intermediate town discourse, Augustin Berque wrote in 1990 that in the final analysis, a simulation of authenticity is part of the awareness of a place in the post-modern era. He claims that the post-modernist Arcadia ideals are giving way to a new mode of perception of the "discovery of the ugly" as a positive aesthetic category[4].

But this kind of perception is circular. It is not about simulation vs. reality or about broken aesthetics, but about the emergence of boredom, which is a kind of patina that destroys the charm of the simulated. The images become redundant and confront one with an aesthetic sequence: from the authentic to the simulated to boredom and finally to an ironic quotation. Barnett Newman wrote in his famous article "the sublime is now", that we change materials the way objects fit in a particular place and then call this sublime – but our actions are only an upgrading of the so called field.

What remains is a constant production of boredom. The collected postcards of Martin Parr show the sudden change from novelty to boredom[5]. They stand for a different, a light perception. They show everyday culture and a predilection for the unnatural and exaggeration. They arouse longings for a landscape which does not always require its meaning to be questioned, which is not exposed to the production of anti-worlds. Martin Parr shows what happens when progressive images coat patina. An outside comes into being inside and an escape into "culture" is withdrawn. The landscape becomes a utopian screen for future purposes and a superabundance of images: places whose representation and complexity render clear classification impossible. Boredom and superabundance complement each other, since neither communicates unequivocal images and hence prompts to future actions.

Fastland Three (Almost Land Already)

Landscapes are fraught with meanings. Gottfried Boehm differentiates between powerful and weak images. A powerful image gives work. It operates metabolism with reality. Powerful images show us something that we would not know without them. They provoke action and are engendered by it.

Images which unequivocally depict familiar intentions and tie them to places are an example of images that are not powerful. They are weak images coming from direct representation of the known. Their weakness is the result of negation of their intrinsic value and of the precedence of the pictorial alignment with the represented[6]. Current discourses will despair in their search for a landscape attuned to the times if future intentions represent an excessively unequivalent and hence weak image.

An image which evokes a corresponding action with a landscape and which goes beyond a mere recording of elements calls for the construction of a melancholic type of design. In response to the loss of cultural landscape, the utopian type of design tried to design a new one – whereas the melancholic one designs places in response to the loss of integrity. In this sense, the design process can no longer rely on the establishment of traditional images. The perception of an unfamiliar landscape requires a view that comprises, in equal measure, the prejudice and the action-oriented comment on the place – in the sense of powerful images, which as landscape units categorise wishes and decisions in terms of that which is given. Thus, overdrawing the context opens the possibility of experimentation.

There is no special demand on the landscape, there are only different players with various inquiries with regard to the landscape. In this connection, landscape must become a disposable product by differentiating landscapes, and potential users can fall back on the landscape of their interest. Landscape attains a value whose reduction would diminish the interest in this landscape. Landscape must be able to instrumentalise a space in the sense of the intended purpose, which is disposal. This would reflect neither the disunity of the landscape, nor attempt harmoniously to cancel it, but continue building.

1 see: Jauss, Hans Robert: Kleine Apologie der ästhetischen Erfahrung. In: Stöhr, Jürgen (publ.): Ästhetische Erfahrung heute. Cologne 1996.

2 see: Huber, Andreas: Heimat in der Postmoderne. Zürich 1999, p. 258-259.

3 see: Fellmann, Ferdinand: Symbolischer Pragmatismus – Hermeneutik nach Dilthey. Hamburg 1991, p. 51.

4 see: 3. Internationales Architekturkolloquium an der Bergischen Universität Wuppertal: Internationales Kolloquium Stadtlandschaft 1997.

5 Collection Martin Parr: Langweilige Postkarten. Berlin 2001.

6 see: Boehm, Gottfried: Zuwachs an Sein – Hermeneutische Reflexion und bildende Kunst. In: Klüser, Bernd (publ.): Hans-Georg Gadamer – Die Moderne und die Grenze der Vergegenständlichung. Munich 1996, p. 95-125.

Resource Region: Regionalpark Mitteldeutschland

Harald Kegler Lutherstadt Wittenberg

Regionalpark Mitteldeutschland (Regional Park of Central Germany) is a creative vision and a political and cultural task. This approach stands for similar tendencies, challenges and action options of regional planning in other urbanised or suburbanised regions in the industrial world. For over a year now, Regional Park Mitteldeutschland has been taking shape – it is being discussed, taken note of, viewed euphorically or sceptically or with disbelief and at the same time driven on with concrete plans and projects at various levels, regional levels and political bodies by initiatives, enterprises and associations. It is "in a state of flux". According to one thesis it is a concrete expression of a "resource of a new kind" for shaping space.

1. Starting Position

The Regional Park is situated between the conurbations of Leipzig/Halle and Berlin/Potsdam. This area has been one of eastern Germany's regions of dynamic transformation since 1989/90: Formerly a central industrial area in this part of Germany referred to as Berlin's "back yard" and characterised by chemical industry, machine building, mining and extensive military areas, but first and foremost by severe environmental pollution, this region initially became a deprived area – economically and socially. But in the meantime it has also become a region with an influx of enormous investments for rehabilitation, infrastructure, major industrial projects and urban and landscape renewal. The cities of Leipzig, Halle, Dessau or Potsdam have become places of radical transformation, of contrasts and of shrinking population. But they are also areas of a variety of new projects, some of which were presented to the world public at EXPO 2000.

2. Components

This corridor with a population of at least two million (not counting Berlin), which measures about 150 by 60 kilometres and touches four states of the Federal Republic, represents an unusual sequence of urban and rural landscapes:

- Formerly dominated by lignite mining, the southern Leipzig area is one of Germany's most extensive and multifaceted landscapes. Extensive rehabilitation and reconstruction work is in progress. Formerly an open-cast mine, Cospudener See is the first newly created lake which has become a popular tourist attraction.
- The northern Leipzig area with its new fair grounds, an international airport and numerous commercial locations, is one of the most dynamic areas, with direct connection to Leipzig and Halle and cross-regional rail and autobahn links connecting Berlin, Northern and Eastern Europe, southern Germany and Southern and Western Europe.
- The urban and suburban settlement landscape between Leipzig and Halle represents an area of concentration of European dimensions: the large modern chemical industry area of Leuna/Merseburg, old town centres and resort towns, old industrial sites, industrial areas stemming from the period of rapid industrial expansion in Germany (1871-73), settlements built between the two world wars, prefab slab settlements – areas of renewal interspersed with signs of decay everywhere, as well as a mix of the most up-to-date commercial sites, shopping facilities and settlements along autobahns. A "new centre" seems to be coming into being between the major cities – a fascinating and at the same time amorphous development space around the centres of Leipzig and Halle, which also form part of the Regional Park.
- Adjoining this area of concentration to the north-east is one of the most easily remembered regions in terms of its cultural historic features; it has become one of the most innovative regions in recent years: The industrial and mining areas of Wolfen/Bitterfeld/Delitzsch, the Gartenreich Dessau-Wörlitz (Garden Realm of Dessau-Wörlitz), Wittenberg (the town put on the map by Luther), the Bauhaus city of Dessau, the Dübener Heide Nature Reserve – a valuable natural and cultural landscape in the centre of the Regional Park offering interesting tourist attractions and novel services. The triangle formed by the cities of Dessau, Bitterfeld and Wittenberg, the industrial garden area, is the centre-piece. "Ferropolis – the city of iron" is the most spectacular project.
- Ferropolis is situated in the centre of the future Regional Park. This internationally known landmark and art centre and the former lignite mines (south-west of Delitzsch via Goitzsche, the Mulde river, Gröbern, Golpa-Nord to Bergwitz) form a multifaceted lake and cultural landscape. It will be a testimonial to creative rehabilitation of a landscape once damaged by man.
- The meadows of the Elbe, the Mulde and Saale rivers form an East-West divide over the entire area. These most sensitive areas are documented as a UNESCO biosphere reserve; they are also most attractive in terms of their ecology. The Dessau-Wörlitz Garden Realm, which in the meantime has been put on the World Heritage List, the reform landscape of the 18th century, stretches along the south bank of the Elbe. Today, it is a Mecca of cultural tourism. But the area is also characterised by a fragmented economic and settlement landscape which poses major planning challenges. The Elbe also links the region with other regions between Hamburg and the Czech Republic – i.e., an ecological refuge of international significance, a string of picturesque towns and a potential chain of future industry (from environmentally compatible shipping to fishing industry and tourism).
- Rising above the northern bank of the Elbe is the Fläming, a sparsely populated range of wooded hills with agriculture as the predominant industry and noted for its recreational value. Another nature reserve in which a proper balance has been struck between agriculture, forestry and individual tourism with a variety of services is situated. The settlement culture is characterised by small picturesque towns.
- Bordering this area is another lowland, the glacial valley of

Baruth – a tranquil zone characterised by open landscape; the accent here is on health tourism. To the north-east, this area abuts the suburbs of Berlin known here as the "Speckgürtel" – literally: "bacon belt" – characterised by an amorphous landscape made up of newly-built commercial facilities and residential areas and diversified infrastructure. Embedded in it is one of the many military conversion areas. Abutting this is the catchment area of the Havel river.

· The cultural landscape of Potsdam, which has also been put on the UNESCO World Cultural Heritage List, the attractive urban area of Potsdam and the media town of Babelsberg form the Regional Park's northern boundary adjoining metropolitan Berlin.

3. Opportunities

"The metropolises touch each other ..."

These landscapes are not simply a succession of "green stretches of land", "spacer areas" or "urban strips", but represent a compact "corridor of potential" in terms of their economy, ecology and cultural history. At the same time they are a space of contradictions uniting historically different cultural landscapes – some of them enjoying the world's highest level of protection –, which represent an almost unique succession of man-made landscapes overformed by Man. In a certain sense they are experimental areas for new landscapes and town sections – in the wake of the age of waste of resources. Although in the short term, the Regional Park should not be seen as a space that is likely to experience rapid population growth, but as one that is likely to show dynamic growth of new branches of industry. Growth and shrinkage are being shaped simultaneously, resulting in an equilibrium, which is the basis of sustainable development.

4. Alternative

The "balanced city": the metropolises keep their distance!

A "balanced city" can be an alternative to an intermediate city, to suburban sprawl. The Regional Park Mitteldeutschland represents the scope of action of players in space to form a synthesis of independence and "globalisation". The landscape is not "adapted" to an anonymous market, but reinforced in its cultural, ecological, economic and social identity when the players embrace this vision and shape it themselves. In this sense, the Regional Park is a cultural programme. The Regional Park acts as a framework for shaping a specific quality in individual parts. It can then become a "resource".

The metropolitan spaces of Leipzig/Halle and Berlin/Potsdam are linked by the Regional Park Mitteldeutschland and at the same time separated by it. A succession of anthropogenic landscapes in which people effectively manage their affairs and shape their environment with due respect for the natural and man-made heritage creates this uniting separation. The Regional Park forms a link between the former leading Central German industrial area and Berlin, establishing a new, balanced connection between the capital and its former "back yard".

5. Learning and Marketing

The Regional Park is also a place of learning in the field of landscape and town planning: it is a network of regional educational institutions with instruction offered in various fields tailored to the area and model places of renewal, i.e., also an "institution" which networks the development of individual towns, cities and landscapes with the region as a whole, co-ordinates research and education in the region and develops international contacts to other regions.

A marketing strategy built around the Regional Park gives the individual regions greater weight within a wider framework. With the Regional Park, the Central German area can attain a position which is perceptible in concert with the major conurbations and which can highlight its special features.

6. Term

The Term Mitteldeutschland (Central Germany) relates to the 1920s. At that time, the Central German industrial district came into being as a planning and economic zone stretching across Leipzig, Erfurt and Magdeburg, with Merseburg as its hub. Today this term is still used colloquially, even though historically, Central Germany no longer exists.

7. Comparison of regional projects

Regional Park Mitteldeutschland is part of a current trend toward regionalisation, which has been going on nationally and internationally since the '90s. A new dimension of regional development was opened with such regional projects as the Emscher Landscape Park.

Similar major regional projects have come into being in the Frankfurt/Main area and include the Landschaftspark Bodensee-Oberschwaben (Landscape Park of Lake Constance, Upper Swabia). Work on another regional park was started recently when the IBA Fürst-Pückler-Land (International Building Exhibition Prince

Pückler Land) was opened in Lusatia. Experience gathered in the IBA Emscher Park and in the Industrial Garden Realm projects is being utilised here. Regional parks are also increasingly being created abroad as a form of integrated development. They include a park in the Po Delta in Italy, a similar park east of London, the "Regional City" in the USA (e.g. Portland, Oregon) and others.

The Regional Park Mitteldeutschland not only follows a trend, but adds a new facet to it by virtue of its characteristic "European dimension" and its specific approach, namely that of a balanced city.

8. Model

The Regional Park can be seen as a model for a new kind of thinking as regards resources. It is no longer simply a matter of clearly definable "available" nature, but of how to shape balanced relations between nature and the people making use of it and of the cultural aspects involved. The towns, cities and landscapes used by people are the most important resource for a future fit for human beings. Now that the entire landscape has been overformed, exploited and reshaped or "done up" by Man, new ways need to be created to make it possible not only to counteract the urge to equalise and adjust to other regions in the sense of increasing one-sided efficiency of utilisation, but to provide a qualitative alternative which arises from the existing stock. Moreover, a new regional policy, reaching beyond administrative confines and parochial politics and yet preserving local independence and armed with the necessary competence for action, must be put in place to carry out this planning and development work. New methods of participation (e.g. "Charrette") and relations of co-operation between the players (networks) are equally necessary. But first and foremost, the hierarchical spatial organisation must be fundamentally reviewed and replaced by a framework policy of spatial development oriented on identity-preserving models.

Literature

Calthorpe, P.; Fulton, W., 2001: The Regional City, Washington.

de la Chevallerie, H., 2001: Landschaft zwischen den Städten – die Idee der Regionalparks, in: Das Bauzentrum/Baukultur 1/2001, pp. 12-18.

Böhme, G., 1997: Gärten und Wüsten, in: Die Gartenkunst 2/97, pp. 235-238.

Dokumentation REK Dübener Heide, ABRAXAS/Labor für Regionalplanung, Weimar, 2000/2001

Forßmann, J.; Kohlbrenner, U., 1999: Nachhaltige Entwicklung – eine Herausforderung an die Regionen Europas, Ergebnisse der 3. Europäischen Planerbiennale, Herne, pp. 265-271.

Freistaat Sachsen, Staatsministerium des Innern (publ.) 2000: Regionalentwicklung durch interkommunale Zusammenarbeit im Freistaat Sachsen, Dresden.

Gratz, R. B.; Mintz, N., 1998: Cities back from the Edge, New York.

Hämer, H.-W., 1999: Lerngemeinschaften, in: Industrielles Gartenreich-2, pp. 222-229, Dessau.

Kegler, H., 2000: Von der "Überlandplanung" zum "innovativen Milieu", in: et – Magazin der Regionen, 2/2000, pp. 58-61.

Kegler, H., 2001: Regionalpark Mitteldeutschland, in: et – Magazin der Regionen, 4/2001, p. 45.

Kegler, H., 2001: Dübener Heide – zwischen Heidemythos und neuen Landschaftsbildern, in: et – Magazin der Regionen, 3/2001, pp. 31-35.

Wessling, Ch., 2000: Verflechtungsstrukturen im Metropolraum Berlin-Brandenburg, Cottbus.

Umlandverband Frankfurt 2001: Tätigkeitsbericht 2000, Frankfurt/Main, pp. 10-13.

www.industrielles-gartenreich.de

Negotiating Margins, Reclaiming Peripheries
The 'Wilderness' Imperative in Architecture and Urban Design

Leigh Woolley Tasmania

Hobart, Tasmania – a city on the edge of 'wilderness'

In the continuum of nature, conventional wisdom suggests wilderness at one pole and the city at the other. Although nature and human perception are indivisible[1], this dality is deeply ingrained in the cultural development of city. Rykwert refers to the process of loss when, in founding the early city, the act of enclosing a piece of ground 'broke the texture of nature by cutting it out of the continuity of landscape'[2].

The remnant landscape of wild nature, though constantly subject to the 'top dressings of modernity'[3], is nonetheless increasingly revered. It is recognised as offering an intrinsic place of encounter or a moment of truth where humans are themselves drawn into its order of being, "Wild nature has a kind of integrity, it is creation itself… we encounter our sources, beyond our resources"[4].

Distinguishing the architecture of culture from that of nature is therefore of particular relevance to the expression of human ecology. Although managed wild nature is a contradiction in terms, those cities nearer the wilderness pole have an important role to play because they focus the urban periphery. The terrain of experience is one where the margin between nature and settlement (be that rural or urban) is not only readily apparent, it is constantly being negotiated.

While the dynamic tension between the inhabited space of the city (inside) and the unbuilt beyond (outside) may be universal[5], the opportunity to locate oneself within such extended landscapes, from the centre of cities is increasingly denied. Maintaining this tension is a challenge to urban design theory and practice because it demands reconciliation of the contemporary with the immutable. It also extends definition of the urban beyond the built.

Those cities where the landscape of nature is strongly present can offer specific insight to this process. These 'edge cities' demand re-evaluation of the scope of urban design beyond its traditional centre focus to an investigation of city region and beyond. An initial analysis of some of these cities, where the natural force of the land is particularly apparent, suggests the following characteristics. They each for example:

- Consider their regional scale to ensure that the 'naturalness' that underpins their culture is not lost.
- Pursue urban design policies which reinforce and build (on) the topography of the place.
- Reinforce the experience of the location and its capacity to accentuate and focus (wild) nature.

They also
· Generate strong civic identities as a result of their extended 'public realm' – an enduring publically shared resource.

As the evolution of the city and its architecture becomes 'The shape of the earth as it is modified by the structures of mankind'[6], so it also expresses the terrain of experience, the landscape of nature. Far from the wilderness construct being the inverse of city, its role as counterpoint should be to assist in appreciating the true scale (particularly its built limits) and the origin of cities. While all cities (and their architecture) demand this counterpoint, very few acknowledge the spatial continuum from city to wild nature in their urban design and planning. Fewer have strategic policies in place based on this conceptual and physical reality.

For most cities, even those with strong topographies, there is at best, a glimpse of protected non-built space. There is, however, a decreasing number of cities where the experience of the undeveloped natural setting is still more dominant than the built, and where from their centres, 'unbuilt' space is still more visually extensive than developed space. Such places raise a number of questions – including how the public realm, as that space managed in the public interest, is defined and what principles should be developed to deal with the planning and design of the extended domain of the city.

To design within and for those regions in which one practices, it is necessary to develop spatial values or constructs capable of operating across different scales, be this at the regional, precinctual or immediate vicinity. Initially this demands an investigation of landform structure and urban morphology of those city regions in which one works. The intent is to ensure that both 'nature' and 'culture' are respected reciprocally as the urban experience, while attention can be brought to sustaining the regional eco-system that lies beneath the existing urban infrastructure. 'A city's natural environment and its urban form, taken together, comprise a record of the interaction between natural processes and human purpose over time'.[7]

But this process is as much an act of mind and spirit as it is of physical planning. In referring to eighteenth century French ports, Claude Levi-Strauss suggests their value lay in a relationship that still existed at the time where a port was 'a human settlement which did not completely destroy, but rather gave pattern to, the natural relationships between geology, geography and vegetation, and thus offered an exceptional kind of reality, a dream-world in which we can find refuge'.[8]

Landscape therefore needs to be considered not as environment but as habitat, as inevitably the interactions between human activities and the natural environment produce an ecosystem different from the one that existed prior to the city. In referring to 'our greatest asset' being damaged by 'an imperialistic ego that believes in the illusion of its own separateness', Australian writer David Tacey suggests the need for Australians at least to develop a deeper, spiritual pact with the land[9], in order that we may become more fully human. There are no better cities from which to address this than through those which can sharpen the distinction between urban and wild landscapes.

The importance of the defined edge, the obvious margin, is not only a proposition at the scale of city, and indeed the cultural scale of region, but also one at the architectural scale of site. The annexation of 'border territories' suggests Leatherbarrow, can confuse identity by disturbing the integrity of a building plan, precisely because it integrates into the surround what one would consider properly excluded from it.[10]

When these spatial values are applied at different site scales and distances, they reflect an interconnected set of processes informing architectural projects. In my own work these include:

'Disclosing terrain':
Responding to movement across the land. Experiencing the force of the land by acknowledging ones natural inclinations, while indicating how topography has been revealed through the process of settlement

'Spatial Prospecting':
Locating individual sites within their regional frame and seeking the freedom of a prospect while demanding the security of a refuge

'Building topography':
Incorporating construction into the process of topography's perpetual becoming. Topography is recognised as the first fact of architecture as well as its fundamental resource.

The role of these values can be further pursued when investigating urban morphology/city structure.

An (abridged) Case Study[11] – Hobart, Tasmania, Australia

Tasmania's capital Hobart, is a city in a 'good place' – it is among only a handful of settlements lying 42 degrees south and beyond. In contrast to its small population and small inventory of built assets, its oceanic context and landscape assets are considerable. Hobart nestles between water's edge and the surrounding abrupt terrain. The sense of scale and proportion provided by nature has a powerful presence here. Ground level views from the central grid of colonial streets stretch one's gaze beyond the built. They connect Hobart's central area streets with the high ground of the Tasman Peninsula some 60 kms away. They confirm how thin the veneer of settlement is and how extended the experience of 'urban' space is within this region. The contrast between topographic containment and urban extension provides Hobart's particular spatial experience.

The diverse topography provides the city with an environmental resource of national significance. The transition from coastal vegetation, through dry lowland and moister upland forests, to alpine scrub and moorland is nowhere in Australia as compressed as here. A microcosm of the state's vegetation and a transect of the nation's botanical wealth provides the constant backdrop and reference to the evolving form and fragile urbanity of the city. The advantages of being a quieter, greener and more reflective place demand to be pursued spatially.

Not surprisingly, it is the unbuilt space of the city which provides its dominant image. The edge between the built and the natural is more obvious in this setting because the natural force of the topography makes it so. The landscape obtrudes, and like many other Australian settings 'insinuates itself against our very will'.[12]

The way the built should contribute to the expression of city, however, has yet to be seriously addressed at the regional scale. Thus far the debate is limited, less by an investigation of form, and more because it has been focussed around saving the bush rather than designing the city. Until this shift occurs the city will be left dealing with the design of pockets of 'found' space, rather than 'structuring' the city as part of the continuous space of the landform. If this can be achieved, it will confirm that most sought after urban asset: a small, compact and legible city within the extended domain of wild nature.

But it is not just as part of the local that these edges deserve critical attention. Tasmanian settlements, particularly Hobart, have a role beyond the local, because they focus the sense of a significant national periphery. Few cities are so sharply defined by their relationship to natural environment values, particularly those of wild nature. Those cities which can maintain some of what was nature's own before human arrival confirm the city as both a sheltering domain and an eternal one.

Truly urbane cities will be those which negotiate their margins by acknowledging their regional and human ecology. Reclaiming ill-defined peripheries from the unconscious extension of city, ensures that 'wilderness' will not be threatened or continue to shrink in the face of an ill-understood or de-humanised city. It is from cities such as these that urban habitation within wild nature can be pursued.

---

1 Simon Schama, Landscape & Memory, Harper Collins, London 1995, p. 6.

2 Joseph Rykwert, The Seduction of Place, Weidenfeld and Nicholson, London 2000, p. 13.

3 For example, the primeval forest as ancient place is now likely to be termed 'wilderness park'. See Schama op. cit. 1995, p.16, and Shepheard, Paul, The Cultivated Wilderness, MIT, Cambridge, Mass. 1995, p. 124.

4 Holmes Roulston, 111, Conserving Natural Value, Columbia University Press, p.15.

5 The history of the city has always focussed a dialogue between inside and outside, particularly in terms of its role as sanctuary and as 'shelter for humanity driven out of the Garden of Eden'. Rykwert op. cit, p. 13.

6 Vincent Scully Architecture, The Natural and the Manmade, New York St Martins Press 1991, p.1.

7 Ann Whiston Spirn, The Granite Garden. Urban Nature and Human Design, Harper Collins, New York 1985, p. 12.

8 Charbonnier, Conversations with Claude Levi Strauss, trans. by John and Doreen Weightman (Johnathan Cape 1969) in: Crowe, Norman. Nature and the Idea of a Man-Made World. MIT Cambridge, 1995, p. 205.

9 David J. Tacey, Edge of the Sacred. Transformation in Australia. Harper Collins. Melbourne 1995, p. 11.

10 David Leatherbarrow, Uncommon Ground. Architecture Technology and Topography, MIT Cambridge, Mass. 2000, p. 61.

11 For a more detailed assessment see: Leigh Woolley, Naturally Urban – The Making of Public Place on the edge of 'wilderness'. Siglo, Hobart Autumn/Winter 1999, pp. 44-50, and Urban Nature and City Design. Island Magazine. Hobart, Spring/Summer 2000-1, pp. 8-14.

12 Tacey 1995 op. cit., p. 6.

Center as Periphery: Studies on Revitalization of Traditional Downtown Areas

Sonia Ricon Baldessarini Berlin

The downtown area around the famous Pelourinho Square in Salvador, Bahia/Brazil, is the oldest intact historical site in Brazil. Tomé de Souza, the first Governor General sent by the Portuguese Kingdom, founded the city on 29 March 1549.

During Portuguese domination in Brazil Salvador was politically and economically the most important city, until 1763 when Rio de Janeiro, situated closest to the iron mines, became the capital. After the middle of the 19th century the traditional inhabitants left the downtown area and moved to the south, around the beaches on the bay area, the Bahia de Todos os Santos.

Since 1985 the Ensemble around the Pelourinho site have become part of UNESCO's World Heritage. The deterioration process of the historical district began in the 1930's. In the 60's effective measures aimed at urban revitalization were launched in this area. The first interventions designed to contain the deterioration of the historical buildings were based on the census of the Pelourinho Area, which was elaborated in 1967. It included the conservation and restoration of the single buildings.

The IPAC (Instituto do Patrimônio Artístico e Cultural da Bahia), which was created 30 years ago, played a central role in this policy. In its different phases of intervention, the IPAC has applied several work philosophies to redevelop the historical center of Salvador. In general they concentrate on the technical conservation and restoration of the buildings, the original purpose of its creation, the IPAC then made the recuperation of the area of the Pelourinho quarter a priority. Time showed that it minimized the effective intervention against the process of decay in the other quarters and ignored the complexity of social structures. This model of intervention reproduced the relationship "Center – Periphery" inside the historical district as an internal cellular problem with the renovated cultural and historical buildings on the one side, while on the other the buildings next to Pelourinho were left out of the process and remained in a "state of ruin". This situation clearly reflects typical dynamic of Brazil's increasing "void" between social sectors and their economies.

So are the studies which Brigitte Thomae developed in the 80s still valid today. They pinpointed the critical problems of the quarters around the Pelourinho/Maciel area as follows: poverty, prostitution, drugs, degradation of living conditions, and development of informal economic activities with low incomes.

It all resulted in a process of urban-architectonic deterioration and social degradation, which made old downtown Salvador a place of "periphery" character – a place to be avoided for many native inhabitants of the town.

Prostitution and informal economic activities are no longer a part of the "private" space in this area, because their demands had changed. It doesn't mean that these practices do not exist, as a form of interchange in the public space where tourists and visitors also circulate.

The revitalization of the historical area of Salvador entered a new phase in 1991, when a new Regulation, the "Termo de Referência", created a list of priorities and steps in the Recuperation Program for the historical center of Salvador. This document introduced an architectural quarter-management model and some strategies to relocate and/or dislocate the inhabitants. Among other forms of appropriation of the private space, the Program made the proposal, for example, to the owner of two buildings the cost-free restoration of one of them. In exchange the State would keep the second one as its property. One of the positive aspects of these new rules has been the expansion of the urban infrastructure and services to the historical area.

Although the interventions in the last decade increased markedly, the critical situation of the area as a whole remains. This could indicate that social problems almost always turn into urban-architectural problems or, in another words, the solution in both cases should be sought in tandem. Like other disciplines, and that is the epistemological reason why they exist, architecture and urban planning are not to be considered only as an aim in itself, but more as a way to deal with human problems and the environmental consequences.

As long as quantitative instead qualitative criteria are used, there will be no improvement in the living conditions and the respective infrastructure, and this point also applies to the urban character of historical traditional living areas.

Bairro da Lapa dos Mercadores in Rio de Janeiro

Alexander Von Humboldt was the first to call Rio de Janeiro, in Brazil, "the most beautiful town in the world", says Gustav Faber in his book "Brasilien: Weltmacht von Morgen". At that time the capital of the former Portuguese Colony already had several thousand inhabitants.

Humboldt was a nature lover and was probably very impressed by the gentle hills and sinuous topography of the old town. He could not possibly imagine what the later physiognomy of the city would look like, but he surely saw one of the most ancient landmarks of the historical central district from Rio de Janeiro: the Bairro da Lapa dos Mercadores and the aqueduct, build in the 18th century to bring water to the neighborhood.

Humboldt's expression became famous. "Rio, Cidade Maravilhosa" is now the second biggest city in Brazil (5,8 million population – 2,7 men; 3,1 women, IBGE 2000).

The City of São Sebastião do Rio de Janeiro was founded on 20 January 1565, at the foot of one of the many hills on the seashore, the Morro do Castelo, which was demolished in the second decade of the last century to give place to the Esplanada do Castelo. The colonial town expanded around the Square with the Residence of the Vice-Kings, today the Praça XV.

The urban modernization of Rio de Janeiro started at the beginning of the 19th century when the Portuguese Real Family arrived in 1808, fleeing from Napoleon's army. In keeping with European needs, the Monarchy build public and private buildings and parks, such as the Botanic Garden near the Rodrigo de Freitas lake.

At the end of the 19th century Rio de Janeiro became the capital of the new republic. The historical center suffered more interventions in order to make place for the new administration buildings and to improve traffic conditions. Large historical sites underwent demolition, while large avenues like Av. Rio Branco and Av. Presidente Vargas were opened. A consequence of this policy, that privileged the commercial and administrative use of the traditional areas and eliminated architectural historical substance, was the transfer of the inhabitants to other expansion areas. The upper classes moved to the beach quarters in the south, while the workers and lower middle classes moved to the suburbs in the north of the town. People with less income and the unemployed remained in the old houses, as well as some immigrant families with their small stores. They remained settled in the fragmented living areas in the surroundings of the traditional downtown area. Those areas form a "social periphery" along the commercial and administrative areas. One of these living forms is the "cortiço", a kind of urban "favela" where the private space of the old homes is reduced to "comodos" (small rooms). These small single rooms are sometimes occupied by a whole family with seven or more persons.

The process of destructive interventions in the urban context of Rio continued until the 70's. During this decade the historical district Lapa lost one of its characteristic hills, Morro de Santo Antonio, and some of the ancient houses.

Lapa quarter is the last bastion with living character just inside the historical downtown area in Rio de Janeiro. Like other examples of inner-city neighborhoods, it bemoans the same problems: poverty, prostitution, drugs, the degradation of living conditions, and the development of informal economic activities with low incomes. In addition to this, the homeless problem is extremely acute in Rio de Janeiro.

However, with the beginning of a revitalization movement in the late 70's, some changes can be observed. According to Lilian Fessler Vaz these changes can be classified into two tendencies: one can be related to the development in the European metropolis and "Global Cities" around the word. This tendency admits the "gentrification" and commercial and cultural use, with the establishment of cultural institutions and show business. The second tendency is associated with "downgrading", the progression of informal economic activities and the concentration of the homeless population. This situation also reproduces the phenomenon of "Center as Periphery".

Over-the-Rhine District in Cincinnati

The district Over-the-Rhine, in the City of Cincinnati (Ohio/USA), is like the other downtown areas cited above, an area that has long been known for its architecture. Several buildings were erected toward the end of the 19th century and provide a typical interpretation of architectural historicism.

The buildings still deserve the attention of the public and specialists. The major problem of the area, however, is social deterioration

In the past the district was mainly inhabited by German immigrants. The District Over-the-Rhine is located on the shores of the Ohio River which inspired them to name the place in honor of the River Rhine. The District occupies the northern section Cincinnati's downtown core. As an inner city district, Over-The-Rhine suffers from classic problems of poverty, conflicting forces of homelessness and gentrification, of housing abandonment and debilitation by absentee landlords.

According to the Over-the-Rhine Community Council the district displays the following features:

· Over the Rhine is Cincinnati's poorest neighborhood.
· of its approximately 12,000 residents, 95% live below the official poverty level according to the last census.
· historically Over-the-Rhine has received many poor forced-migrants from the Appalachians and the rural South
· 5,000 of the 7,500 apartment units in Over-the-Rhine are damaged or abandoned.

Besides these structural problems the neighborhood developed a strong civic community that struggles to remain living there.

Because of its special character Over-The-Rhine can be compared to neighborhoods in other American cities, such as the French Quarter, Greenwich Village, and Beacon Hill. Like these neighborhoods, Over-The-Rhine has suffered much urban decay in the last thirty to forty years.

In addition to some other measures directed toward urban revitalization in Over-the-Rhine, the district will receive a new impulse to strengthen its social and cultural identity: it is precisely in this eruptive downtown area that the Contemporary Arts Center is erecting its new building in Cincinnati, designed by the architect Zaha Hadid. It will be the first American art museum to be designed by a woman, and it is her first project in America.

Bairro de São Vítor in Porto

Apart from the classic forms of intervention in urban sites, Portuguese architects and authorities used the favorable climate of social and political change after the Cloves Revolution to try and implement a new model.

The SAAL (Serviço de Apoio Voluntário Local) was an official initiative to improve the living conditions and take care of the acute home deficits of the population. Although the initiative existed for only a brief period (1974 until 1976), it still managed to register some successful work.

Concentrating their activities on projects with residential character, the work groups (brigadas) tried to operate on the basis of social communication with the inhabitants and potential neighbors.

Some of the projects were carried out in the historical areas of cities like Lisbon, Porto, and Guimarães. The planners made critical revisions and interpretations of the old site structure in order to answer to the needs of the present.

In dealing with the complexity of the formal, social and urban character of the city landscape, it is possible to achieve a new specific identity for building, that will be reflected in the qualities of its use.

In his interventions in the housing complex of São Vítor's Quarter in Porto (1974-1977), for example, the Architect Álvaro Siza Vieira systematically worked with these conceptions.

References:

Alves Costa, Alexandre. Portugal Architektur im 20. Jahrhundert. DAM.

Faber, Gustav. Brasilien. Weltmacht von Morgen.

Prefeitura da Cidade do Rio de Janeiro. "Corredor Cultural".

Thomae, Brigitte. Programme u. Problematik der Altstadtsanierung in Salvador.

Vaz, Lilian Fessler. "O Bairro da Lapa", in: Viver en el Centro, Ludena, Pfeiffer ed.

Wilhelmy, Herbert. Südamerika in Spiegel seiner Städte.

Zane L. Miller and Bruce Tucker. Changing Plans for America's Inner Cities.

Cincinnati's Over-the-Rhine and its implications for city planning in Ohio.

Land and Democracy

Don Grant **Lisa Ting** **Ian Williamson** Sydney

Land, and the interaction of human societies with it, has resulted in many economic, social, political and environmental outcomes and concerns. The dialogue between these competing and overlapping factors requires a land administration system that is able to support the ever-changing relationship between humankind and land, to facilitate complex decision-making and to support the implementation of those decisions. Therefore, appropriate and effective land administration is of crucial importance for sustainable development.

Fortunately, there has been a growing awareness of these issues and, as a result, world attitudes are changing. Nations are now generally more conscious of the fact that the actions of individuals can have global consequences. This awareness has arisen as a result of the greater availability of information. This in turn has re-ignited the debate about how land can best be administered for the good, not only of individual landowners and users, but also for the community as a whole. The imperative to re-examine land administration systems in the context of sustainable development is now overwhelming.

It is likely that, as the concept of obligations in land grows and as the complexity and scope of interests increases and becomes, not only parcel relevant, but also areal, zonal, catchment or regional based, a new approach to the recording of these interests will emerge. At the global level, it is clear that ecological systems treat jurisdictional boundaries with contempt. As the reality of these new concepts becomes widespread there will be a need for a change of individual and group behaviour in relation to land use as the societal rights take on a new significance. Spatial information infrastructures that embrace such diversities are a necessary component to be built into any meaningful re-engineering of land administration systems.

Whatever form it takes, a public recording of these rights and responsibilities means that members of society will be made aware of the behaviour of others and the impact of their own behaviour on members of the state, national and perhaps global community. The consequences of local actions therefore will be seen and considered in the framework of a global cadastre by adopting the concepts of custom, territoriality and the links between humankind and land, which encompass the symbiotic relationship between terrestrial space and the behaviour of human society. A capacity to record the landscapes of the mind, which holds the key to customary land rights is necessary and should be developed. As citizens of the world we should direct our energies to explain and instruct society at large in the means of comprehensively showing, and making public, individual and group rights and obligations to the land.

The Art of the Feasible
The Role of Construction in Architecture

Jörg Schlaich Stuttgart

After the industrial revolution had made it possible to manufacture iron and steel structures with systematically planned geometric proportions and strengths, the profession of the all-round master-builder was superseded by two occupations concerned with planning and designing buildings: that of the architect and that of the engineer. The report prepared by Poloni in 1748 on the dome at St. Peter's in Rome is considered to mark the birth of the civil engineer. Strictly speaking, Poloni was "only" a structural engineer, because he merely analysed what existed, rather like the natural scientist who simply discovers what nature holds in store for us. This is very different to the work of the structural designer who invents objects that are completely new and hitherto unheard of! In their designs for the magnificent Galérie des Machines for the 1889 Paris Exhibition, structural designers brilliantly used their knowledge and imaginations to design totally new forms that broke with all that had existed. The hinged joints that supported the structures of the hall overthrew all the old notions and conceptions cherished by architects. The cross-sections of the supports were narrowed down at that very point where the forces are greatest: at the base, just above the ground, right where the structure ought to have the greatest mass! – Isn't this what nature teaches us? This novel design swept the ground away from under the feet of those – in other words: the architects – who had learned the art of construction without studying statics and had – at best – a sense of the flow of forces, yet lacked the experience. Wisely, and with long faces, they left these structures to the engineers, turning instead to decorating the engineers' elegant filigree railway stations and bridges with magnificent facades and bombastic towers.

When we, as engineers, discuss our contribution to building culture, wouldn't it be more appropriate if we spoke of works of engineering? And don't we all – architects and engineers – basically want the same thing: namely, good buildings – buildings where the question of who has created what is ultimately irrelevant? In any case, the division of labour generally decides this on a rational basis. The relationship between the two professions evidently suffers from a congenital defect: offended, the architect withdrew because the engineer, confident in his newly discovered creativity, rejected the autocratic rule of the architect. The architect, instead of taking up these ideas and drawing inspiration from the engineer, failed to recognise the signs of the time; he reached out for whatever he could find – for all those things in which the engineer showed no interest – and dug into the construction box of history (how appropriate!). The engineer, in turn, exploited his new knowledge as an instrument of power: "This is how it is done. Full stop!" And, in so doing, he degraded his role to that of engineering structures, instead of conveying his knowledge in alternative designs and co-operating with the architect to produce holistic solutions.

When reinforced concrete came into use during the first half of the 20th century, their roles changed again. But this time the architect won out. The architect was now in a position to design structural forms which, owing to the calculations demanded, confronted the engineer with insoluble problems. Whereas, for example, the layman has no difficulty imagining a concrete-poured ceiling as an alternative to a wooden joist floor in his home, this is a gigantic challenge for the engineer. And although the lever principle will suffice for computing statics, nothing less than a fourth order partial differential equation is needed to design a flat slab, not to mention the complex mathematics involved in designing a curved shell.

In his Space, Time and Architecture, Siegfried Giedion welcomed ferroconcrete as a material that had emancipated architecture from its fetters. No longer tied to the engineer, the architect was now free to design plastic forms. The forces could now flow freely (in correspondence with the unlimited democratic ideas of the 20th century) in all directions. Giedion argued that architecture had taken the lead, and frequently demanded more of the engineer than the later was able to deliver.

The fact that architects were far more adept at using "their" concrete than engineers were, sent shock waves right through the engineering profession, which was still mainly working with reinforced steel. And to this very day, engineers have apparently still not recovered from this shock. After a brief revolt, which produced some beautiful, suspended reinforced concrete shells (Nervi, Candela, Isler) and a few outstanding bridges (Maillart, Menn, Manterola), they simply gave up. From this moment on, engineers became the sycophantic slaves of architects, calculating everything they could with their 6B pencils without realising that a composed architect is only too pleased to submit to the "discipline" of a good structure, and actually seeks the assistance of the engineer as a partner in design. The comment made by the architect Michael Graves, whom nobody would surely suspect of being a constructivist, typifies this development: "now that I can erect the columns wherever I want, I no longer know where I ought to put them." And so, form and structure went their separate ways, despite the fact that they belong together like wine and bottles, music and timing, dance and rhythm.

However, the crazy notion gained currency that the architect was responsible for the overall design and the engineer for the statics and the structure. However, what really distinguishes them are the complex social design parameters on the one hand, and the non-functional technical parameters on the other. But they are both responsible for the design within their respective fields. Just as the engineer has never learned to design a semi-detached house in its totality, the architect lacks the necessary know-how to design a bridge. They can only co-operate successfully if they can listen to and inspire one another.

We engineers obviously welcome the way architects have changed their understanding of, and relationship to, construction in recent years. These high-tech structures constructed of steel and cables make use of technological developments in materials and computers in a variety of ways: from determining the technical design, to static and dynamic calculations, to CNC production and (in the near future) to GPS and robot assembly. These developments owe their origins and popularity – in the early days at least, let's say in the 70s and 80s – to the tendency to render the flow of forces visible in order to win favour with designs that made load-bearing behaviour intelligible. People like things they understand. The result is distinguished by its simplicity and transparency. Compressive stress flows into tubes, tensile stress into rods and cables made with (the greatest possible number of) fork heads, fittings and turnbuckles – preferably of special steel. We, as engineers, obviously welcome this approach because it gives us an opportunity to make an original contribution to architecture. Consequently, we do not see – or do not want to see – when we exaggerate, or allow ourselves to be pressured into exaggeration by architects carried away with forces, who unnecessarily maximise the number of building components. Such developments become ends in themselves, and display exhibitionistic traits. High-tech elitism turns into a game of maximising effects for effects' sake as, for example, a comparison of the John Hancock Center in Chicago with the Hong Kong Bank demonstrates.

Solar chimney power station Architects: Jörg Schlaich and gmp von Gerkan, Marg and partners

In recent years, the high-tech approach that once justified itself with the argument of lightness and efficiency alone has been enriched by the ideas of the ecology movement. Its ideas provided a welcome pretext to clad high-tech structures in glass, thereby creating bright, beautiful, sun-flooded rooms, courtyards and atriums. Here too, of course, high-tech can rapidly come to mean maximum effect. In the process, reason and efficiency were transformed into pomposity and wastefulness; for example, when glass areas are maximised to the point of absurdity and all available technical means are employed to ensure that these buildings can be used at the height of summer and in the depths of winter. People talk of "energy-saving" buildings using "intelligent technology" (as if moronic technology existed), and of "intelligent systems". In so doing, they hide – or refuse to admit – the fact that the energy used to construct these buildings and operate their technology exceeds that saved by the passive or active use of solar energy. Both sides are to blame here: the design architects as well as the engineers who maximise glass areas regardless of where a building is located and which direction it is facing. Then they use sunshades and antiglare shields as a welcome excuse to erect double facades wherever possible, rather than where it might make sense. In addition, there are building physicists who believe that concert-hall quality acoustics and air-conditioning are necessary all day long and all the year round, regardless of location, the length of time people stay in these buildings, and whether people can protect themselves with natural means. Why aren't winter-garden conditions considered sufficient in a railway station hall or a covered courtyard for those few days a year when people shiver with cold, sweat, are dazzled, leave on their coats, or use the newspaper to fan themselves or cover their faces? Human beings have seven senses. And, fortunately, it is not possible to satisfy all these senses continually or simultaneously. And surely most people are willing to make a "sacrifice" for a beautiful, inspiring ambience as they would do in the countryside, where nobody is to blame when things go wrong.

Shouldn't it give us food for thought knowing that we can only sell this high-tech-effect architecture to "the rich" here and in Arabia (but not in India and Ethiopia) – in other words: wherever money counts for less than representation and show: among the banks and insurance companies, at airports, fairs, etc., but not at housing projects, universities – or at best, perhaps, in satirised form on commercial estates and – occasionally – in industry. Hence, apart from asking whether this architecture really is ecological (one only has to think of all the hullabaloo about photovoltaics) we ought to admit that it can be also very asocial: both with respect to the projects to in which it is used and the means employed. We need to display greater social responsibility and a general awareness of ecological issues, in other words: show more restraint and discipline, and pay greater attention to inner than to external values.

Alongside the creative work they have done on "intelligent buildings" designed by architects, engineers have, over the past few years, made a few innovative contributions of their own to building culture, thanks to the technological advances mentioned earlier. Engineers design and construct wide-spanned cable-net roofs and membranes for halls and stadiums, filigree glass domes for courtyards and swimming pools, communications and lookout towers, not to mention bridges, dams, watercourses, airports, etc. The civil engineer is responsible for the entire technical infrastructure, and fully aware of the cultural obligations this implies. He knows that the only way he can adequately respond to spoiling the natural environment is to offer culture in return, and that an infrastructure without culture is a technocracy, but that infrastructure with culture means civilisation. As far as this last question is concerned, he also knows, incidentally, that he still has to make an immense effort when it comes to designing bridges, especially railway bridges.

Natural science and intuition come together indivisibly in the design of civil engineering works. Consequently, we engineers can always justify what we do, rely on our know-how and – when we are in doubt and at a loss for ideas – simply do the right thing.

Until recently, statics and manufacturing engineering had imposed painful constraints on the engineer. All this has changed thanks to the growing use of computers. What a nice thought: this high-tech tool has enabled us to rediscover simple calculation methods, such as the brilliantly transparent, distribution-of-forces calculation spread over two pages. And now we can design again without a care in the world, because big brother is there in the background, keeping a watchful eye on us. This development could lead to a new diversity of forms. Computers are stupid but hard-working devices. To a computer-controlled saw it is irrelevant whether it cuts lots of bars of equal length, or countless bars of different lengths down to a fraction of a millimetre. If a tensegrity dome is to be constructed, a CNC mill will cut the complex nodes for the same price, no matter whether they are of equal length or different lengths. Consequently, much of the life-work of icons like Konrad Wachsmann, Buckminster Fuller, Max Mengeringhausen and Frei Otto now seems obsolete, despite all the years of intense effort they put into systematising space-frame structures. Their buildings benefited not only in terms of manufacturing engineering but also visually from the displine imposed by both rules and the serenity of uniformity, as found in the European brick house and the Japanese house with its tatami mats. But isn't there now a danger that the boundless freedoms offered by modern manufacturing engineering will produce arbitrary forms and even visual chaos, because everyone can make anything they like given the right software?

In fact, experience teaches us that when we are designing, we always – and indeed must – come up with new ideas when we face uncommonly difficult tasks and marginal constraints. On the other hand, nothing is more certain to deaden the imagination than a trivial design environment. In other words, when – thanks to the computer – "everything is possible" from the point of view of statics and manufacture, we may one day have artificially force ourselves to produce special and striking solutions for competition entries, for example. A very unpleasant thought, at least for engineers. Take the Wild Ones, for instance: they outdo one another with inflated and crumpled blobs, often satirising and caricaturing misunderstood "idols" in works lacking all relationship between form, function and the flow of forces. Let's hope this is a fleeting fad in our "good-time society". At the same time, however, there is no way that we, as engineers, can seriously oppose progress by busily opposing the very automation that takes the drudgery off our hands.

The hope remains that, in designing our buildings, we will be able to pass on those things that we repeatedly experience in ourselves "in real life": i.e. that self-discipline and a liberal environment do more to promote self-responsibility and creativity in the long run than warding off external constraints. (Just as wars may well give sudden boosts to developments, but can never be constructive in the long run.)

I would like to conclude with a recent example from my own field of work, because it is still so fresh that I cannot find any development of comparable scale, nor one applied so consciously. I was in Berlin, looking at the glass roofs of the DG Bank with Frank O. Gehry, the architect. One of these six glass structures alone needs 2,490 bars of various length and 826 nodes with 14,940 angles milled from special steel. And what did I think? I liked it anyway!

Working at Three Scales

Newton Harrison Helen Mayer Harrison San Diego

This presentation will consider planning at three scales – city scale, regional scale, and sub-continental scale. I should explain that over thirty years ago we, Helen Mayer Harrison and I, as an artist team, took the decision to do no work that did not variously comment on, support, restore and value the ecosystem in which the place we found ourselves invited to work was embedded. Thus we found our way into the domain of ecologically based planning through the lens of the artist's eye. We hold that every place is telling the story of its own becoming, creating and destroying its own history moment by moment. We also hold that the artist is, in as many, although different, ways than the writer, the storyteller of the culture. And, of course, no story can function as a story without embodying existing metaphors and out of this process of creation, new metaphors emerge. The narrative properties of our work take a variety of forms, models, diverse visual images, and diverse linguistic approaches. The outcomes tend to be installations that propose new elements designed to first create new space for the mind and thereafter engage the built history of place.

Frankfurt: The Citywide Scale

When Rolf Toyka, Director of Education for the Chamber of Architects of Hessen, and Manuel Cuadra, Chairman of our Committee, invited us to be part of a workshop that would make proposals for the city of Frankfurt, we posed to ourselves the same questions we always do:

1. How big is here?
2. What are the boundary conditions of here?
3. What are the social and economic conditions that operate?
4. What is the state of ecological wellbeing of this place?
5. What is endangered here? And if so, how?
6. Are there new patterns emerging? And if so, What are they?
7. Can we perceive other patterns that might possibly emerge?

At this point, based upon our past experience, intuition, and research, as well as the information that flows to us from the direct experience of place, we search for what appears important to say, and how we may add value to the story of the becoming of this place.

In Frankfurt, we soon grasped the fiscal difficulties of the city: based on

8. a reduction of its tax base due to the exodus of the wealthy
9. a rather large debt to the banks as a result of developing the cultural infrastructure
10. an unduly large administrative structure
11. the amount of monies variously required by unemployment and
12. a general reduction of funds that came of the flow of monies to the east.

The major remedies for this state of affairs put forward by internal planners represented two opposing views. The first was for the city to expand, to annex some of the encircling towns that were attracting Frankfurt's population and thereby to keep its tax base. The other was to not expand but to grow upward and to improve quality within. We chose the latter as the better possibility. The former was fraught with difficulties that we enlarged upon in our Frankfurt proposal, which has been reproduced for those present at this conference. The concept that we developed is based on valuing the farms within and bordering the city, of which there are many, as well as the forests, the meadows, and the rivers. Frankfurt is blessed with a rich potentially biodiverse surround and the City of Frankfurt could profit by acting as a figure within this potentially valuable field. The headlines for the proposals are as follows:

On Valuing the Farms, the Forests, the Rivers, the Meadows and the City of Frankfurt within its own Boarders: The Great Encircling Walk

On an Attentuated Botanical Garden that addresses the Future of Farm Life and Botanical Life for the City of Frankfurt

On Farming in the City of Frankfurt: On the Production of Certain Foods for itself within its own Boarders

On a New River for Frankfurt

More on the Great Circling Walk

Fundamentally, the concepts that flow from these headlines are as follows: A botanical boundary running through the farmlands which are so apparent along one edge of Frankfurt, the farms themselves valued differently, and green farming subsidized by the recycling of the waste of the city, enhancing the topsoil. The forested lands encouraged to go into succession. The center of the city grows upward and the neighborhoods refurbished and valued. Finally, a new edge is created by the construction of a stream dropping from the Main at the upper border, swinging through the forest, and coming back into the river at Frankfurt's lower border. The waters themselves are purified and space is located for some high-end development at the new river edge. If the concept were accepted, Frankfurt would become a figure within a diverse field, partly ecologically driven, part botanically driven, part farm – in fact, a new eco-urban edge would be created to form a diverse boundary for the city. An amenity, an adventure, a limitation that would encourage the creation of value within its borders!

We believe that every city will in due course need to grasp and make a virtue of its own boundary conditions. If we are thinking about the increase of population, and the concentration of population, if we are thinking about boundary conditions for cities, particularly those advantaged by existing within larger farm and forest domains, then we need, we believe, to address these boundary conditions and limitations so that the boundary conditions themselves feed value back into the communities within them.

The Regional Scale: Holland

The second scale is regional in nature, the approximately 3,500 square kilometers of the Green Heart of Holland in relationship to the Randstad. The Harrison Studio, at the request of the Cultural Council of South Holland, undertook a consideration of the Green Heart. Beginning with an appreciation of the appropriateness of the metaphor embedded in the naming "Green Heart"; they searched for its meaning and its identity. The Green Heart can be seen as a vast farm centering a ring of cities – Rotterdam, Den Haag, Amsterdam, Utrecht, et al. Its landforms represent the history of Holland. It is a landmark by which the rest of the world recognizes Holland. It can be seen as a great Central Park for this ensemble of cities, the Randstad, which in turn are an economic and cultural center in Europe. But the Green Heart was facing encroachment from without and development from within. Its value at that time was being debated, its identity questioned, its very existence threatened. It was being threatened by the belief that 6 million plus houses, or a $150 billion economic engine was the inevitable use for these lands. The question was, could we redirect this force by a simple, clear and convincing vision?

In responding to this, the Studio constructed a series of images – part ceramic tile, part map, part drawing, part video and part text that propose a new vision for the Green Heart and the Randstad. And then possible new boundary conditions would clarify the roles of the urban, the ecological and the farming communities. These altogether comprise a major part of the Dutch landscape.

The Green Heart Vision is a single form that encodes a complex array of functions, processes and concerns. The form itself is a ring around the Green Heart, approximately 140 kilometers long and one to two kilometers wide, that we call the Ring of Biodiversity. The inside edge facing farm and polder, the outside edge seen as the eco-urban edge, posits that a new form of design is needed where the urban interfaces with the reconstructed ecology. Modeled on Nieuwkoopsche Plassen and de Kerkweg along the Meije with their spontaneous succession ecologies, the ring offers the possibility of presenting the full range of biodiversity available to the polder landscape.

The outreach arms of the Bio-diversity Ring take up the issue of maintaining cultural diversity and minimizing urban sprawl by separating the major cities of the Randstad. Every outreach arm can be seen as a many kilometer linear park, varying between one and two kilometers in width. Every outreach arm terminates in an ecologically vital area – forests, hills, and dunes, the Ijsselmeer, et al. Every outreach arm prevents two major urban forms from joining and forming the alienated space that results when large cities overbuild their perimeters. Every outreach arm would be designed as multiple use park space, with housing at the perimeter whenever possible.

The outreach arms as a whole define sufficient space outside of the new Green Heart border for the presently planned for 600,000 houses and infrastructure. The Harrison Studio, however, suggested that without clear limits to growth the Randstad community as a whole would suffer cultural and environmental stress in the near term coupled with economic stress in the long term.

Both the Green Heart and the Randstad are expressed imagistically and poetically as one place, a reciprocating figure-ground relationship, a yin-yang totality. In contradistinction to normal mapping operations, the biodiversity argument and the cultural and economic arguments are wedded to structure here. We have been told specifically that the principles of this argument have been accepted and are in the plan. That is to say, the region is now being planned as a whole rather than in fragments. Cities are being kept separate by parkland, the Green Heart will not be developed. However, the Biodiversity Ring, the most original of our ideas, well probably not be done.

The Sub-Continental Scale: "Peninsula Europe"

In 1998, we were invited by the Schweisfurth-Stiftung to write one of the ten books produced by Hannover 2000 on the future of the European Landscape. To define the terrain, we looked at a map of the Eurasian Continent with the European Peninsula jutting out into the Ocean. We took our pencils and we emphasized the rivers, defined the mountain ranges and whited out the roads and a surprising image emerged. It appears that the Vistula drops off the Carpathians flowing north and ends in the Baltic Sea. The Dnestr drops off from the Carpathians flowing south and lands in the Black Sea. In the Carpathians, the beginnings of these rivers, their first tributaries, are only about 30 kilometers apart. Where these rivers hit the vast Russian plain, a large wetland edges them. Thus the European Peninsula has an almost complete water boundary, an astonishing geophysical revelation, as this means that the European peninsula is connected to the Russian plain by only 30 kilometers.

Seeing

The geophysical heartland of Europe
As a peninsula
Extending from the continent of Eurasia
With ocean boundaries
Cojoined by the Dnestr and Vistula Rivers
Separating it from the Russian plain
Making it almost an island

Peninsula Europe
Contemplating its domain
The high grounds emerge
As both pattern and icon

Thinking

Contemplating the pattern
Reflecting on the properties of its
domain
I said,
"It's an array of drain basins cradled by the mountains
formed by the pouring forth of the rivers
that begin in the high grounds."
You said,
"Most of Europe's water begins there."
I said,
"If the forests were left to re-grow
and the grasslands released from overgrazing
Then the resulting bio-mass
Could help to purify
The outpouring of water."
And you said,
"Then the bio-diversity ribbons again could grow
across the high grounds from the Pyrenees
to the Carpathians"
I asked,
"Where would you begin?"
and you said,
"Where the terrain permits and the will exists
Choose your mountain."

The Schweisfurth-Stiftung called a small conference on the concept of Peninsula Europe to test its viability to see if it made sense to seek funding from the European Union and the Deutsche Stiftung Umwelt and diverse museums to develop and present the concept "Peninsula Europe – The High Ground". This meeting was attended by a diverse group of people. Several ecologists with a particular focus on transmontaine systems, an economist, a high level bureaucrat from the European Union involved in environmental protection, a person from an independent environmental organization from Hamburg whose focus was on environmental economics and generating systems of production that tuned to natural systems. With the usual maps on the wall we made the following presentation. We said that we saw Peninsula Europe: the High Ground as a trans-European eco-cultural form at great scale wanting to happen. The trans-European bio-diversity ribbons could re-establish the biomass and redundancy necessary for the well being of large ecosystems, such as existed formerly, but are no longer part of the current cultural landscape. We believe this new forest/meadow system has the potential of protecting the top end of every river within its domain. Therefore, most of the countries of Europe would gain a pure water downhill flow of great magnitude when the water crisis across Europe will begin to become acute in the next fifty or sixty years, perhaps sooner. All agreed but the EU representative. He stopped us here, asking where the money would come from and how would EU interests be served and how would such a system be sustained over time? You said, "Think of the waters. About 1430 billion cubic meters of water fall on the high ground. At least a thousand billion flow downhill. The rest remains in the mountains because of percolation and evapo-transpiration. So we are looking at about a thousand billion cubic meters of water yearly." Then I said, "Everybody pays at least on average one Euro and sometimes two or three times that downstream for clean water. So we are looking at a trans-peninsula expense pattern of perhaps 1 trillion Euro yearly assuming all downhill flow is used at least once. As the gross transnational product appears to be about 8 trillion Euro, a rough calculation suggests that a modest water tax of about 1/2 per cent would yield about 50 billion Euro a year. Projecting this over a 15 year period that amount of money would go a long way to putting this system in place. Once in place the system would move to self-maintenance and therefore costs would drop significantly. Thereafter the gentleman from the EU said, "Go for an EU grant."

Operating in scales of the dimension of "Peninsula Europe" is normal for our highway systems. Almost normal for our military systems such as NATO. Certainly normal in our monetary systems, thinking about the euro. And obviously normal if we look at "globalization". "Peninsula Europe" operates in that context. However, we see planning imperative on the 50- to 100-year scale. It takes the following shape. The high grounds of Europe take up almost 40% of the land mass, but the population of the high grounds is only 40 million. The lower lands, which are about 60% of the landmass, have a population of about 280 million. Now when the greenhouse effect raises temperature levels, the tree line will rise, farming will change its nature, some flooding will take place in the lowlands, while population increases. We predict that the high grounds will perhaps double their population. Who will conceive the new cities and towns and future eco-urban edges in ways that protect the high ground, the cultural landscapes and ecosystems, but those who share the disciplines of people in this conference?

INNOVATION AND TRADITION
DIALOGUE BETWEEN CULTURES

Introduction and Abstracts

INNOVATION UND TRADITION
DIALOG DER KULTUREN

Einführung und Abstracts

Peter Niehaus Munich

Thomas Herzog Munich

Elizabeth Sikiaridi Cologne

Moshe Zarhy Tel Aviv

Gabor Bojár Budapest

Chris Butters Oslo

Ma Hang Shenzhen

Yury Gnedovski Moscow

Klaus Unger Dublin

Barbara Engel Cottbus

Ugur Tanyeli Istanbul

David Sutherland Melbourne

Introduction
Einführung

What we are looking at here are not just visions of what will be technically feasible; we are also concerned with giving the concept of innovation a cultural safeguard and integrating it into the "horizon of the earth". In other words, innovative architecture and urban development should be in peaceful coexistence with natural conditions. In terms of basic assumptions at least. 'Innovation' is a process of renewal and optimisation aimed at bringing about well-researched, planned and controlled changes in functional relationships, in structural and visual forms of architecture and in the urban environment. New ideas, perspectives, technologies and techniques facilitate transitions to new and more efficient systems. The guiding principle is not the technical interpretation alone, but the assimilation of architectural developments into the "horizon of the earth".

Context

Innovation as an "intellectual resource" transforms architecture and urban development into key cultural media that can help to reveal the structure of future developments, thus presaging "anticipated reality" (Walter Benjamin). Even so, coexistence with natural factors, the use of environmental energy sources, especially the unlimited energy potential of the sun, are indispensable preconditions for shaping our built environment. Innovation through resource-saving construction methods includes the kind of urban development that saves space and protects landscapes in all forms of society.

Innovation as a Guiding Principle in Resource-saving Construction

The concept of innovation is often linked one-sidedly to the improvement of technological conditions. Changed conditions of communication and work, changed social structures and greater mobility within a global framework seem to generate innovation, thus posing new challenges for architecture and cities. New typologies for homes and jobs, new communication facilities for conference centres, fairs and exhibitions, hybrid functional programmes and new typologies for airports, tourist and leisure facilities as well as museums are being developed in response to changes in overall conditions. In the transformation of cities into entertainment centres, changed forms of production and consumption within the information and media society reflect innovation patterns that deserve a critical response.

The development of the concept of innovation must be multi-dimensional, embracing not only the globalisation of the economy, but also technology, the Internet and pop culture. Architecture that takes due account of natural settings can develop into a resource of innovation which can become effective as an

Es geht nicht allein um Perspektiven des technisch Machbaren – sondern, als reflexive Folgerung, um eine Orientierung des Innovationsbegriffs an einer kulturellen Rückversicherung, an der Einordnung in den „Horizont der Erde". Innovative Architektur und Stadtentwicklung in friedlicher Koexistenz mit den natürlichen Gegebenheiten. Wenigstens in den Grundannahmen. „Innovation" ist der Erneuerungs- und Optimierungsprozess, der recherchierte, geplante und kontrollierte Veränderungen von Funktionsbeziehungen, Strukturen und Erscheinungsformen von Architektur und städtebaulicher Umwelt zum Ziel hat. Unter Anwendung neuer Ideen, Perspektiven und Techniken werden Übergänge in neue oder leistungsfähigere Systeme geschaffen. Leitvorstellung ist nicht die allein technische Interpretation, sondern die Einordnung der Entwicklung der Architektur in den „Horizont der Erde".

Kontext

Innovation als „geistige Ressource" lässt Architektur und Städtebau als kulturelle Leitmedien erscheinen, die in der Lage sind, Zukünftiges in ihren Strukturen aufscheinen zu lassen und „antizipierte Realität" vorwegzunehmen (Walter Benjamin). Dabei sind die Koexistenz mit den natürlichen Faktoren, die Nutzung der Umweltenergien, insbesondere die Nutzung des unerschöpflichen Energiepotenzials der Sonne unverzichtbare Grundvoraussetzungen für die Gestaltung der gebauten Umwelt. Innovation durch ressourcenschonendes Bauen schließt flächensparende und landschaftsschonende Stadtentwicklung in allen Gesellschaftsformen ein.

Innovation als Leitvorstellung des ressourcensparenden Bauens

Die Vorstellung von Innovation ist häufig einseitig an die Verbesserung technischer Konditionen gebunden. Veränderte Bedingungen von Kommunikation und Arbeit, veränderte Sozialstrukturen und gewachsene Mobilität im globalen Rahmen scheinen Innovation zu schaffen – neue Herausforderungen für die Entwicklung von Architektur und Stadt. Neue Typologien für Wohnen und Arbeiten, neue kommunikative Einrichtungen für Konferenzzentren und Messen, hybride Funktionsprogramme und neue Typologien für Flughäfen, Tourismus-, Freizeiteinrichtungen und Museen werden entwickelt und antworten auf die veränderten Rahmenbedingungen. Bei der Umwandlung von Citys zu Entertainment Centres spiegeln veränderte Produktions- und Konsumformen der Informations- und Mediengesellschaft kritisch einzuschätzende Muster von Innovation.

Die Entwicklung des Innovationsbegriffs hat weitere Dimensionen einzuschließen als die der Globalisierung von Ökonomie,

"intellectual resource". It relates innovative developments in architecture and urban development to the "horizon of the earth" rather than to an abstract concept of globalisation. Architecture and the city as projects, as projected ideas of the social development of the city, can be compared with an "endlessly extended and improved narrow walkway that analytical reason has constructed above an abyss". (Claude Lévi-Strauss)

Responses

Architecture is developing its inner power of vision and has a responsibility to be forward-looking in its actions and to actively influence history. Hence, innovation does not simply mean more technology, but more critical reflection, more sustainable and experimental concepts encouraging a symbiosis of technical and natural, historical and cultural, social and economic solutions – a symbiosis of nature and architecture, of nature and the city.

Proposition

The comprehensive quality of architecture as a resource for innovation reflects and anticipates aspects of social change. Resource-saving and solar techniques in construction must be considered as a challenge that harbours significant, unexplored potential (not least for developing countries).

Innovation means relating the various dimensions of architecture to one another in an appropriate fashion, as the potential of a "societal identity", as a future-oriented, visionary challenge. It has to deal with new materials, new technologies and new production methods (CAM).

Technik, Internet- und Pop-Kultur. Architektur, die sich ihrer natürlichen Kontexte versichert, kann sich zu einer Ressource der Innovation entwickeln, die als „geistige Ressource" wirksam zu werden vermag. Sie bezieht innovative Entwicklungen in der Architektur und Innovation in der Stadtentwicklung weniger auf einen abstrakten Begriff der Globalisierung als auf eine Einordnung in den „Horizont der Erde". Architektur und Stadt als Projekte, als projizierte, vorausgeworfene Ideen der Gesellschaftsentwicklung in der Stadt, sind vergleichbar einem „unaufhörlich verlängerten und verbesserten schmalen Steg, den die analytische Vernunft über einen Abgrund baut". (Claude Lévi-Strauss)

Erwiderungen

Architektur entwickelt die innere Kraft der Vision und steht in der Verantwortung, zukunftsweisend zu agieren und Geschichte aktiv zu beeinflussen. Innovation bedeutet deshalb nicht einfach ein Mehr an Technik, sondern ein Mehr an kritischer Reflexion, eine Zunahme an nachhaltigen und experimentellen Konzepten, welche eine Symbiose von technischen und natürlichen, von historischen, kulturellen, sozialen und ökonomischen Lösungen, eine Symbiose von Natur und Architektur, von Natur und Stadt befördert.

Thesis

Architektur als Ressource der Innovation spiegelt und nimmt, in ihrer umfassenden Qualität, Aspekte des Wandels der Gesellschaft vorweg. Das ressourcenschonende und das solare Bauen sind als Herausforderung und als bedeutsames, unerschlossenes Potenzial (nicht zuletzt für die Entwicklungsländer) wahrzunehmen und zu entwickeln.

Innovation heißt, die verschiedenen Dimensionen von Architektur angemessen aufeinander zu beziehen, als Potenzial einer „gesellschaftlichen Identität", als zukunftsweisende, visionäre Herausforderung. Sie hat sich mit neuen Materialien, neuen Technologien und neuen Produktionsmethoden (CAM) auseinander zu setzen.

Reinhart Wustlich Hennef

The City of 2030 – Network and Processes

Peter Niehaus Munich

1. Fixed urban-design models and prognoses on the future of the city have often missed out the reality of urban developments, as any review of the past thirty years will show. Economic and planning cycles are becoming more short-lived; new players have entered the scene. Instead of the completed 'urban product', it is networks and processes that are becoming more and more important in urban development. Where do we find innovative urban clusters? Who are the promoters of processes, and who the 'mere partakers'? How are urban development processes structured, and what phases of such processes are critical?

2. In recent years Siemens has initiated large development projects in Berlin, Frankfurt/Main, Munich and in the United States, while at the same time stimulating the debate on urban design with new ideas and methods. This is exemplified by the urban renewal project for the world's largest Siemens premises in Munich, the Siemens City 'Isar South'. Established in 1927 as a factory for telephones, at present the complex is being redeveloped into an open urban quarter and the core location of Siemens's IT and communications technology. We know that we can only win innovative employees on a long-term basis if we provide them with an attractive working and living environment. Siemens Real Estate, as the proprietor and real estate administrator of the Siemens Group, aims to create modern work spaces, offer flexible working methods to the brains of the next IT generations and to reactivate unused property potential. The planners envisage urbanising the area and creating c.775,000 sq.m of floor space of offices, residential space and related facilities, hotels, catering outlets and leisure amenities.

   A large project of this nature will only be successful if the proprietor, the users and the city as the authority in charge of urban planning join to form a threesome of creative energy which will ensure its win-win potentials and define common goals. In Munich, to this end, we discovered and developed an informal, co-operative planning procedure, teaming up with the municipal planning authority. The first milestones were the positive benchmark decision by the city council and an urban-design and landscaping ideas competition. The functional structures and zoning of the area are currently being defined in weekly sessions by the project group of Siemens managers and representatives of the city's planning department, at times supported by sub-groups specialised in particular fields. A steering committee of high-profile municipal and corporate representatives decides on any basic questions remaining open or matters of conflict.

   Planning is based on the winning design by the young Cologne architects of Jaspert & Steffens–JSWD who proposed a somewhat unspectacular, but 'elastic', urban street-block grid with a central green area. Unlike individual buildings and urban estates, the design of urban layouts requires less 'exalted', timeless basic structures.

   Apart from consultations with neighbours, citizens and the district committee, the project benefits from important innovations and best-practice-oriented ideas proposed by expert consultants with a background of science and/or practice who accompany the innovatory process of collecting and applying information. Time-management models are of particular interest in the world of work, as are new workplace concepts, alternative forms of living and working, space-saving and environmental traffic management. It goes without saying that a technology producer like Siemens also applies – in an experiment on itself – its own guiding IT, building and traffic technology.

   In today's construction industry, the developers' economic performance goals are of increasing importance. By developers we mean both the real-estate owner (e.g. Siemens) and subsequent investors of capital. This speaks for including 'investors' as early as possible in a consultant function. After all, paying developers' bills is the duty of every client. This is not to say that profit is the only priority, but only that the larger a development project, the more balanced the assessment of its economic and social aspects must be. Do all the social groups concerned assume their part in this process?

3. What we need on the road to urban development up to the year 2030, is high professionality in process design and an 'elastic' project management. The planning steps from the initial stage through to marketing, letting, moving in and aftercare may not be new, but are rarely assessed and determined for the entire life of a completed project, beyond a ten-year period. It is crucial that processes evolve less hierarchically, i.e. more 'from the bottom up' than 'from the top down' and are understood as a value-adding chain of private and public initiatives. Experience tells us that proprietors, users and enterprisingly minded city representatives are the motors behind them. They possess clear goals, planning tools, economic thinking and social competence, and this enables them to establish a confidence-based informal consensus long before council decisions and contractual agreements are made. Siemens has introduced key data programming to the initial phase of development and construction projects. We should expect preliminary planning measures, even with large-scale urban development projects – i.e. the careful conception of planning goals, of alternative architectural, technical and economic strategies as well as time, committee-work and process structures – to take up 40 per cent of the entire planning time and work.

   The control of planning processes is, however, only one half of urban development. The other is a sensitive taking stock of the long-term economic and social change in structures and values. As far as we know today, the city of 2030 should be read as a 'basic expectation' of its inhabitants, as 'urban life on demand', i.e. a combination of a great number of amenities. It resembles a car rental company: instead of having their own car, citizens discover the attractions of permanent availability of service and of various paradigms. Once the city of 2030 has become – in terms of society and buildings – the 'elastic' field of action for constantly changing designs of living, we will also have to reckon with the related revision of zoning and use structures. Who will be able to afford these then?

Innovation and Tradition

Thomas Herzog Munich

On the one hand, we face a wealth of global phenomena like climate changes, population growth, worldwide migration movements, food scarcity, energy supplies ...;

with world championships, Olympic Games, world trade and worldwide communications ...;

with scientific communities, the UN and UNESCO, the WHO, with GPS and our own UIA ...

On the other hand, there are thousands of different languages and dialects spoken around the globe, thousands of different everyday and public holiday traditions and we perceive differences in our environment over short distances. In mountainous regions densely populated east-west valleys run into north-south valleys with another topography and a distinctly different climate. This generates cultural characteristics, even in relatively small areas, so that individuals and communities who live there develop a sense of belonging, of homeland, from their specific features.

Architecture has to deal with the one and the other, with local dwellings built of local materials, with creating shelters and living space in different climates and regions: in deserts, on sea shores, in the mountains, on the plains and on hillsides, in the natural jungles and those of the megacities.

Yet architecture is also concerned with internationally marketed high technology. The buildings on the Nile, built of hand-formed unfired loam bricks – an age-old traditional construction method – rise up and finally crumble again next to modern industrial plants manufacturing flat glass or anodise huge steel sections for assembly lines which produce prefab prestressed RC trusses. Alternatively such products are imported.

The results are often material gains as well as cultural losses. After thousands of years of balance between the great river of Egypt and the people living from and on it, there are now other 'crops' growing up on its banks that change the proportion of things: multi-storey skeleton structures with fired brick infills that sprawl out from Cairo in their thousands.

The production of entire non-structural facades and their details, of solar collectors for roof installation, of built-in luminaries, suspended ceilings, movable partitions, floor coverings and media units matches the technologies used in other areas of industrial production in terms of engineering and logistics, conversion technology, surface finishing, high-end precision and control.

Yet there is an additional problem arising from the fact that, similar to cars, houses are being built the world over to almost the same typological design, with architects and builders pursuing only the technically and economically feasible without considering either the ridiculously, irresponsibly high energy input required to operate them or the parallel destruction of the values which influence and generate architectural identity.

Talking of cars, i.e. the means of mobility, in the same breath as buildings, by nature stationary, i.e. immobile, is highly questionable. Just as living creatures are able to live in certain regions because they are exactly suited to the conditions prevailing there, buildings, too, must relate to their surroundings to function efficiently and to use the potentials offered there.

The polar bear cannot subsist near the equator, despite all its strength, and the gazelle has no chance of surviving in the Arctic Ocean, despite its speed.

Innovation should neither only differ from what has existed before, nor make us just expect partial success in certain separate areas. It should lead us to new holistic qualities in an interplay between all the material and immaterial givens. The criteria for such qualities must first and foremost be human criteria.

Human existence, in its material aspects, depends on nothing else so much as on the availability of energy. Where energy is available in sufficient usable quantity, civilisations can develop: water for drinking and watering crops, for heating and cooling, trade and transport, health services and education – in short, everything that makes the building of a city possible as a rich and complex space for life.

In the past, people and communities left a region when natural forces proved uncontrollable by man and made subsistence impossible. This phenomenon has been changed radically in crucial areas by the technological developments of the past decades, especially by the new potentials for using solar energy which is available in immeasurable abundance every day all over the world. This creates the prerequisite for the development of civilisations.

Architects like to talk of their responsibility towards society. In this country this has even been encoded in our professional ethos.

Yet neither our universities nor professionals take note of this responsibility and the activities of architects which are life-giving for humanity as a whole. Migration movements towards the industrialised countries and the unspeakable daily sufferings of people in Third-World countries due to lack of food and safe drinking water can only be stopped if and when

1. technical and functional knowledge is made available to them, and

2. the architectural models followed in developing countries are suited to the respective region instead of being reproductions of the fun-society, of abundance and utterly inappropriate symbols, as well as an often extremely arrogant overestimation of our own abilities. (The politicians, self-declared democrats, who demand a royal palace for this city – if not the whole, then at least the aristocratic symbol of the facade – even though it may cost many hundreds of millions, are in fact declaring our cultural bankrupcy. Considering the above, it is the exact opposite of what should be demanded and would be right and proper in view of public impact.)

It is only when we accept the challenge and opportunity, the artistic stimulus of new means and, at the same time, aim to create a subtly adapted, materially and aesthetically varied architecture, that we will generate a modernism that deserves this designation as a cultural goal.

The attached graph presents some of the important elements in building today. It shows how much more diverse the technological repertoire has become over the past few decades.

Gebäudespezifische Nutzungen von Solarenergie | **Building related use of solar energy**

Idensity®: Public Space in the Information/Communication Age

Elizabeth Sikiaridi **Frans Vogelaar** Cologne

"The new city presupposes that the cables of interhuman relations are switched reversibly, not in bundles as with television, but in real networks, respons(e)ibly, as in the telephone network. These are technical questions; and they are to be solved by urbanists and architects."

Vilém Flusser 1990.

Urbanism in the Communication Age

To reinforce the significance of public space we have to deal with at least two "publics", the global and the local public, by creating spheres where local and global public space can fuse and interchange.

Bridging the gap and connecting the global media spheres with local content and place, an architecture of communication spaces proposes a combined analog-digital infrastructure: publicly accessible interfaces between the global media space and the local urban place. 'Public Media Urban Interfaces' is an alternative scenario for the interplay of mass media in order to reinforce the function of public (urban) space. This project develops a hybrid urban network-space, a fusion of media space and urban space. It emphasises the role of the public in an increasingly privatised society and occupies the vacuum in between the local and the global. The products of this alliance of urban and media networks are "hybrid" spaces that are at the same time analog and digital, virtual and material, local and global.

This project represents a prototype for a new interdisciplinary field of design and planning ('Soft Urbanism'), researching the transformations of architectural/urban space of the emerging "information/communication age", exploring the dynamic interaction of urbanism and the space of mass media and communication networks. 'Soft Urbanism', dealing with the "soft" aspects of the city, not only intervenes in the realm of infrastructures, but also adopts their concept and paradigm: by supplying networks, 'Soft Urbanism' creates new fields of possibilities and frameworks for self-organisational processes.

Today, the communicational paradigm, with its "network-cities", "nodes" and "terminal architectures" is infiltrating and transforming both architectural/urban discourse and practice. Within this framework, 'idensity®' is proposed as a conceptual tool for developing space in the information/communication age. This composite term consists of the combination/fusion of the word "density" of real/urban and "virtual"/media communication spaces (density of connections) and of the word "identity".

Public Media Urban Interfaces

The local-based public 'tele-feeder facility (at your neighbourhood's laundrette)', the primary unit of Public Media Urban Interfaces, enables the public to produce messages and to narrow-broadcast and receive them in a dynamic communication environment. Creating a locally-based dynamic media network from the bottom up, local events can be accelerated and reinforced to temporarily invade the glocal media space.

This link between global media space and local place with its interfaces in public space makes it possible to broadcast, access and influence the global media environment from the (urban) local neighbourhood.

A demo project, exploiting London's urban tensions and structure unfolds strategies and visualises aspects of these investigations, confronting a working hypothesis with the idiosyncrasies of a specific urban situation.

Media Babies

128 feeder houses (Media Babies) distributed evenly over London's sprawling conurbation and interconnected by means of a digital network supply eight Bridge Clubs located on the Thames with a continuous stream of (non-)events. The Media Baby at your neighbourhood launderette consists of a Catching Gallery, two Intro Booths, a Debutantes' Booth, a Connector Platform and a Microwave Transmitter. The Catching Gallery is the area where the public can view the narrow/broadcasting activities of eight other Media Babies and one Bridge Club. Interactive technology enables the public to intervene in those narrow/broadcasts but also creates the possibility to establish direct contacts, thus forming endless smaller networks within the larger framework of Public Media Urban Interfaces.

Bridge Clubs

In providing the space for public events on an urban scale, The Bridge Club bridges the gap between programs meant for local distribution and those that deserve a larger audience. Using the larger broadcast facilities available to the Club, the selected programs are experienced and transformed to suit a mass audience. The Bridge Club, being a node in the net of translocalities, also serves the function of bridging programmatic events related to the site where the club is located.

Replacing the Right to Vote with the Right to Broadcast

The publicly distributed 'Air Time for All' Smart Card allows you to produce and narrow/broadcast and also gives you the opportunity to adopt a message (not your own) by giving it extra Air Time. At the Media Baby in the neighbourhood, you will find the necessary programming facilities to make your program and the means to monitor it as it goes on the air. You can also accelerate messages (not your own) by giving them extra broadcasting time with the help of the special Smart-Card. And as a message gains strength, its chances of reaching a much larger audience increase, reaching more Media Babies, a Bridge Club, the city or even the whole country, Europe and the rest of the world. Replacing the right to vote, a right to narrow/broadcast is established.

Mobile Containers

A fleet of container-boats, caravans, riksha's, taxi's, trucks and limousines (equipped with transmitters/receivers and interactive life jackets) form a transportation/communication infrastructure servicing the users of the network and also commuters, nomads, migrants and tourists. The traditional translocal (mobility/communication) networks are thus knitted to the new glocal media networks (Internet/TV).

These capsules containing (from rudimentary to more sophisticated) media units are mobile nodes in the translocal networked environments, "vessels" within the complex multilayerings of the space of flows. They serve as spaces of exchange (export/import trade), as laboratories of glocal cultural bastardisation. With these containers new hybrid, media and real spaces emerge, that are no longer tied to any one specific location but rather are the result of their interconnection.

Soft Urbanism

In architecture's role of defining and materialising the spaces for social interaction, designing the relationship between the physical and digital public domain is becoming more and more of a challenge: investigating the relation and interconnection of the "soft" city with its finite material counterpart, the living environment, speculating about interfaces between the "virtual" and the material urban world and designing hybrid (analog-digital) communicational spaces.

Soft Urbanism deals with information/communication processes in public space, the soft aspects overlying the urban sprawl and modifying it: the invisible networks acting as attractors, transforming the traditional urban structure, interweaving, ripping open and cutting through the urban tissue, demanding interfaces.

Soft Urbanism not only intervenes in the realm of infrastructures, but also adopts their concept and follows their paradigm. It brings an inherently flexible approach by expanding the field of possibilities of social interaction and opening new paths of urban development. It is therefore not about determining places, but about creating frameworks for processes of self-organisation. Not accepting being powerless in the face of the forces of the market, Soft Urbanism rethinks the strategies of interventions to reintroduce programmatic speculations about the public domain in urbanism.

The interventions will not be about control and determination, but about expanding infrastructures, frameworks for processes of self-organisation. "Soft" strategies will be "bottom-up" strategies: rather than defining first the global result of the interaction and then determining the necessary relation between the elements in order to produce that interaction (which would be a "top-down" approach), simple rules for a set of independent elements will be developed and what emerges from the interaction of these elements is aleatory. According to biological models, these fields of interaction of plural forces could serve as a reservoir for the selection processes needed for the urban transformations.

Idensity®

In the contradictory dynamics of today's urban environment with its antithetical tendencies of concentration and decentralisation, of functional mix and segregation, traditional terms of spatial distinction lose their validity. In this fragmented urban landscape, categories such as "functional zoning" (living, working, recreation), "centre" versus "periphery", "landscape" versus "city" are becoming obsolete.

The polarity between private and public space is disintegrating. Public and private environments are becoming intermingled and blurring in the fusion of media and "real" space. We see this in the hybrid spaces of the publicly broadcasted (inverted) privacies of reality TV and the "Big Brothers", in the media presence of war intruding on our living rooms and in the private communication space of mobile telephony within public urban space.

To understand the fusions, this superimposition and the interactions of media and "real" architectural/urban spaces, the new term 'idensity®' is introduced, replacing the obsolete conventional terms of spatial distinction. It does not differentiate between media networks and urban/architectural environments and it offers an integrated model for dealing with "hybrid" (media and "real") space in the information/communication age.

This model can incorporate the widest range of (future) spaces:
· from the 'tele-feeder unit at your neighbourhood's laundrette', a public infrastructure for teleshopping, telelearning or tele-democracy, see 'Public Media Urban Interfaces',
· to new "club" facilities, providing the space for "hybrid" (media and "real" space) events on a larger urban scale, see 'Bridge Clubs',
· and the networked 'Mobile Containers', see for example, 'ReBoot-NRW.NL', a networked boat with 80 participants on board, descending the Rhine in 1999, connecting the network and the nodes (harbour-cities) of the river with media networks (Internet/TV),
· or the combined media and "real" space of your bank, presenting itself in its telebanking application with the corporate identity of its "real" architectural building while fusing in the representational entry of its headquarters a high-touch architectural space with the media spaces of its net presence, in the form of monitors, projections, etc. (just visit your bank).

'Idensity®' is a composite term consisting of the combination of the word "density" of real (urban) and "virtual" (media) communication spaces (density of connections) and of the word "identity".

It integrates the concept of "density" (density of connections, density of physical and digital infrastructure, density of communication-spaces, etc.) with the concept of "identity" (image policies, urban brands, etc.). It can therefore, for example, help in understanding the processes of spatial segregation and distinction between urban fragments that have qualities of global performance and that can be seen as part of a "global urban condition" and those other, sometimes neighbouring (parts of) cities that lose in relevance and disappear from (global) mental maps. 'Idensity®' can be implemented as an operative tool to steer the processes of urban development.

But it is not a mere summation of the two concepts of "density" and "identity". It is rather a fusion, as it inverts "identity", linking it to communication: "identity" being defined by connectivity.

Therefore, it does not just address the "clear-cut identity, the particularity, the individuality of the traditional places or cites (like centres and monuments)" but also the layered 'idensity®' of the "non-lieux" ["non-places"] of today's generic cities, which are to be found especially in the realms of mobility and consumption (airports, hotels, shopping malls, motorway rest areas, etc.). It does not refer only to object-qualities but describes a field of superimposed communication spaces: the branded space of the chain-shop, the symbolic space of the traditional building the shop is housed in, the media space of teleshopping, the communication space of the GSM...

This new term is implemented to describe and analyse the communication spaces of the coming "network society", a society not so much based on the traditional, relatively static structures of belonging in the family, the corporation or the state, but on flexible, dynamic, ever-changing networks of exchange and communication. It carries the discussion on the urban from the morphological level of a formal description of the network patterns of the "network city" to a more integrated structural understanding of the networks of spaces for social communication.

The term 'idensity®' is a conceptual tool for researching and developing space in the information/communication age.

MEATCO

Architecture and the Information Era

Moshe Zarhy Tel Aviv

The last quarter of the Twentieth Century has been characterized by rapid technological development in all fields. The most significant changes that we have witnessed at the turn of the century relating to architecture and the role of the architect are the momentous developments in the exchange of information.

Accessibility, speed of transference and immediate communication have a deep influence on ways of life and creative possibilities.

As we enter the new century, high-tech industries and services will become one of the main sources of employment worldwide. Buildings where science and high-tech activities can be carried out optimally are a necessity, and there is a growing demand for better and more suitable facilities. This new kind of built environment will naturally change life styles and behavior patterns. The history of architecture teaches us that, at times, architects were slow to grasp the potential embodied in new technologies. The leaders of progress were other than architects.

The change is not always tangible. Just as developments in building material brought new challenges, new technologies in industry and communications open new horizons in architectural thought. We have to study and comprehend what is happening in front of us, draw the right conclusions, understand our place and our role as architects in the Era we live in, and place ourselves at the spearhead of progress.

We are aware of the classic perception of the role of the architect and his relationship with his building team. We should define the changes in this perceived role vis a vis the developments in communications and modern industry. It is indeed true that there is some measure of architectural specialization depending on the designated purpose of the building, but we are not addressing the issue of one more kind of building. We are dealing with substantial alterations in the role of architecture in this new context.

What is the impact of the Information Era on Architecture?

There are many questions concerning this subject but no one decisive answer. Nevertheless, this is an important issue that deserves deep investigation and discussion. Let us examine some of its aspects:

One of the challenges facing architects in the new Era is how to settle the contradictions between "one world" born from the newest communication techniques and the yearning for local identity.

On one hand – the swiftness of information exchange between different parts of the world, the standardization of interpersonal relationships and business management and overexposure to foreign cultures – all these push for architectural globalization and blur borders and cultural characterization between different countries.

On the other hand, we are anxious witnesses to the loss of local architectural heritage, and crave the preservation and development of national architectural identity. Each nation clings to its cultural heritage and yearns for continuity. Our challenge is to settle these two conflicting tendencies. The modern means of communication are the outcome of decades of research that created applications in consumer goods and a revolution in the industry that produces these goods. Most of them are produced in high-tech industries.

Sophisticated industries deal with diverse topics: design and manufacture of computer hardware, software for diverse purposes, different fields of electronics, such as micro-electronics and electro-optics, biotechnology and genetic engineering, pharmaceutics, rare chemicals, agro-technics, robotics, bio-engineering, development of office equipment and many other fields.

The end products of high-tech facilities are those that enable the existence of the "Information Generation". They have become cheaper and available to all. It is hard to imagine the society, economy and culture of the Twenty-First Century generations without the constant flow of information made possible by these products. This is a direct concern to the architectural community – we have to strive to deal with the unique problems connected with the Information Era and Architecture.

As mentioned before, we foresee that, as we progress into the new century, we shall see more and more built-up areas that provide working places for many people. These built-up areas are of interest and of concern to the architectural community.

Technology-rich industry has a special nature. It is innovative, its scope and its goals adapt rapidly to new ideas and demands and it attracts people of a unique kind. We have to keep in mind that the driving force behind high-tech industries are the entrepreneurs, the inventors, the men of ideas, the thinkers and the innovators. Although money is a big factor – it is not the main element.

All this, of course, has to be taken into consideration when planning facilities to house high-tech industries and scientific research.

The location of sophisticated industries is of great importance. In many cases, such as in software development industries, we could argue that the only requirements would be comfortable and efficient office stations. This presumption is erroneous – the dynamics of the field, the teamwork, the constant interchange of personnel, the ability to expand or diminish according to present needs, the linkage of hi-tech industries to other fields, require planning a flexible facility, that includes a variety of systems.

Most high-tech industries can be planned so as to be environment friendly – therefore there is nothing to prevent their incorporation in the urban fabric. In other words, most modern industry campuses are returning to the cities and can be optimally located within urban housing schemes.

The urban character of these clusters of buildings is directed by their local surroundings and by the wish to impart to the campus a unique image. It is neither a university campus, nor is it an industrial area. The leaders of these high-technology campuses and the high-tech industry managers want the campus to broadcast "success". In high-tech centers, environmental design and landscaping enhance the quality of life and give a prestigious image to the whole campus. A high-tech campus built on the right site has a better chance of success. Proximity to an upper-level population and an area of high-quality of life, enough reserve area for expansion and achievement of a "critical mass", affinity to a university or research institute – all these factors contribute to the success of a high-tech campus.

Besides choosing the right site for a high-tech campus, it is extremely important to develop a Master Plan and define its integration within the surrounding urban neighborhood and the access routes that connect it to the main urban and national road system. The establishment of a high-tech industrial campus has an impact on the neighboring environment. There is a constant interaction between the campus and it neighboring town fabric. The consequences of this interaction are many, but our time is too short to go into this most important subject.

Some architects may think that when planning a high-tech facility they have to express modernism and emphasize futuristic architecture. It is not so. We have seen incredible failures that followed this line of thought. We have to put architecture in the service of the subject and not the other way round. We have to attend to the needs of the clients to the best of our visual and conceptual abilities – and I believe that this cannot be accomplished by any member of the building team, other than the architect.

The theme of the twenty-first UIA Congress "Resource Architecture" has been open to many interpretations. In the context of our paper we can say that architecture is founded on two resources: one material, the other spiritual.

The Information Era was born, and continues to develop and grow mainly on inexhaustible spiritual resources. In the course of history, radical changes in architecture occurred as a result of population migration, social changes, technological discoveries, and new building materials. All these are material resources. Today, we are dealing mainly with spiritual resources and changes in lifestyle. How will they influence modern and future architecture?

Summing up our paper we can say that accessibility, speed of transference and immediate communication have a deep influence on ways of life and creative possibilities. They have, and they will continue to have their impact on Architecture and Town Planning.

The ease in communications, the incredible progress in technology, the abundance of consumer products and services – all these factors necessarily change human society in all of its aspects. The present generations do not live, interact or work in the same manner as former generations – and therefore need a habitat suited to their needs. We architects are required to see the whole picture – to be able to learn from the past, see the present, look into the future and be the pioneers in creating the environment that will suit the needs of this Era.

Back in the Driver's Seat
How Architects can Regain their Renaissance Role

Gabor Bojár Budapest

When weighing the dangers which threaten the architectural profession today, I believe one of the most alarming trends is that architects retain ever-decreasing influence over the final product. In other words, the finished building is not always the one you dreamed up.

A few hundred years ago, the architect, as a genuine Renaissance personality, acted as "master builder" and controlled all aspects of the entire construction project, down to the processing of the very last piece of stone. In contrast, in today's technological environment, on the one hand, skills are highly specialized and, on the other, time is accelerated and has become the dominating constraint on the construction project. The client will pay most attention to those who have the greatest impact on deadlines and budgets. The architect often becomes just one of the players among many other contributors to the design, such as structural engineers, HVAC and electrical engineers, interior designers, civil engineers and others. In principle the architect is still responsible for co-ordinating the work of all these participants, but the reality is different. The information flow among the participants takes place on paper, with all the inherent limitations of paper, and under the pressure of deadline and budget constraints, in most cases it is the general contractor who ends up making the important decisions. Who loses? The loser is the finished building, our built environment, and consequently the entire society.

How can we return to our roots? How can we regain the control architects enjoyed during the Renaissance? How can we climb back into the driver's seat of the construction industry?

We Hungarians have a proverb: "Take a hair of the dog that bit you!" Technological developments deprived the architectural profession of its leading role; but current technological developments have the potential to return architecture to its former stature. Technology made building information so complex that without the proper tools, the architect could no longer remain in charge of all this information. But another technological development, namely information technology, can fix the mess and put things back into their natural place.

Information technology (IT) has long been perceived as a means to automate tedious manual work. The first generation of Computer Aided Design (CAD) workstations were used as automated drawing boards, to improve the efficiency of the drawingprocess. This was the case some 20-30 years ago among nearly all computer applications, but in the last 10 years "computing technology" has been re-named "information technology" and has gained a completely new interpretation. Computers are not just a means of automation, but first and foremost a means of information control. The IT revolution is not a continuation of the industrial revolution, not just a further development of machine automation. Rather, it is the beginning of another revolution, with much more profound impact on the destiny of the human race. The IT revolution of our age was preceded in human history by two other revolutions related to "information technology". The first was the invention of the spoken language, some 40,000 years ago, enabling us to communicate detailed information. The second was the invention of writing, enabling us to store and accumulate information throughout generations. The invention of writing may be dubbed the "Paper Age". But Gutenberg's marvelous invention inherently forecast an end to the very age it created: The amount of knowledge collected in millions of books in thousands of libraries is now so huge that it has become practically inaccessible.

The third IT revolution, which characterizes the current era, addresses this challenge and is as profound as the invention of speech and writing. While the spoken language enabled us to formulate information, and the invention of writing enabled us to store information, computers now enable us to organize, process and control information – in other words, to find among this ocean of information the exact piece we need. It marks the beginning of a new era of the human race, called the "Information Age".

I believe that the pulleys and levers of ancient times were merely secondary factors in the building of pyramids. The spoken language, which made it possible to co-ordinate the work of thousands of people, and the gradually increasing know-how of generations preserved in written form, were the real keys to the spectacular achievements of the ancient world.

Similarly, the improvement of drawing efficiency, offered by today's CAD workstations, is just secondary to the ability to control complex building information.

The rules for communicating building design on paper were first developed some 400 years ago when architectural drawing became the lingua franca of the profession. But paper, as a medium for communicating building information, has three major limitations:

1. Limits on the amount of information that can be stored on paper. As building technologies and buildings themselves become ever more complex, the size and number of drawings grows exponentially, and the management of hundreds of huge DIN A0 size drawings becomes a real challenge.

2. Limits on parallel access to the same information. We all know how many revisions are performed on the building plans during the design phase and even later, during the construction phase. These revisions often lead to conflicts among the different disciplines, such as structural engineers or building services, or to budget conflicts. It is the architect who should

be in charge of resolving these conflicts. But as there is no parallel and concurrent access to the paper-based drawings, participants do not see the effect of changes made by others on the same building. This limitation slows down the information exchange and consumes the most critical element of any construction project, TIME.

3. Lack of ability to represent relations among the different aspects of the same information. Paper has only two dimensions, while the real building has at least three, but in fact far more. There are several functional and logical relations among the different aspects of the building, representing the fourth, fifth and further dimensions. For example, how HVAC requirements will be affected if a glass surface is enlarged, not to mention the effect on construction expenses, or operational cost. Paper drawings do not reflect these dynamic relations, although they represent the most critical building information.

The first and second limitations (the amount of information and concurrent access to it) have been partly resolved by the first generation of architectural CAD systems, the "automated drawing boards". There are no physical limits to the electronic version of the drawings, they are much more flexible for revisions, are much easier to manage and organize, and the synchronized access to them by several players has recently been solved using intelligently networked computer systems.

However, as long as computers just imitate paper, ruler and pencil, they will never offer any solution to the third and most important challenge, namely the limits on the depth of the building information on paper. For this we have to leave the "Paper Age" and really enter the "Information Age". It is not the paper that should be simulated on computer, but the building itself. The digital prototype of the real building (we call it "Virtual Building") contains much more than just floor plans, elevations and sections, even more than a visually attractive 3D model of the building. It comprises all information about the real physical building and all the inherent logical relationships within this complex information, on a level that could never be perceived and communicated on paper. The architectural design software should go far beyond being a mere productivity enhancement tool. It is an authoring and reporting tool about the Virtual Building and it is the key to control the information about it.

The original author of the Virtual Building is the architect, who owns most information about the actual building. But most importantly the architect becomes the one person in charge of coordinating all the information coming from different contributors, such as structural and building services engineers, interior designers, or from the demand side, i.e. the general contractor, the facility manager and the owner.

The ability to simulate a building's behavior, both before it is built and throughout its life cycle, will naturally affect the architects' design process, but it will also change his relationship to the client, the contractor, the community and to the entire construction industry. For example, as the owners and caretakers of the Virtual Building, architects will enjoy competitive advantages in procuring all future work associated with the same building, including maintenance and management, where much more financial spending is concentrated than in the design phase. The architectural firm will offer comprehensive services by integrating all design disciplines – engineering, civil, facilities – and become a lucrative consulting business as well.

In other words, IT will put the profession of architecture back in the driver's seat of the construction industry and allow architects to maintain full control over the building project, just as master builders did hundreds of years ago.

Architects in Time and Space: Understanding Sustainability

Chris Butters Oslo

Three Facets of Sustainability

People in different ages and cultures understand the world differently. In our age, the environmental imperative is recognised by and unites all cultures.

Though it is not easy to define, sustainability as understood today always has three basic dimensions: ecological, economic, and social. No society can exist without a healthy natural base, hence environmental awareness and efficiency are crucial. However, the ultimate goal of sustainability is not just healthy ecosystems, but healthy community.

Planning, urban development and buildings are responsible for 50% of the total energy use, resource flows and emissions in our societies. They also represent over 50% of all real capital. Our profession thus has a major responsibility. In our age architecture is not complete, and is certainly not good architecture, if it is not reasonably sustainable.

So the goal of sustainability has far reaching effects for architecture. It is not an empty cliché; nor is it just about technical fixes for environmental problems. We have the three keywords:

Ecology: the "ecological footprint" resulting from our energy and resource consumption is of critical importance, and has been a main focus for ecological design. But eco-technology is only the material dimension of sustainability.

Economics: in its original sense economy refers not just to financial systems, but to the overall organisation and management of society. Economic sustainability thus includes issues of inter-generational planning, equity and participation.

Society: similarly, sustainable development implies a healthy relationship between nature and culture. The dimensions of social and cultural sustainability are therefore just as important.

Tradition, Modernism and Ecology

Traditional societies often had a wholistic view of their place in the cosmos and their interventions in nature, and their architecture was well adapted to local environment. However their technology was often far from sustainable, in part due to a lack of the global awareness and scientific knowledge which we have today. Their sociopolitical systems – tribal, feudal, dictatorships or whatever – were not always something to get very romantic about either.

The modernist world replaced the gods of old with rationality and science. Modernism sought a new, universal vision, but in doing so discarded both tradition-based wisdom, and design based on local context. In addition, the specialisation of knowledge and functions led to disorientation and fragmentation of the world, both physically and spiritually.

The ecological vision combines global view and a local context. And it integrates the dimension of time into architecture, so that both the global and the local, both the past and the future, become our informants in the act of design.

The Whole of Space and Time

The ecological paradigm places architecture in a new framework which encompasses the whole of space – and the whole of time.

As regards space: at the macro end of the scale, we know that our choices – our buildings, energy systems and materials – may have an effect on the ozone layer over Sydney or the rainforests of Brazil. At the micro-level, Sick Building Syndrome and the new science of building ecology have made us aware that we architects must consider the murky secrets of industrial chemistry, of formaldehyde and radon, PCBs and PVC.

And as regards time: we know that our cities and our buildings will have effects for many generations. Life cycle based production is becoming a requirement in modern economies; and architecture must consider the long term future, lifetime maintenance, deconstruction and recycling as well as immediate needs. This affects the way buildings are to be designed and assembled.

In a sense, this perception is not new. Pioneers such as the sociologist Le Play and biologist-planner Patrick Geddes developed wholistic models. These are reinforced by insights from thermodynamics and modern systems theory. The new view thus connects age-old wisdom in the forming and management of human settlements, with new scientific insights.

High-Tech versus Low-Tech

The old polarity between high-tech versus low-tech is no longer relevant. In many cases these were little more than architectural metaphors anyway. An "organic" form was seen as being ecological even if it was built of polyurethane foam – or a house with "natural" timber facades, even if they were poorly designed and needed to be replaced every ten years. A high-tech building in stainless steel was seen as being energy intensive and not ecological, even though it might last five times as long and without any maintenance. Sustainable solutions may be both low-tech and high-tech – when seen in a lifecycle context and on condition that they are used in the right ways.

Debis Building, Potsdamer Platz, Berlin Architect: Renzo Piano Building Workshop Architects

The polarity of traditional versus modern also loses its sense. Many modern eco-buildings "rediscover" principles, such as for thermal zoning, which are identical to those we find in traditional building. Whilst modernism wished to cast off history, sustainable building is a conscious reintegration of the two forces of continuity and change, of old principles and new awareness. It is an architecture of evolution.

Architecture – Elitism and Abstraction?

Architecture can only be a resource in relation to the world as architects understand the world. And our world is a real world. Much of today's architectural practice, and discourse, is in danger of becoming increasingly irrelevant. It retreats into elitism, or into graphical abstractions.

Architects sometimes seem more interested in being exciting than in being relevant.

Some propose that the answer is a sort of enlarged modernism which includes ecological aspects. This attempt to "save" modernism doesn't take account of the fact that today's world is no longer the industrial-era world of the 20th century. Modernism does not necessarily provide modern architecture. The tools and methods of modernism are powerful, but they must be applied to an up-to-date vision of the world.

Others suggest that regional development and urban design are now subject to the so-called greater forces of globalisation and capital development. This is an abdication of responsibility, for cities are still real places. They are about infrastructures, energy, streets, buildings and people. They embody real environment, real economy, real community. So they must be designed – either well or badly – either sustainably or not.

Sustainability is not a Style

The modernist approach is no longer enough: nor is a purely traditional approach. Above all, we are still far too concerned with aesthetics and trends in architecture. Style, by definition, is not sustainable; therefore, sustainability can't be about style!

Architecture is often still seen as a linear progression of historical styles and technologies. This overlooks both the underlying premisses, and the ecological content. Evolution is not about surfaces. If we ask, what will tomorrow's architecture look like, then we are asking the wrong kind of question.

The grand idea of functionalism, that form follows function, has little meaning. Many modernist buildings were never more than metaphors of functionalism, or of technologism, whereas many extremely fine, functional buildings have non-specific if not entirely circumstantial form. There is not necessarily an explicit relationship between content, function and form.

The idea of ecology as leading to some kind of new architecture needs to be discarded. Sustainability concerns processes and resource cycles, not buildings as objects. It concerns all the "invisible" contents of architecture which lie under the surface – site impact, energy systems, unseen effects of components and building materials. If ecological buildings can be high-tech or low-tech, traditional or modernistic, then this means that they can take many forms. On the other hand, when we work from the starting point of local climate, site, solar energy, natural ventilation, then sometimes these parameters in themselves are the generators of new, exciting formal solutions. So ecology can lead to some new forms, too!

No Easy Soundbites

Recent Nordic examples show interesting links between tradition and modern sustainable design; as well as a humanist approach which combines high quality in design with a sense of global responsibility. Similar signs can be seen in other regional responses to the need for sustainable building. Tools and discourses for evaluating sustainability are however as yet not well developed. Sustainability is becoming fashionable, even with some famous architects, but there is not enough critical evaluation of high profile projects. There are many traps in so-called sustainable architecture. The message is getting confused with the meaning, and the content with the form.

If we could make some nice, clear, concise soundbite statement about what sustainable design is, then that statement would certainly be wrong. Because reality is more complex – and so is architecture. It is not this, not that, both, and, neither, nor. Avoiding the tendency of rational western thought to categorise things, may be our biggest challenge. The imperative of sustainability opens new, inspiring perspectives for design. Global solidarity becomes as important as local context; the past becomes as important as the future. The care for attractive, high-quality detail which our profession loves, takes on a more complete meaning. Ethics and aesthetics need not be in contradiction. There arises an aesthetics which embraces both product and process, both form and content. Both must be beautiful – the surface of architecture, as well as what is under the surface, and the processes that go into making it, where it comes from, and where it is going.

Things must change – but first of all we must change our way of seeing.

Tradition and Innovation
The Evolution of Regional Architecture in China

Ma Hang **Chen Jiawei** Shenzhen

The built environment is generally believed to be the product of technological, political, economic and socio-cultural processes. After presenting the basic characters of Chinese traditional architecture and their evolution since 1950, the thesis focuses on five aspects – cultural and living pattern, new exertion of traditional structural technology, historical and original, technological, and city style – in the course of pursuing and retaining the traditional features of Chinese architecture, which include not only the new exploring conventional approaches of Chinese architectures, but also experimental attempts in new directions, as well as new ideas of how to treat traditional architecture. All of these make Chinese contemporary architecture evolve in multi-directions, in terms of forms, contents and theories of Chinese architecture, and will continue to dictate the future development of Chinese architecture.

Research Background

In countries of abundant traditional legacies, people sense a strong crisis. On the one hand, such crises come from increasingly efficient technological equipment and the requirement of such; on the other hand, it is generated by a sincere longing for traditional form endowed with rich glamour.

In future, our society is to wit lifestyle globalization and cultural localization. Although people's life style get nearer, as a matter of fact, people hope to retain profound and distinguishing legacy, including language, history, culture and region, etc. Therefore, the nations which have been nourished by traditional culture for a long time, whose psychological quality, view of life and aesthetic sense would be inseparable from the tradition because of the changing of modern lives furthermore, the tradition combining with the modern culture will still retain their respective character during the developing phase.

Architectural tradition is a composite of multi-level and multi-direction, the definition of "The Characters of Chinese Architecture"[1] was summarized by Mr. Liang Sicheng in 1954 as followed:

1. Single architecture is composed of three parts: "down part" – platform, "middle part" – house, "up part" – roof;
2. Grouped architecture form yard, which becomes "the outdoor living room";
3. It is wood frames not walls up to support pressure, and wood structure is a mainly structure;
4. The brackets are used to support overhanging and reduce the shearing forces situating in the intersecting points of columns and beams;
5. The bending roof is constituted by raising-and-depression (a method of defining the slope of pitched trusses in Song Dynasty) and raising-and-truss (a method of defining the slope of pitched trusses in Qing Dynasty);
6. The curved roofs help to outstand the architecture's exterior forms;
7. The colorful pigment and decorative paintings are bravely used in the ornamentation;
8. The intersecting points of components are mostly bald, and most of the projecting parts had been aesthetic treated;
9. The gazed tiles and sculptures made by brick, stone or wood are used in a large quantity.

The Process of Innovation

There are mainly three phrases during the evolution of traditional architecture in China.

1. By the beginning of 1950's, Chinese architects had begun to show solicitude for the design of the regional architecture and drawn fully on the experiences of folk dwellings' designing methods that were mainly involved in traditional details.

The Cultural Palace of Nationalities[2] designed by Zhang Bo in 1959 is a good example. The designer strived to build high-rise architecture with national forms. Located at north of the Fuxingfu Street, it looks fresh and bright, elegant and cheerful in color. Its central tower is crowned with a traditional double pyramid roof echoed by similar roofs around it. Moreover, the traditional peacock blue glazed roofing contrasts sharply with the white flat finish wall, so it is renowned for its rich local flavor.

2. At the start of 1970's, architects focused on the exploring the traditional methods relating to topography, weather conditions and plan layout.

The Enlightenment of Fragrant Hill Hotel

Fragrant Hill Hotel designed by I.M.Pei adopts the traditional decorative elements and pursues the expressions of traditional spirit and artistic conceptions; in addition, it breaks through some rigid patterns existing in the minds of Chinese people.

The features of the hotel are as followed:

- Respecting the nature – the building preserves many valuable old trees and original old walls, which are harmony with the new building;
- Derived from Chinese courtyard layout and classical Chinese gardens, eleven courtyards in the hotel and eighteen elaborately designed landscapes reflect traditional gardening beauty.
- The hotel has not a traditional cured roof, and the elevations

implicate the features of the folk houses of South China in the Tang and Song Dynasties. Moreover, it appears more charming when the sunshine casts the shadow of black diamond terrazzo window's frames on the white walls.

3. Since the early 1980's, some regional architecture had broken through the rigid forms of tradition and strived to widen the design ranges of regional buildings. There are five aspects to embody the trend:

a. The thinking of "making famous into strange"[3] – from cultural and living pattern's aspect

The architects emphasize not only the idea of the continuity of cultural structure, but also its inheritance and development. In addition, they utilize the thinking of "making famous into strange", which means attaching importance to the existed things in the cultural structure and conceiving existing ones familiarized by people. Most of important, grafting or combining local architectural culture with external one forms new concepts, theories and design methods.

The Mausoleum of Prince Nanyue of the Western Han Dynasty[4] stresses the blending the modern architectural approaches with classical principles. It consists of three parts: Displaying Hall, Coffin Chamber and Treasure Hall, the 3-story Displaying Hall is for ordinary exhibits. The Coffin Chamber is enclosed with verandah and covered with a metal glass roof in geometric shape that is the tomb style of the Qin and Han Dynasties, and the 2-story Treasure Hall exhibits all the funeral valuables in the tomb. The architects designed the building in compliance with the features of the low hill at its back after taking into account of the environment. Furthermore, the three parts of the Mausoleum are also harmoniously linked to create an ambiance of pleasing wholeness.

b. The emergence of criticizing region – from new exertion of traditional structural technology's aspect

By means of studying the constructive forms, character and materials of local buildings, the architects intend to give people a good feeling of traditional character harmonizing with the modern technology. Meantime, combining with the local climate and specific terrain are main points of departure in design.

The Contemporary Art Gallery of Quanzhou[5] designed by Mr. Zhang Yonghe is an appropriate example. It used the old building material and "brick out and stone in", which is a constructional way existing for a long period in Quanzhou. Moreover, different roof frames of old buildings were renewed in order to reduce the cost of construction, which is also a conservation of vernacular dwellings in the demolition period. Specifically, changing the dimensions and directions of roof frames weaken the traditional rigid form.

c. The regeneration of new original patterns[6] – from historical and original aspect

The architects mainly stress on utilizing the ceremony character and ancient-style of the primitive buildings in order to be in harmony with the nature. Meantime, they think that the typological elements endowed with historical and original character should be reappeared, transformed, transplanted and regenerated to form a countryside scene. Since 1994, Mr.Liu Jiakun had designed a series of "Artist Working Studios"[4], he tried to create higher artistic quality with low cost and low technology and explored an architectural way which can be applied in the countries of short of funds and having a long history.

Concretely, Luo Zhongli Working Studio originates from the gray cave dwellings in flatlands, and Dan Hong Working Studio and He Duocen Working Studio adopt the traditional elements – courtyard. The architects think that the basic constructional elements – wall, curved surface, ramp, and gallery have great effect on the form and the lasting sense of the architecture can result from the simple, unadorned and succinct ways such as coarse stucco, cobblestone, tilted roof decorated by gray shale and black iron window frame. In a word, they select originated way to manifest the mind of site, but sometimes the application ranges of the buildings are limited to some degree.

d. The thought of technological worship – from technological aspect

The regional buildings stress the distinguishing feature and consistency; especially rational technological core becomes a starting point of design. Gradually architects import the simple ecological thought into the design and advocate that architects should analyze the regional forms which are effected greatly by local climate and geography environment, which actually be the main basis for designing. Furthermore, Chinese architects learn much from the western buildings.

e. The emergence of city renew – from city style aspect
The architects advocate that drawing on the experience of single house's form, retaining the identity of environment and regarding the city as an entity.

Ju'er Alley Residential Quarters explores the ways of solving some important problems in reconstructing the old city of Beijing. First, the new courtyard housing system not only affords high floor area ratio, but also suits well to the traditional city style, thus the protection of the famous historical city goes down to residential quarters – the grass roots of a city. Secondly, the project is perhaps the best way available at present to renew the old city of Beijing. Thirdly, the project puts forth a fresh, simple outlook, which combines the tradition with modern function and technology. Lastly, it explores the way of creating a condominium form with funds pooled by the inhabitants.

Vanke Wonderland[7] helps to widen the fringe dwelling district theory of contemporary cities in China. Especially, the new dwelling conduct model has a great significance during the urbanizing and suburbanizing.

There are Four Aspects to Manifest it

- The vernacular architectures in KeJia and neighborhood blocks – they both share the character of "the floating space is enclosing not sealing". Every neighborhood block is consisted of two or three dwelling buildings. According to building orientation and climate, the architect adopted the plan layout of enclosing not sealing to strengthen the fluidity of space and ventilation in summer.
- Memorial archway with balcony – traditional buildings in China always use the arcades to embody the transitional space between interiors and outdoors space. The traditional elements largely are used in the blocks and entrances to pedestrian paths, and the balcony can be led to main bedrooms, so it is named as "the memorial gateway with balcony".
- Commercial space under the sotto portico with big depth – an architectural language of vernacular dwellings in LingNan. The architect paid attention to the function of sotto portico, moreover, he widened the depth from 2 to 3.3 m in order to intense its traditional sense.
- The confliction and amalgamation between the modern style and tradition one – there are two facets to expound it. Firstly, there is strong comparison between the Community Center with modern style and the residential buildings whose white-pitched roof and dormer are signs of classic style. Secondly, the dwelling buildings with modern constructivism have a good harmony with those with white walls, gray tiles and two-tilt roofs.

Conclusion

While monotonous architecture without cultural connotation and the loss of historic locales and contexts to similarity and oversimplification are equally undesirable, quite a few developing nations that oscillating between modernity and tradition have successfully revived traditional culture with the help of their growing economic wealth, even though modern architecture seems to be an obvious choice for fast constructions at first glance. Chinese architecture since 1980 features richer traditional elements and cultural connotation compared to those in previous decades. Although in the above cases, our analysis only concentrates on five aspects, it should be pointed out here that any successful architecture is not limited to such and should entail creative architectural conception, sound structural design, good plan composition, exiting modeling, and elaborated detailing. A major development since 1980 is a criticism of simple-minded imitation and reproduction of the external appearance of the traditional architecture and a concern for the overall environment and the underpinning cultural ideals. This development reflects a deeper understanding of the true content of Chinese traditional architecture and opens new ways in sustaining and developing this tradition.

---

1 Liang Sicheng, The Character of Chinese Architecture, Architectural Journal, 1954 (1).

2 Zhu Yongchun, Zhao Wei, Sun Zhongwei, Jin Lei, Yang Yongsheng, Chinese Architecture, China Architecture & Building Press, 1999.

3 Wang Shaosen, The Cultural Tendency of Architecture Design in Contemporary China, New Architecture, 1990 (3).

4 Contemporary Chinese Architects-He Jingtang, China Architecture & Building Press, 2000.

5 Zhang Yonghe, The Contemporary Art Galley of Quanzhou, Architecture, 89.

6 Liu Jiakun, Low Voice Recount and Low Technology, Architecture, 78.

7 Chen Fang, Finge Dwelling District – Vanke Wonderland, Architectural Journal, 2001 (1).

Architecture of Russia in the Epoch of Globalization

Yury P. Gnedovski Moscow

1. The 21st century – new challenges:
   globalization and anti-globalists. Cultural identity as antithesis of cultural unification.

2. Architecture of the 20th century:
   an attempt to create universal town planning systems and artistic concepts. Change of styles and variety of local schools. Masterpieces of architecture of the 20th century, their individual and general values.

3. Russia of the 20th century:
   political and artistic revolutions. Architecture of totalitarianism and that of mass-scale social programs. Loss of prestige in architectural creation and rehabilitation attempts.

4. Anti-communist revolt. Rehabilitation of market economy. Freedom of creation and the role of new clients:
   businessmen and officials. Commercial architecture and commercial architects. Decentralization, new role of regions.

5. The new times – time of Putin.
   An attempt to rehabilitate the State organization and Federal programs of development. The law on land and effect of the cities' master plans. Participation of architects in investment processes.

6. Vector of architecture's movement.
   Rehabilitation of original architectural and artistic values, rehabilitation of the avant-garde and modernism. Placing an order for "accordant to national history" architecture – "Neo-Russian style", "Neo-Baroque" and "Neo-Classicism". Modernism and de-constructivism are the languages of new generations. Protection of historical values throughout Russia, preservation and rehabilitation of historical centres, prohibition on the invasion of sky-scrapers. The new architecture is mainly for "sleeping districts". "Moscow-City" is the Moscow "Defense" and "The Russian cultural centre". It is a tribute to historicism. St. Petersburg, architectural school of Nizny Novgorod, work in the regions.

7. Conclusions. Each country takes part in globalization of the economy.
   Russia is on the threshold of entering the WTO. The peculiarities of geography, economy, culture, mentality are reflected in contemporary architecture. "High-tech", without high technologies in the construction process, is only a superficial tribute to a fashion. Organic architecture today is architecture that is adequate to the country considering its aspirations for the future.

   Are high-rise "down-towns" and cottages surrounding them still an existing inevitability? Or will the history of Europe and Russia find another form of organizing labor and recreation in the structure of towns with century-old histories?

   Russia is entering the WTO and is opening up to the world for discussions and international competitions, which we hope will attract the best architects in the world.

   We want our country to be revived. The increase of competitions for entering Architectural Schools confirms our optimistic forecasts for the future of architecture in Russia.

The Modernisation of Old Buildings and the Preservation of Historic Monuments – The Irish Experience

Klaus Unger Dublin

The world was at war, when in 1916, a rebellion against the Union with Great Britain broke out in Dublin, as Ireland made its bid for freedom and independence. This event led eventually to the formation of the Irish State in 1922, but with the country damaged and impoverished. The urgent need to create unity, reconstruction and vision for the future, was instrumental in seeing available accommodation provided for government and parliament.

Two hundred years earlier, throughout the 18th century, Ireland had its own parliament, separate from London, and housed in a purposefully designed building. That Neo-Palladian "Parliament House" consisted of a two-chamber legislature, the Upper and Lower Houses placed in an informal relationship to one another, with a corridor surrounding the Lower House on three sides, ante-rooms protecting both chambers, committee rooms clustering the periphery, and a grand colonnaded atrium by way of screen and frontispiece. The success of the American and French Revolutions had inspired an early revolution in Ireland in 1798 which failed, and the Act of Union with Britain which followed in 1801 ended the then Irish Parliament. The redundant building, still regarded architecturally as one of the most important buildings in Dublin, was sold to the Bank of Ireland in 1803 and remains its headquarters to this date.

A brief breakaway parliament formed in 1919 and met on a few occasions in the Round Room of the Mansion House (the Lord Mayor's premises). But it was the availability of the lecture theatre at the Royal Dublin Society's building – known as Leinster House – that the new Irish Government saw fit to rent in 1922, to provide for parliamentary meetings, on a temporary basis. Concurrently, the government took possession of the newly completed complex of the Local Government Board of British Administration and the Royal College of Science and State Offices. It founded its Ministerial Offices, Cabinet Room, and the senior Department of the Taoiseach (Prime Minister) and of Finance, within the complex, sharing space for many years with the Faculty of Engineering from the University. That complex eventually became known as Government Buildings.

Both Leinster House and Government Buildings occupy sites on a city block of 8.2 hectares, 1 km south of the city centre.

Leinster House was originally a large private house built in 1745 in a then unfashionable part of the city. In 1850 it was acquired by the Royal Dublin Society as their headquarters to promote and develop interests in farming, husbandry, and arts and crafts. The Society organised architectural competitions throughout the remaining decades of the nineteenth century to build new cultural institutions in its grounds around Leinster House. Here, beginning in 1859, they built a Natural History Museum, a National Art Gallery, a National Library and a National Museum, after being stimulated earlier by the success of the 1853 Dublin Exhibition which was visited by over a million people in six months.

The 19th century cultural buildings flanking Leinster House, within the city block, were prior to the time of their construction, surrounded by 18th century terraced family houses and parks. These were part of the elegant Georgian architectural development of Dublin that now saw this once unfashionable part of the city become stylish.

It was this cultural enclave that then became the first home of the new establishment and legislature in 1922. In 1924 constrained public finances determined that parliament should remain in Leinster House. This decision triggered a programme of consolidation of accommodation within the city block through further investments and the acquisition of adjoining properties by successive governments. One of the first actions taken was to complete the final section of the recently constructed National Library as a symbolic gesture towards learning and the future. Leinster House was also modified to better suit the initial requirements. The complex of buildings which commenced life as a private home with wonderful spaces, decorative plaster ceilings, ornamental fireplaces, and other domestic scale details, settled comfortably over the following decades into a successful parliamentary accommodation, by careful conservation of works and by sensitive additions and alterations, as resources allowed. Throughout the 1950s, 60s and 70s the State-owned buildings within the block were maintained and enhanced where possible. A major new commercial office block was built tangential to the Leinster House precinct in 1972, and was acquired by government to house the then important Department of Agriculture. It was, however, the successful adaptation and conversion of older buildings throughout the 1980s, that helped to underline the potential of using old buildings for new, modern purposes.

The accession to the European Economic Community in 1973 placed an additional burden on the government to provide suitable accommodation for Ireland's hosting of the European Presidency, which provoked a debate over new-build conference facilities versus adaptation and extension. The final outcome was the modernisation and extension of structures at Dublin Castle in 1986.

In 1981 the development of the National Concert Hall in the Aula Maxima of the 19th century Catholic University showed that, when finances are scarce, solutions can be found to successfully exploit existing resources which can be justified in terms of sustainable urban planning. Likewise, with the founding of the Museum of Modern Art at the restored 17th century Royal Hospital in 1983, the preservation of a historical building and its continued occupation and use for appropriate new purposes is seen as justified at a number of levels, not least, in maintaining the urban fabric of the surrounding city.

The 1990s have seen considerable additional investments made to all the State-owned buildings on the block. Upgrading, extensions, alterations, restorations and conservation have all been employed to further consolidate and reinforce the success which the State and cultural institutions have together enjoyed, sharing not only the same urban site, but also common facilities. This physical closeness, has served at both the practical and philosophical levels to ensure that there is no great remove between public representatives, State officials, and the custodians of heritage and culture.

The Irish Government's Policy of Architecture 1999 commits the State to encourage high standards of architecture and to give a good example to society, by leading when it acts as its own developer. The government has taken significant steps towards strengthening the protection of our architectural heritage through the recent introduction of a comprehensive and systematic legislative mechanism for protecting this heritage as part of the planning code.

Encouragement and facilitation by government at local government level has also assisted in the successful regeneration of run-down areas in the city centre – most notably at Temple Bar, south of the River Liffey, the HARP area to the North, and the more recent O'Connell Street rejuvenation plan launched in 2002.

In this regard it is also important to note the Resolution of the European council of 12 February 2001, on "architectural quality in urban and rural environments" which expressed its attachment to

a) the common characteristics shared by European towns and cities, such as the importance of historical continuity, the quality of public areas, the social mix and the richness of urban diversity;

b) the fact that good quality architecture, by improving the living context and the relationship between citizens and their environment, whether rural or urban, can contribute effectively towards social cohesion and job creation, the promotion of cultural tourism and regional economic development.

In the past great Irish literary figures have also had associations with parts of the block:

James Joyce made reference to the museum and library buildings through his character Leopold Bloom in Ulysses.

Oscar Wilde lived within metres of the National Art Gallery, and George Bernard Shaw was so indebted to the delights of the National Art Gallery that he bequeathed half his estate to it upon his death.

The Irish poet William Butler Yeats, writing of the events of the Easter Rising in 1916 and the subsequent difficult years before the establishment of the Free State, penned the line to his poem commemorative of the event: "All changed, changed utterly: a terrible beauty is born".

The inference was fear of the unknown, the future, yet hope too, that the outcome would be beneficial. Now, after eighty years of successful self-government it is possible to confirm the maturity of architectural responsibility and commitment to quality architecture at government level.

Present, and planned future developments within the block are optimistically viewed as stamps of approval for the course of actions taken in 1922.

In this regard, the example set at the very centre, by the on-site commitment to "resource architecture" and by the attendant levels of architectural excellence, is worthy of note beyond the shores of Ireland.

Processes of Transformation in the "New Cities" of Siberia The City Angarsk

Barbara Engel Cottbus

Processes of transformation in the New Cities of Siberia. Requirements and chances for a sustainable development in the city Angarsk.

The Blue Cities[1] of the Soviet Union

In the former Soviet Union there are cities, which were founded in a time of big industrialization between 1955 and 1975 in the northern part of Siberia. They are located near to new transport lines and development axes, which lead from important cites along the Trans-Siberian railway toward the North. Those cities were founded because of the industrial production nearby, big resources of raw materials, or as scientist cities.

The word "blue cities" comes from the sixties and means new cities from Siberia and the Far East. The word 'blue' means 'unreal', 'dreamlike', first because the inhabitants, the founders of the cities transformed their wishes and dreams into those cities, and second because they were secret, closed cites, and many Russians did not know them at all.

Changes since Perestroika

Because of the vast political changes over more than ten years, big changes also affected those cities at many levels, depending on their social, economic and spatial structure. In the Communist era those cities were supported by central subsidies of the Russian government, which is now no longer possible because of the contemporary economic situation. It takes time to change from a planned economy to a liberal market. The results are low economic growth and a high rate of unemployment.

Requirements for Sustainable City Development

The New Cities in Russia have – depending on the regional and specific situation – enormous potential, which could be further developed. A high level of education, good technological knowledge and a developed infrastructure, decreasing population rates and a young population structure provide a good basis for the future of the cities.

Development of the City Angarsk

Angarsk, a mono-functional city, is located in East Siberia, 50 km from the regional capital Irkutsk with an area of 1,143.78 sq km, between the two rivers Angara and Kitoi. The main idea and motivation for the foundation of the city (1951) was industry and the need of housing areas for the workers of the chemical enterprises. The areas of industrial production with adverse health effects need big buffer zones to the housing areas: Angarsk develops with linear city structure and strictly divided functional zones, living, leisure and working areas. The industrial zone is located near to the water and prevents the access of the city to the river Angarsk, which gave the city its name.

The example for the first houses, which were built in the fifties, was St. Petersburg. Little districts (2–3 hectares) with two- or three-, later four-story-buildings, are reminiscent of their "big sister" in the country, not just because of their ensemble but also because of the lovely details. They are followed by the industrial housing areas of the era under Khrushchev, which have a well-designed scale and open space. Today those houses are very popular housing districts. In the 70s and 80s the density of the housing area grew from six to eight hectares with nine- and ten-level-buildings. Huge areas of settlements developed within the open space.

Demographic and Economic Tendencies

In the year 2000 264,500 inhabitants lived in the city Angarsk, tendency constant. People are not leaving the city at the moment. Neither are people coming from other parts of the region to Angarsk; just a very few, who lived in provisional cities in the north, come to stay in Angarsk.

The statistics from this year make believe, that the bottom of economic development has been reached, even though those statistics do not obscure the facts about the still dramatic economic situation in the country. The economic situation causes other problems in demographic development: A low birth rate, a high mortality rate and low life expectancy, because of decreasing health facilities provided by the state and because of the worsening ecological situation.

With the decline of industrial production more and more space lies fallow. The industrial enterprises try to sell those areas to the city government in order not to pay taxes for the unused areas. For financial reasons there are no developing perspectives for those areas.

In the meantime the city is trying to change its mono-functional profile. Further new enterprises have already been established. The activities of small trade companies and crafts enterprises are growing, such as furniture production or car-repairs.

Housing

Even though the number of inhabitants is nor growing in the city of Angarsk, the city has an enormous demand for new dwellings. The demand to enlarge the area of the city has several reasons: About 30,000 people live in dormitories or they stay in so-called kommunalkas, where the people share the kitchens and bathrooms or live with many others in one room. On average the people live in a standard living space, stipulated in Soviet times as 9 sq m, but which is now 20 sq m. People need inexpensive dwellings that they can pay for, because the state cannot afford enormous subsidies like in Soviet times.

People who can afford it occupy more space. Many flats are made into one, which were three in former times. One big problem is the increasing demand for single housing. Just recently, in 1998, the rule from Soviet times, not permitting the building of single houses, was repealed.

But because it was not possible to integrate these new individual demands within a whole concept, we now see the "chaotic" development of many new dwellings of the so called "New Russians". It is a big problem within the urban structure without any basic planning law. Low density and individual dwellings were not envisaged by the still existing general plan from the seventies. So one important task for urban development in the future will be to think about alternative housing concepts for both low-income and wealthier people.

Transformed Use and Wrong Use of Public Space

The effects of new economic developments resulting from political change are also visible in public spaces. The growing trade-market and the beginning of a liberal market system led initially to the emergence of temporary buildings – at bus stops, train stations and squares. But now more and more steel-glass buildings are being built to replace the former kiosks. Former kindergardens, dormitories and cultural clubs are being converted to offices. Also some dwellings have been transformed into shops and stores. On many facades of the dwelling houses you can see signs "remont" and "office" indicating that rooms in the flats are being put to new uses. In the initial phase emerging independent enterprises start up in the backyards of the dwellings.

The increasing demand for provisional constructions, such as gas stations, parking spaces and stores encroaches on the territory of the well-designed green public spaces. Spaces, which were designed as leisure space or for sports facilities are now being used as parking lots or even being built on over night.

The city center has not been built yet. It is one of the city's most prestigious districts, which should connect the two parts of the city, the northeastern and the southwestern parts. At the moment there are two sub centers. The central street, which in the new center should become the Angarsker Prospekt. In the general plan, and in some alternative results of a competition as well, there are some huge and crazy ideas for the center of former times. We are lucky, that these ideas have not been realized. At the moment the Angarsker Prospekt is like a bypass: in the one direction it ends with a housing complex opposite which there are more than 1,000 garages, which were built by the inhabitants on their private initiative. This street is a central axis when defined by its geometrical location, without any beginning and without any end.

Possible Ideas for the Future Development of Angarsk: the Public Space

Angarsk has many problems, mainly of an ecological or economic nature. The socialist ideas of Soviet times, of planning a city oriented exclusively toward industrial enterprise and providing equal conditions for all of the inhabitants, has now lost its legal foundations.

But a great deal of additional potential exists, such as the location, the technical and social infrastructure, the landscape and a well-educated, young population. With its existing cultural activities and opportunities Angarsk, which its inhabitants also call the "Siberian St. Petersburg", could become the cultural capital of the region.

One special characteristic of Angarsk is the building structure of the first construction stage from the fifties. We see very valuable typologies of buildings with high-quality open spaces, streets and squares. The variety of types of streets and their shape is a specific feature of this city compared to others in Siberia.

One particular quality of the Siberian cities is the surrounding environment, the landscape. The integration of the landscape within the city structure has been one of Siberia's central parameters in city planning. The green public spaces – the taiga as little decorative forests in the dwelling yards and the parks connecting the districts – are used not only as leisure space for the inhabitants but also represent the specific image and identity of the city. This is also an important aspect for the future development of tourist enterprises.

The public spaces offer the chance to try out new forms of functional uses. The city government cannot pay for running all the cultural facilities anymore. For long-term development new kinds of public-private partnerships will have to be found.

By converting the large industrial area of Angarsk, a former private area could become a public space. This is good chance to develop further interests and satisfy different needs in the city. Like in the projects of the IBA in Germany (IBA Emscher Park and IBA Lausitz), which were also established to support the transformation process of big industrial spaces, the initiative for a sustainable development in this East Siberian city could also be started. The regeneration of nature on the embankment of the river Angara and the improvement of the ecological situation of the environment must become one of the priorities of any planning issue. The existing ecological problems are moving toward an ecological disaster, and this should be addressed as soon as possible. An IBA-Projekt could also help sensitize people to those problems and to begin starting real, effective projects.

Sustainable development of the public spaces might make it necessary to reduce or privatize open spaces, if they are no longer needed because of the continually decreasing population rate or new functional requirements. For the development of Angarsk city center this could mean that it does not make sense to build a public cultural center on the currently empty space, but to use this area instead for building new dwelling houses and so prevent further urban sprawl in the periphery of Angarsk.

Conclusion

With its enormous open and public spaces Angarsk owns a potential for the development of the city, which still has to be realized in order to optimize its possibilities according to their social and economic value. In contrast to often general conclusions that Siberian cities have no future, the example of Angarsk shows that the beginnings for a sustainable city development already exist, and it is worth continuing to cultivate them. Clearly, sustainable city development has first to develop the political and economic requirements, incentives for investments and really practicable laws have to be made.

But in any case, the public space, is a resource for the city's development. It can stimulate and provide impulses for launching the transformation process of the cities of Siberia on the path to becoming sustainable cities in the future.

1 cf. the song from the feature film "Zwei Sonntage"

DZ Bank Berlin Branch Architect: Frank O. Gehry

Architectural Politics of Belated Modernity: Re-Traditionalizing the Islamic World

Ugur Tanyeli Istanbul

Belated modernity is not only a sequence of transformations merely chronologically different from the "timely", or "punctual", modernities which characterize the luckier parts of the world, but is a situation structurally different from the "other" modernities. I do not want to go so far as to propose that the modernities other than the "timely" ones are "failed" modernities. No parameter, or criterion, can convincingly prove a situation of failed modernity which is unable to change a society structurally. On the other hand, what is vital for all the modernities are not their achievements and failures, which are frequently confused with the value judgements of their observers, but their destructive energy in face of the "ancien regimes" in which they were born. Beginning with the early Turkish experiments in the 18th century, the whole Islamic world was destroyed, and its centuries-old social structures are still being energetically destroyed. However, the problem is to create new social and cultural structures and practices to cope with the needs of a changing society. In this respect, a belatedly modernizing society radically differs from the other modernities. It is not able to contend with the new practices and structures substituting the old ones which are considered incomparably precious, and more important than the indigenous, while the new ones are regarded as alien imported features.

What I intend here is to define one of the architectural consequences of belated modernity in the context of an anxiety that makes such societies extremely sensitive towards the loss of traditions. A by-product of this sensitivity is the aesthetic instrumentalization of cultural differences. At first sight, it seems that the anxiety of the loss of traditions and, consequently, the identity mania have nothing to do peculiarly with the Third World. As rightly posited by Thomas Meyer[1], the identity mania and the variations of fundamentalism are not only the socio-political problems of the Third World, but also of the economically developed countries in Europe and America. But, as far as the recent architectural developments in the Third World are concerned, the need to aestheticize cultural differientation seems vital only for the belated modernities. Although the loss of traditions was observed in 19th century Europe with the same anxiety as now in the belated modernities, the responses and reactions seem to be completely different.

Why is the need to aestheticize the cultural differences so instrumental in such a belatedly modernizing society? I only attempt to answer this question in the sphere of architecture which forms a component of the general anxiety of those countries.

As those are experienced elsewhere, traditional cultural practices gradually become extinct in the non-western world along with the processes of modernization. Apart from this, centuries-old architectural practices and the traditional patterns of demand and supply are also decaying. Along with the death of traditional cultural-artistic-architectural practices, the cultural dichotomy which was described by Adorno for the western modernities, is being experienced in the non-western world[2], though its consequences are inherently different. The dichotomy of popular (or mass) and high cultures appears also in the belatedly modernizing societies. New architectural practices were to take their proper place within this cultural system. Architectural production under the aegis of popular culture, the first component of the dichotomy, is not thought by the high culture half of the system as an adequate means to realize the project of modernization, intending to create an indigenous modern architectural culture. Pop-cultural architecture has its own practices, criteria and parameters, all of which are posited by an attitude defined by the commodification of culture. This means that none of them is related to, or depends on, aesthetic and cultural discourses. In this respect, they slightly differ from the pop-cultural practices of the western world. The differences have to be traced mainly within the formal aspects of its products. In contrast to this, high cultures in the western and non-western worlds do not have even a slight structural resemblance. On the other hand, in the belatedly modernizing societies, the architecture under the aegis of high culture finds itself an extremely small area of existence. Almost all of the architectural production of such a society is contained in the sphere of the pop-culture area. From the "architecture without architects" in the illegal squatter housing settlements, to the identical apartment blocks of the inner city, pop-cultural architecture is characteristically similar in every belatedly modernizing society.

Huyssen, who discusses the situation in the "timely" modernities, argues that mass culture has always been the hidden subtext of modernism[3]. High culture in the belated modernities has, however, three subtexts, one of which is hidden while the others are openly expressed. The first, explicit, subtext is what is defined as "western culture". The newly emerging high culture of belatedly modernizing society legitimizes itself as a challenge to the cultural practices which are declared "western". On the other hand, it has to deal with the other subtexts: traditional cultural practices which block the way towards modernity, and the new but prolific mass culture of the majority. These are the heavy burden of a high culture which is demanded and practiced by a tiny minority. Besides, this is a rather contradictory burden to be borne. The need to aestheticize cultural differences, which are frequently expressed by the intelligentsia in the non-western societies, is an indirect derivative of the triple functions, or subtexts, of their high cultures.

The "raison d'être" of the first subtext, of "western culture", seems perfectly clear. "Timely" modernity of the West gives the impression of swallowing up culturally the non-western world. Its productive energy economically and culturally are so great as to be incomparable with that of the latter. All the means and channels of information and communication are controlled by it. Consequently, it has such an impressive image of powerfulness that no non-western people seems to have an alternative in the future, other than consuming its products forever. They pathetically fear they will never be able to create authentic cultural products and feel obliged to use the languages, industrial and cultural products of the western world. These pessimistic discourses, the results of this atmosphere of desperation, can only draw reactionary itineraries for modernity in the non-western world. The projects to create indigenous and necessarily non-western modernisms are consequently the axis of these reactionary attitudes. This means that, for any modernizing non-western society, cultural opposition towards the West is a vital necessity. In this sense, the "western culture" as a subtext is nothing but an empty category symbolizing everything the "other" contains culturally. Put differently, the creators and "consumers" of the high cultures, who describe themselves as symbolically identical with the totality of their own societies, can only legitimize their existence and their efforts in such a reactionary manner against the fictive cultural totality of the "other".

The empty category, entitled "western culture", is not the only subtext of the high cultures in the belatedly modernizing societies. Pre-modern cultural practices and structures are also the implicit objects of challenge. However, the nature of the challenge is a problematic, even controversial one. It is almost an affair of love and hate. On the one hand, the high culture has to overcome the obstacles of the traditions on the way proceeding towards modernity. On the other hand, the very same traditions have to be idealized and sublimated as to become the "essence" upon which the modern cultural foundations of that society will be built. For example, the practices of "hammam" have to be exalted, while traditional hygiene is being condemned, because of its obviously non-scientific basis. The traditional family has to be declared a paradise of humanistic relationships, while traditional patterns of gender relations are being gradually abolished. The solidarism of the traditional community structures has to be sublimated, while the escape from the "mahalle" is the only way to be free from the suffocating atmosphere of the pre-modern structures of "Gemeinschaft" (community). The resulting contradictions can only be avoided again by making the traditional culture an empty category. In other words, the pre-modern cultural practices and traditions are expected to transcend their own historicity, and become non-temporal and omnipresent realities devoid of their original contents.

The newly emerging area of mass culture has generally been the implicit subtext of any non-western high culture, like that in the "timely" modernities. For example, in Turkey, high culture continually opposes the indigenous pop music called "arabesk", which began in the 1960s. In architecture, the pop-cultural morphology of the apartment blocks, and the system of "quid pro quo", which provided several generations of urban dwellers with modern homes, are the objects of constant criticism by the Turkish high culture. Like the first and second ones, the third subtext of the high culture is also an empty category. It has potentialities to create cultural consciousness by defining actual counter-positions, though it seldom occurs. In the following paragraph, what makes this subtext an empty category is going to be discussed within the attitude of high culture, which ideologically accuses the "mass" who "misunderstands" modernity.

In a belatedly modernizing society, the high culture has always been in a struggle with inner and outer enemies. As those are in the other high culture areas, in architecture too, the inner and outer struggles are paradoxically interrelated. In whole cultural practices, this interrelationship is the main factor which defines the scope and content of the discourses. The discourses I called "re-tradionalizing" are the results of a contradiction which stems, on the one hand, from the search for a cultural identity to be declared to the world, that can only be defined within the traditions, and on the other hand, from the intention to destroy the very same traditions inside which block the path of modernity. Thus, traditional cultural structures have first to be destroyed (because modernity can only be built on this levelled area), and then they have to be reconstructed. Generally, these two have to be done simultaneously. The resulting re-traditionalized cultural structures have to be not plural but singular, i.e. national, because belated modernity is always simultaneous with the nation-building processes of a country. What makes the practices of popular culture a target is this need to singularize the cultural atmosphere, which is vital for a project of nation-building. Mass cultural practices, however, cannot be singular. They are inherently plural, though seldom pluralistic. Because, when the capitalistic process of commodification becomes the guiding principle, cultural practices and artifacts have to become tautologically diversified. Thus, they can easily become a threat to any disciplinary singularizing attempt, such the project of nation-building. For this reason, high culture has to pretend not to see mass culture, or simply ignore it. Or in some cases, the people, who experiences the practices of mass culture, and its products are generally condemned by the agents of the high culture: they are the "others" who misunderstand modernization, as if there is an orthodox modernity to be understood properly. As a result, they have to be reeducated. However, it is a futile disciplinary attempt to rectify the situation which is thought to be mis-shapen by the mass who creates the popular culture. In reality, the aim is to restore the long-lost singularity and inner consistency of the traditional systems beneath the cloak of modernity. Everything that was solid has long since melted into thin air, but the discourses of the belatedly modernizing society desperately seek to solidify that which was vaporized.

At this point, the aestheticization of the cultural differences becomes instrumental for making architectural discourses work. Within the high cultures of the belatedly modernizing societies, cultural differences, which are ordinary and natural elsewhere, become invariables valuable "per se". The fact that the disresemblances between the cultures are natural characteristics for any society tends to be ignored by the newly emerging nation-states. Thus, they try to produce architectural images of disresemblance. Without doubt, the most important virtue of those images has to be their disresemblance to the western models and images. The products of popular culture, obviously dissimilar to the western models, however, are not placed in such an exalted position. Because, what makes the disresemblances aestheticized are not the morphologies but the discourses formulated by the three empty categories defined by the subtexts of the high culture. In these circumstances, the extremely limited extent of the high cultural production is thought by those who create it to represent the whole belatedly modernizing society with all its complexities. They agonize over it, as if the whole cultural burden of the very same country is on their shoulders. However, it is just an illusion generated by an over-estimation of its own real extent.

As a result, the argument I expressed at the beginning, that the cultural dichotomy, theorized by Adorno and then by Huyssen for "timely" modernities, has to be inherently different in the belatedly modernizing countries. For an in-depth understanding of architecture in the Third World, we have to go behind the mere morphological analysis and study this specific dichotomy in its own context.

1 Thomas Meyer, Fundamentalism and the Politicization of Cultural Differences, Zed Books, London & New York, 2001.

2 cf. Theodor W. Adorno and Max Horkheimer, Dialectic of Enlightenment, translated by J. Cumming, Verso, London & New York, 1979, pp. 120 – 67; Theodor W. Adorno, Aesthetic Theory, translated, edited with an Introduction by R. Hullot-Kentor, University of Minnesota Press, Minneapolis, 1997, p. 16.

3 Andreas Huyssen, "Mass Culture as Woman: Modernism's Other", in A. Huyssen, After the Great Divide – Modernism, Mass Culture, Postmodernism, Indiana University Press, Bloomington & Indianapolis, 1986, p. 47. For a detailed discussion of the relationship of high and mass cultures: Andreas Huyssen, "Adorno in Reverse: From Hollywood to Richard Wagner", in ibid., pp. 16 – 43.

The Eureka Tower

David Sutherland Melbourne

Situated in the Southbank district of Melbourne, Australia, south of the Yarra River which bisects the city centre, there will be, when completed, the tallest residential building in the world. The site itself is in a prime location, although prior to Eureka it had a history of failed development aspirations. What is the logic of such a tall building? The underlying premise is very simple. At the time of purchase the site had current planning permission for two tall buildings adjacent each other along City Road, a major arterial road which borders the site to the south. Constructing the buildings in that manner would have exacerbated the already poorly considered location of a series of tallish buildings, thereby creating a wall of buildings to City Road. Nonda Katsalidis, a Principal of Fender Katsalidis Architects had the idea of, simply, putting one approved building envelope on top of the other, thereby creating a taller building that would provide a much slimmer profile, allowing light and space to flow into City Road, and creating a far more iconic solution.

When we first started work on the development, our approach was differentiated from that on other projects only by its size; at that time the building was a little under 90 storeys high (it has since grown to just over 90 storeys). In other words, we commenced our work on Eureka using traditional approaches familiar to architects since the birth of the profession.[1]

As we progressed through the design and documentation processes, however, our delivery systems changed dramatically. How and why did that happen? That is the subject of this presentation. To explain that we will initially examine the changing environment within which architects in Australia and other countries are operating and then describe the place of our adopted methodology within that milieu.

We find ourselves working in a professional environment where we are expected to provide a greater level of information than in the past, but for a lesser cost. Other participants in the building procurement process are developing the tools and expertise to carry out some of the tasks traditionally associated with architects. That approach is being encouraged by large construction companies who, because of their financial base, are underwriting aspects of projects and, therefore, gaining a greater influence over the whole of the building procurement process. We are experiencing more and more pressure on the time available for us to do our work, yet the nature of the problems we are dealing with at a design level are complex, and require due consideration so as to arrive at an optimum solution. While we accept those pressures on us to do our work quicker, we also understand the underlying conundrum that, to derive the greatest value from our designs, our design methodology must be based on an iterative process and, indeed, we should be considering several strands of enquiry simultaneously. We know that the best way of doing so is by working in parallel with other participants, but we find that the time taken with the tools generally used in the construction industry means that that approach is very difficult to achieve. While we understand that the changing nature of this industry means that we will inevitably be concentrating on our core skills of design and management, and allowing other participants to provide much of the construction information traditionally associated with architects, in the back of our minds we are not necessarily convinced that we are able to appropriately test or prototype our designs. In fact we find that studies have shown that the greatest proportion of building failures result from design errors.

Clearly we need to work differently from our traditional approaches, and we need the appropriate tools to allow us to do so. And if we can achieve that, we have the potential to reposition the architectural profession back to a central role in building procurement processes.

Our success with Eureka Tower has shown an approach which will help us achieve that aim.

The Eureka Tower project commenced its design development and construction documentation phase at a time when we were seriously questioning prevailing attitudes and methodologies inherent in the construction industry, and the resultant demands placed on the architectural profession.

The revolution in information technology has until recently been of little benefit to architects. At our core activity of design, computer-based tools have generally replicated traditional methods of building analysis or description at the cost of a huge increase in complexity. That complexity has been allied with a loss of transparency and overview of the design activities and has brought with it an increased fragmentation and decreased cohesion of project teams.

The underlying basis of our approach to Eureka Tower (and now all projects within the practice) is to use 3D object-oriented software that allows sharing of information.[2] That approach means that we simulate the building in three dimensions throughout the design and documentation phases. Interestingly, those phases are now distinguished from each other by the end objective of each phase, rather than the processes adopted during them. As well as being able to design more effectively, team members can now collaborate cohesively as they are co-operating to craft the one virtual building rather than working on separate files or pieces of information. The architectural process can now be one of unified finalisation of design intent, rather than that of in-creasing subdivision into design and documentation units. Rather than an office of segregated tribes, we have a practice of collaborating teams.

That approach means that, instead of exchanging two dimensional data which is essentially limited to dimensional and locational information, we are able to share rich three dimensional information that not only has dimensions and locations, but can include descriptions of materiality and quality, provide specification data, allow analysis of environmental, financial and constructional aspects, and provide a visualisation of the proposed building. All of those benefits are derived from a single project database of information.

Therefore those outcomes are a by-product of that project database. The term "by-product" has become a mantra within our practice. We direct our primary efforts towards the activity of design, be it broad conceptual design, or the detailed design that is construction detailing, using our 3D software as our primary tool. From that database can be derived through automated processes the required 2D, 3D or text- based information.

If we consider that under traditional processes those who were involved in the design of buildings were generally directing their energies towards arriving at a design solution, and that those who were working at the description of that building for construction purposes were always working to produce a drawing, we can see that a process oriented around the idea of by-product means that now everyone is designing.

That approach is advantageous at several levels. Working on the 3D virtual database means that the underlying structure of information assists co-ordination in three dimensions. It also means that the co-ordination is ongoing and accumulative. By contrast, in traditional processes, that co-ordination was carried out by people who would compare disparate pieces of 2D information and then stitch that information together in their minds to create a three dimensional whole. Unfortunately, whenever a drawing was changed, that process had to start all over again – it was not accumulative. Our approach means that whenever a design is changed, the visual manifestation is updated automatically, whether that be design image or construction drawing. Critical to these processes is working through a single database of information.[3] Under traditional processes visualisations of designs tended to be aspirational in nature, produced in parallel with the documentation team who in the meantime potentially were travelling down a different path entirely; by contrast our visualisations show the building as it will actually be constructed. They are visual images of the prototype building.

Using a 3D approach allows us to understand the nature of the problem earlier, to be able to understand and evaluate our design responses quicker in the overall process, and to therefore be able to venture further into the resolution of the design without having to fundamentally reconsider our solutions. Using the 3D virtual building we are able to reduce interruptions between activities. In traditional modes of working, those interruptions were the exchange of 2D information, the need for each recipient to work through his or her understanding of how the received information formed part of a complex three dimensional jigsaw puzzle, and the need to wait for long periods of time for each participant to finish that aspect of work before again sharing their 2D information with others. Working in 3D facilitates the simple and immediate sharing of rich information, thereby reducing interruptions to the flow of ongoing design resolution. The construction of the podium levels surrounding Eureka Tower has been completed, and construction on the Tower itself commenced earlier this year.

---

1 If we define such methods as being essentially defined by the use of lines and text intended to symbolise the desired building, and the co-ordination of that information is carried by manual techniques, then we can understand that 2D CAD falls squarely into the definition of traditional techniques.

2 The practice commenced that process using ArchiCAD version 6.0 by Graphisoft. Currently we are using ArchiCAD version 7.0. That is our core software tool for assisting us with design and construction documentation. The basic methodology adopted within the practice is to allow the ArchiCAD software to contain the underlying design information, and to use other software to represent that information in various ways, be they 3D visualisations, 2D documentation, or text-based data. When the underlying design is updated, so automatically are the representations of that information.

3 In this approach, the 3D database, or 3D virtual building is generally created and maintained by the architect. The 3D model stays active for the whole of the life of the building. It provides information for planning and cost modelling, and allows for creation and enrichment of data during the design, construction and operational phases by all parties involved in the project – simultaneously. The value of the 3D model, under this process, is increased and maintained throughout the life of the building; it becomes the central control document. This presents a key opportunity to reposition the architectural profession. By maintaining control of the information we revalue our position in the building procurement process.

SPACE AND IDENTITY
ARCHITECTURE AS THE FOCUS OF SOCIAL VALUE SYSTEMS

Introduction and Abstracts

RAUM UND IDENTITÄT
ARCHITEKTUR ALS FOKUS DER GESELLSCHAFTLICHEN WERTORIENTIERUNG

Einführung und Abstracts

Milica Bajic Brkovic Belgrade

Werner Durth Darmstadt

Georgi Stoilov Sofia

Janet Ward Boulder

Hou Hanru Paris

Dieter Hassenpflug Weimar

Ábalos & Herreros Madrid

Mohamed Ayman Deaf Assiut

Michael Burt Haifa

Krzysztof Chwalibóg Warsaw

Thomas Spiegelhalter Pittsburgh

Tay Kheng Soon Singapore

Llewellyn van Wyk Philadelphia

Richard Kroeker Halifax

Bruno Stagno San José

Introduction
Einführung

The decline of aesthetic conventions, the disappearance of local identities and the dominance of non-places in urban contexts (Marc Augé) is undermining the traditional qualities of architecture. The penetration of architecture as well as of media-occupied and virtual space and the dominance of visual surfaces has led to the dissolution of traditionally communicated forms of perception. The globalisation of economic relations has reduced urban 'places' to 'business locations' devoid of any identity. The universalism of international companies and the new economies is leading to the spread of standardised prototypes. The new electronic media are fostering the consumption of "images" of architecture that lack any recognisable content, cultural or social significance. Forms of architectural aesthetics, which – in the traditional sense – are communicated culturally and have their roots in society, are coming under ever greater pressure from without. Whilst "building culture" is a call for integration, for the commitment to proceed down a common path in the interests of quality and to subordinate other goals to this commitment to quality, the identity of urban centres has been called into question in many different parts of the world.

Context

Traditions in architectural aesthetics and the structures of urban planning involving an exchange with nature and landscapes are being forced to compete with the aesthetic arbitrariness of a "one-world architecture", which recognises neither ties to a specific place nor to cultural individuality. This leaves us with anonymous, uniform works of architecture that lack – within the context of the respective local culture – any "inner meaning" and cultural necessity. Given the growing opportunities to consume architectural aesthetics in the form of fancy historical and contemporary buildings all over the world, to design architecture with arbitrarily chosen symbols and surfaces and to construct towns and cities with no overall inner conception and no resource-saving relationship with their environment, it would appear that new images of the city are being generated together with images of people that leave virtually no room for urban man – for the zoon politicon, the zoon symbolicon – who has been able to relate to his own unmistakable cities and architecture in so many regions across the globe.

Der Verlust ästhetischer Konventionen, das Verschwinden lokaler Identitäten und die Dominanz von Nicht-Orten in städtischen Kontexten (Marc Augé) schwächt traditionelle Qualitäten der Architektur. Die Durchdringung von Architektur, medienbesetztem und virtuellem Raum und die Dominanz visueller Oberflächen führt zur Auflösung traditionell vermittelter Wahrnehmungsformen. Die weltweite Verflechtung der Wirtschaft reduziert urbane „Orte" zu „Standorten", denen es an Identität mangelt. Der Universalismus der Weltkonzerne und der neuen Ökonomien verbreitet standardisierte Prototypen. Die neuen elektronischen Medien fördern den Konsum von „Bildern" der Architektur ohne erkennbare inhaltliche, kulturelle oder soziale Bedeutungen. Der Druck auf im traditionellen Sinne kulturell vermittelte, sozial begründete Formen baulicher Ästhetik wächst. Während „Baukultur" der Aufruf ist zur Integration, zur Verpflichtung, einen gemeinsamen Weg im Interesse der Qualität zu gehen und dieser Qualitätsverpflichtung andere Ziele unterzuordnen, ist städtebauliche Identität in unterschiedlichen Regionen der Welt in Frage gestellt.

Kontext

Traditionen der Architekturästhetik und der städtebaulichen Ordnung im Austausch mit Natur und Landschaft werden in Konkurrenz versetzt zur ästhetischen Beliebigkeit einer „One-World-Architecture", die weder Ortsbindung noch kulturelle Individualität würdigt: anonyme, uniforme Architekturen, denen es – im Kontext der jeweils lokalen Kulturen – an „innerem Sinn" und kultureller Notwendigkeit mangelt. Mit der Zunahme der Möglichkeiten, Architekturästhetik in Gestalt historischer oder zeitgenössischer Versatzstücke weltweit zu konsumieren, Architektur unter Verwendung beliebiger Zeichen und Oberflächen zu entwerfen, Städte ohne innere Gesamtkonzeption und ressourcenschonenden Umweltbezug zu bauen, scheinen nicht nur neue Stadtbilder zu entstehen, sondern auch Menschenbilder, in denen der urbane Mensch, das zoon politicon, das zoon symbolicon, der Mensch, der sich seit Jahrhunderten auf seine ureigenen, unverwechselbaren Städte in den Regionen der Welt und auf deren Architektur beziehen konnte, kaum noch Raum hat.

Responses

Architectural and urban developments, committed to long development cycles and shaped by the cultural history of humanity on all continents, must be preserved. Aesthetics as an expression of social and cultural identity that might well find structure and form in a culturally communicated architecture must be revived. The development from architecture to building culture implies progress towards expression, from form to content, from the building to the surroundings, from the outcome to a basic attitude. With design becoming a pure marketing instrument, an instrumental part of aesthetics has developed a life of its own. Minimalist and symbolist architectures are creating alternatives to the visual chaos of the metropolises and the megacities.

Proposition

Ortega y Gasset reflects the opposing position in his statement that man alone is gifted with the ability to plan. It remains to be seen, he says, whether the ability to plan is a scientifically proven competence that cannot be rejected. Or whether "being" is directly expressive, i.e. without any expert communication, in the sense of constructive thought that also incorporates planning in a context which includes architecture and has identity as its goal: "Architecture is the art of building that produces magical objects". Contemporary aesthetics as a criterion of environmental quality and sustainable architecture must be redefined in connection with the creative use of new materials and technologies.

Erwiderungen

Architektonische und städtebauliche Entwicklungen, langen Entwicklungszyklen verpflichtet und durch die Kulturgeschichte der Menschheit auf den Kontinenten geprägt, sind zu bewahren. Ästhetik als Ausdruck sozialer und kultureller Identität, die in einer kulturell vermittelten Architektur Struktur und Gestalt finden könnten, ist neu zu beleben. Die Entwicklung von der Architektur zur Baukultur meint ein Fortschreiten zum Ausdruck, von der Form zum Inhalt, vom Objekt zur Umgebung, vom Ergebnis zur Grundhaltung. Mit der Bedeutung des Designs als Marketinginstrument hat sich ein instrumenteller Teil der Ästhetik verselbständigt. Gegen das visuelle Chaos der Metropolen und Megacitys setzen Architekturen des Minimalismus und des Symbolismus Alternativen.

Thesis

Für die Gegenposition steht Ortega y Gassets Feststellung, allein der Mensch sei begabt mit der Fähigkeit zur Planung. Zwar sei es dahingestellt, ob die Fähigkeit zur Planung eine wissenschaftlich begründete, nicht abweisbare Kompetenz sei. Oder ob das „Sein" unmittelbar – direkt, und nicht wissenschaftlich vermittelt – ausdrucksstark im Sinne eines konstruktiven Denkens sei, das auch Planung in einem Sinn umfasse, der Baukunst einschließe, Identität zum Ziel habe: „Architektur ist die Baukunst, die magische Dinge produziert." Ästhetik, zeitgenössische Ästhetik als Kriterium von Umweltqualität und nachhaltiger Architektur, ist in Verbindung mit dem kreativen Gebrauch neuer Materialien und Technologien neu zu bestimmen.

Reinhart Wustlich Hennef

Global Movements and New Resources: a Planner's Perspective

Milica Bajic Brkovic Belgrade

During the last twenty years the planning profession has undergone a tremendous transformation. Not only have the visions, ideas and tools we use changed, but the overall framework for our actions as well. The planning world of today is a different world, and a new culture of planning is on the horizon. The New World environment affects not only the way planners work, but also the very meaning and the role of the profession as well. It is the spectrum of these questions that calls for our attention today. In this presentation some aspects of the relationship: the world environment – the planning professions, will be emphasized and brought into light.

The New World Environment

Globalization – Regionalization – Localization

Often described as a process of integrating national economies into the world market, or as a transitional passage from a high volume economy to the high value economy, globalization is amongst the key phenomena that mark the turn of the century worldwide. The phenomenon per se and its real effects appears to be widely articulated both at geographical and political level. Much emphasis has been placed on the relevant social, technological, political and cultural structures and processes it embraces. It is said that globalization modifies deeply the structural framework of rational choice and as such it is relevant for planning and the work of planners.

In the 1990's the economic aspects of globalization were in focus. From then on a pendulum swung towards other facets and our whole vision has changed. The key feature that attracts our attention today is the heterogeneity of the world. The differences matter, sometimes more than it appears, and we recognize the other two foci as equally important for our work: regionalization and localization.

The triangle globalization – regionalization – localization creates a new framework for today's planning. It influences the way our cities and regions develop, but also places a new agenda before the planning community to investigate the practical implications.

Sustainability and Diversity

Following the 1987 Brundtland report – that humanity has the ability to make development sustainable – to ensure that it meets the needs of the present without compromising the ability of future generations to meet their own needs, the concept of sustainability grew into a much more complex system of values and gradually advanced into a worldwide development strategy. The notion of diversity which is embedded in it and presents one of the key features of sustainability, is of particular relevance to the planning profession. The world is too complex, and countries and regions often differ too much. Diversity is not a question of culture only. Simplified and culture related interpretations belong to the past. Today, the notion of diversity is closely associated with (1) development potentials and development options, and (2) community and civic participation. Diversity in planning and planning for diversity thus become a new motto of the planning world.

www as a New Resource

The mobility of information today is an important factor in the world of international competition. A new generation of information and communication technologies creates new opportunities and potentials of unprecedented values. The World Wide Web supports mobility in space and time. Yesterday of a limited use, tomorrow it will become the intrinsic part of professional routine. The global network is not only the place where people meet. It is also the economic place, and a powerful economic resource that fully participates in shaping development of many regions and cities worldwide, and as such is of tremendous importance for the planning profession. Transparency, efficiency and economy, the key words of the www, support a call for information networking on the Internet to channel organizational and operational resources.

Environment

Environment is in the very focus of any discussion on development, regardless of region, country, or city. Providing a sound environment is one of the first steps in a long process of regaining the vitality and pivotal roles that urban areas play. Many countries and governments today pursue a whole range of different policies with a consistent environmental orientation. An undamaged environment contributes not only to healthy living but a high quality of life. High environmental standards are becoming an increasingly important location factor. The market for environmental goods, clean processes and production technologies, as well as ecological services is expanding. Companies that open up to this new market today, will be creating jobs tomorrow. Local and state authorities have a clear interest in supporting such companies as part of a broader industrial strategy.

Emerging Paradigm of the Planning Practice

It is the intersection and overlay of these dimensions that creates a new framework for planning. Their interdependence has to be fully understood while developing a coherent and sustainable planning practice. In the early days, planning was concerned with creating "grand plans" for building and construction. This is no longer the case. Today's planning practice has a different and new role. It is aimed at creating flexible frameworks and orientation, while cities/areas continue to evolve.

Major Changes and Trends

From

· Cities as enlarged architecture
· Physical space
· Quality of physical space
· Planning is aimed at creating and controlling whole environments
· Planning provides the "end state" solution and control
· Top-down planning
· Planners' product as a result of design rules
· Control by "plot ratio"
· Separation of activities/land uses for health reasons
Confidence in the computer and quantitative methods
· Public acceptance
· Planning as the only discipline involved in planning

Through

· Cities as complex social, economic and physical systems
· Physical, social, economic space
· Quality of physical, social, economic space
· Planning is aimed at creating responsive environments
· Planning provides development strategies and schemes
· Encouraging self-help initiatives and public participation
· Planners' product as a result of professional standards and norms
· Development impact assessment; Environmental impact assessment
· Mixture of activities/uses for social diversity
· Mistrust of model based planning
· Public participation
· Interdisciplinarity and cooperation of sectorial knowledge

To

· Cities as complex structures composed of resources, processes and effects of processes
· Environment as a common denominator
· Quality of life
· Planning is aimed at creating flexible framework and orientation
· The "framework" planning is attuned to implementation and action
· Developing partnership and public involvement
· Equity of access and involvement by all stakeholders
· Planners product as a result of evaluated opportunities and risks
· Strategic environmental assessment
Sustainability impact assessment
· Different models in accordance with local needs and potentials
· Renaissance of computer-aided work; Telemmatics; Internet
· Public cooperation and partnership
· Transdisciplinarity and cooperation of professional and lay knowledge

A major transition from traditional and rule-book practice has already taken place worldwide. Current efforts are focused towards a flexible and innovative planning practice that creates new opportunities but resources too. The ideal pattern is not only one that prescribes but one that provides and supports. "Do universally applicable principles lead to universally applicable solutions? The answer is yes and no: there is a broad good practice solution that is applicable in virtually every city, North and South, rich and poor; but this solution will need to be tweaked in particular ways to suit local conditions, local constraints and local opportunities" (Hall, 2000).

Globalisation and Regionality

Werner Durth Darmstadt

Since the end of the political separation of the world into a western and an eastern hemisphere, the simplified geographics of which corresponded to some forms of the competition between the two systems of capitalism and socialism, a new partition of the world in the shape of a splitting of our social reality into two layers is becoming apparent.

People are continuing to go about their businesses with both feet on the ground, living their daily lives in a sensorially perceivable environment, whose structural and spatial form is also marked by its topography and history, depending on its regional peculiarities. Over this layer of lived daily life with its day/night rhythms and acceleration and deceleration of human activities, another, immaterial layer of global information streams and decision-making processes is expanding and densifying with increasing rapidity. These processes are controlled from command centres in the large agglomerations and have a direct influence on life in the regions, the distribution of wealth and the conflicts between states, the local consequences of which are seemingly unforeseeable.

These information and exploitation processes are, in the last analysis, decisive for people's everyday lives and are leading the world over to a hitherto unknown flexibility in regard to lifestyles and to an existential uncertainty. The need for self-reassurance in a concrete form and collective identity formation is associated with their becoming lost to view. History having been forgotten for decades and as traditional attachments become universally dissolved, in architecture and urban planning this disposition is leading to a positive obsession with history, with a view not only to conserving what has survived, but also to regaining what has historically been lost by reproducing it. Traditions are revived when they fall into disuse, regionality becomes a subject for discussion when it can no longer be lived, and waves of modernisation are cloaked as folklore.

In the area of tension between the processes of modernisation, which are generally experienced as threatening, on the one hand, and the compensatory cultivation of tradition on the other, together the two terms "globalisation and regionalisation" define a dialectic in the structuring of our spatial environment which has been resulting in the most widely varying trends and currents in architecture and urban planning for over a hundred years. Allow me to simplify the analysis by presenting a brief review.

A future global culture was already anticipated in the period leading up to the second industrial revolution, around 1850. It was to have a "cosmopolitan structure" spanning all regional and national peculiarities, as we can read in the Communist Manifesto of 1848. On the one hand the simultaneous expansion and densification of the networks of economic relations between the countries and continents was associated with the spreading of industrial mass production and urban civilisation, including all the new opportunities afforded by telecommunications and electronic engineering, while on the other the emergence of the industrial society intensified national romantic trends which were translated into concrete form in many of the important public buildings of the 19th century, in a bid to safeguard collective identities.

As a move to counter the claims to economic power of Britain and the USA, from 1900 onwards the impetus for an extensive modernisation of the world of objects in terms of the design of industrial products came from Berlin, which had become a new centre of innovation. In defiance of all nationalistic aims and objectives, the foundation of the German Association of Architects, Designers and Industrialists (Deutscher Werkbund) set the course for the development of an International Style and cosmopolitan urban planning. In contrast, the movement supporting garden cities and the "Movement for the Protection of the Heimat" (Heimatschutzbund) created a balance for the costs of modernisation.

In regard to the first half of the 20th century in the other highly industrialised countries, it can also in retrospect be said that in contrast to the widely held vision of a modern era triumphantly gaining ground against antiquated and stifling traditions, it was the very interaction between the ideas developed by the Neues Bauen movement for a completely new style of building and the attempts to make amends by means of tradition-conscious innovation that characterised the building culture of the last century.

In the totalitarian systems of Fascism and Stalinism this interplay was frankly enacted in artistic form, for example in Germany, where the "Gleichschaltung" (total standardisation) of lifestyles and the modernisation of communications in the Reich achieved by the ubiquitous propaganda that was broadcast with the aid of the "People's Receivers" and mass mobility achieved by building motorways were accompanied by regional variations in architecture created in accordance with "Reich Landscape Standards" (Reichslandschaftsnormen) and an impressive redesigning of the main cities of the administrative districts (Gaus). At the same time, Stalin was linking his policy of total subjugation of the republics of the Soviet Union under the centralist rule of Moscow with a demonstrative cultivation of national architectural traditions which was also decreed and implemented in the same manner in the eastern European states after 1945. This trend even reached Berlin, where in the new capital of the German Democratic Republic the entire Stalinallee ensemble was intended to document the continuity of the history of local urban architecture, in a direct attempt to create a contrast to the much decried "cosmopolitanism" of American "cultural imperialism" in the West.

As a result of the political and cultural devaluation of tradition-bound models of architecture and urban planning in the decades after the end of the war the interplay between globalisation and regionality and modernisation and preservation of traditions was to a great extent abrogated by expectations that the building and construction sector would be industrialised ... until, not by coincidence, as a result of the global reflection on the "limits of growth" and the emergence of the ecology movement, from 1970 onwards in the West a style of building oriented towards preserving resources and the existing fabric led to new forms of cautious urban redevelopment, which, after the phase of rediscovery, preservation, re-use and supplementing of the fabric of historical buildings

- on the one hand provided the rationale for forward-looking technologies of sustainable construction methods and some high-tech architectures and
- on the other hand paved the way for aesthetic strategies for adding to existing buildings and erecting new buildings along the lines of historical models.

The reversion to the designs of the vaguely generalised "European Town" of the 19th century with the intended dialectic between technological functionality and a nostalgic historical atmosphere is increasingly being combined with references to the global New Urbanism movement which promises the happiness of having local roots and a locally referenced identity formation in a world controlled by streams of locationless information.

In contrast to a revival of regionality reduced to an emphasis on folklore achieved by reproducing specific forms of building or tight design rules, under the conditions of globalisation understood as a development of regional design potentials, building culture should be mainly oriented towards

- topographical features such as landscape relief, the courses of rivers, the borders of woods, etc.: "Urban development is the improvement of landscapes" (Ernst May 1929), and also towards
- historical development phases and characteristics (the ground plans and morphologies of towns, landmarks and building typologies)
- the urban context, interweaving open spaces that can be used and experienced in many different ways to strengthen socio-cultural qualities.

Only when towns and their landscapes, centres and peripheries, historical centres, new service institutions and industrial areas as well as their technical infrastructures are considered together can the characteristics of a region be strengthened by simultaneous constructional innovation and conservation of the cultural heritage. New landmarks of globalised architectural production can also contribute to this, as demonstrated, for example, in Bilbao.

Finally, in an attempt to evaluate current trends in architecture and urban planning, the polyvalence of new forms of regionality in the area of tension between

- the political positions of a nationalist right-wing populism,
- strategies of economic exploitation with the creation of images suitable for upmarket sites and
- urban cultural resistance movements,

is being discussed, taking into account the widely varying reactions to the consequences of globalisation processes, which, along with the third technological revolution of micro-electronics and the new communication technologies, are in any case beginning to effect a radical change in people's sensory perception and physical experience of space in this age of information.

Globalization and Identity in Architecture

Georgi Stoilov Haifa

I. Globalization and Society

1. Architecture is a supreme manifestation of the creative power of man. It's main target is the creation of optimal conditions, under which a personality would fully develop and express itself. Architecture follows the pattern of the evolution of human society.

The nature and the core values of modern architecture could not be appreciated without a deep understanding of the nature and the core values of the modern age.

Having gone through a number of well-known socio-historical stages of development, the world has entered a new age – the age of

The Global Society

2. Although the political, economic and military aspects of the problem, quite naturally, have a decisive impact on the development of globalization, we are not going to discuss these here.

The technological, social, ecological and cultural aspects of globalization are of major importance to architecture, and they are the issues discussed in this presentation.

3. High technologies are a key factor for the development of the global society. They have led to radical changes in the system of communication, computer science, energy, biology, etc. This is an unprecedented success. They are an enormous tracing force for the formation of an integrated interdependable world.

4. Deep changes have taken place in the social sphere. Urbanization has spread on a worldwide scale and has become a universal phenomenon. The right technologies and urbanization have altered the style and the standard of living of society. Even the most distinctive ethno-cultural communities, such as the Japanese, the Chinese, the Arabic, etc., have been influenced by this process.

5. The role and functions of the national state are changing. The old, isolated state governmental structures are opening and are being integrated into larger megastructures – the European Union, the Union of the Sovereign States, the North American Union, etc.

Presumably, this process is going to continue in the Arabic world, in Black Africa, etc. A concurring process could also be observed as an opposition to the tendency of megaintegration, the old nations are being split into minute communities (the Basques, the Flemish, the Montenegrins, etc.) and there is yet another phenomenon to be taken into account – a strong impulse has been given to the development of individualism and the appreciation of the significance of the personality. All this has deeply changed social psychology, the traditions, the existing moral and social values.

6. At the end of the 20th century ecological problems became greatly aggravated.

As a consequence of the unthinkably rapacious exploitation and covetously wasteful overconsumption of 10% of the planet's population, humanity has been faced with the danger of extinction, caused by an ecological disaster.

The destruction of forests, the soil, the extermination of biological species, the ozone layer, the greenhouse effect – this is our common doom. These problems are of global significance and are globalizing the world even more intensely than our celebrated achievements.

7. And, finally, we come to the question of culture. Although culture has a major Impact on personal and national identity, it also undergoes an intensive process of globalization.

Blue Jeans, Macdonald's, TV action movies and series dominate and turn the world's cultural variety into a mass monoculture. A tendency towards globalization is forced onto the standard of living by the distribution of products of multinational companies: Television, Internet, the car, household electrical appliances, etc. All this predetermines global socio-psychology and the style of living in the new age.

II. Identity and Specificality

8. Identity is the most important characteristic of culture and art. It also applies to architecture, as it is a great socially orientated spatial art. There is no major work of architecture, let alone a masterpiece, neither in the past, nor in the present without a distinct identity of its own. The opposite phenomenon would be anonymity, typicality, lack of self-expression, conventionality and fullness, which could be anything but Art, spelt with a capital letter.

Here we come to the basic questions on the problem:

1. What are the essential characteristics and the fundamental substance of a work of architecture ?

2. Where are the sources and the roots of this identity to be found ?

3. Could the features of an architectural identity and the masterpiece be pre-planned and programmed ?

To answer these questions means to reveal the contents and the mechanism of the process of creation by considering its fundamental characteristics.

9. So as not to plunge into the obscurities of mysticism we should stick to reality.

The Parthenon, St. Sofia, the Palace of the Doges, the Notre Dame Cathedral, The Kremlin, Gur Emir, the Sultan Selim Mosque, The Golden Temple in Kyoto, the Temple of the Sky in Peking, the Sagrada Familia – these are all masterpieces of architecture.

The survey, carried out by the International Academy of Architecture, ranked a number of 20th century works among the masterpieces of architecture. Some of them are the Pavilion of Miss Van der Roe in Barcelona, the Ronchan Chapel by Le Corbusier, the House above the Waterfall by F. L. Right, the Sydney Opera House by I. Utzon, the Pompidou Centre by R. Piano and R. Rogers , the Olympic Complex in Tokyo by K. Tange, The Town of Brasilia by O. Nimeyer, etc.

All these works of the Golden Fund of the history of architecture possess a clearly displayed identity and a brilliant architectural image.

It is the expression of the spirit and the nature of their civilization that makes them significant.

At the same time we observe a principal difference between the substantial characteristics of the 20th century works of architecture and the ones of all other centuries.

And while all masterpieces till the 20th century are a result of an accumulation of typology and style, through centuries and generations, the 20th century ones are the personal achievement of the authors themselves. They are so varied in their stylistic characteristics and means of expression, that it is hard to consider them as works of a homogenous architectural style. That is why architectural critics have desperately been trying to find a single common denominator and have given many different names: modernism, rationalism, brutalism, minimalism, neomodernism, etc. And if we add to that a number of eclectic phenomena and names, such as: neoclassicism, postmodernism, deconstructivism, etc., the overall picture will become rather vague.

10. Where does the truth lie ?

It lies in the different character of 20th century civilization as compared with all previous ages. All preceding ages have formed their identity of spirit and culture through many centuries, resulting in a specific pattern or style in the way of life and art. Due to the high speed of development of informatics in the 20th century and the resulting dynamic changes, we can expect developments in the 21st century to span only 5 years in comparison to 100 years by the old standards.

Therefore as it is declared in the IAA "Horizon 200": "The age of the homogenous great styles and 'isms' in architecture has come to an end and a new approach is implemented in the creation of specific personified and identical intransigent architectural values". Each attempt of imitation of the past or of something already created to form a national or another type of style is a complete failure. None of the masterpieces of the 20th century is a result of such an approach.

11. Does that mean that we disapprove of national values and approve of globalism as a basis of new values in culture? No, that's not true.

We consider globalism to be a monoculture, primitivism, unification and destruction of the cultural heritage and the specific intrinsic characteristics of each nation.

In the old times ethnic identification was the crucial factor for the formation of spiritual values and their identity. Buildings used to be constructed according to the requirements of strict religious doctrines and aesthetic stereotypes, and they represented the views of a given ethnic group or a society rather than the views of the designer.

But in the 21st century it is the individual personality that creates the values and their identity. This is their main source.

The national characteristics are intrinsically woven into the personal perspective of the architect.

The international contests and exhibitions, the designs made for other countries, the constant exchange of information on the Internet and all other activities on a global scale intensify this trend.

And there is something even more important to be expressed through art – the supreme values of the Universe – harmony, vitality, progress, grandeur, infinity, etc. All of them are embodied in every masterpiece of architecture. That is why in the past the greatest masters of arts created works which crossed the boundaries of their native countries and age, and we treat them as our contemporaries nowadays. How could we say that Shakespeare, Goethe, Tolstoy, Beethoven, Tchaikovsky, etc. are merely artists of a single national culture. They are authors who have managed to express the cultural values of the whole of mankind and the whole Universe – they have conveyed the universal values.

In conclusion, we appreciate both the universal and personal values and their artistic identity expressed through their imagery, whereas monoculture and primitivism cannot be considered as art.

The specific features of the national spirit could also be found in the works of contemporary masters, i.e. the Mexicans: Luis Baragan, Pedro Ramirez Vaskez, Theodoro Gonzalez de Leon, Abraham Zabludovsky, Augustin Hernandez and others. This also applies to the Japanese masters: Kenzo Tange, Kisho Kikutake, Tadao Ando, etc. It could also be traced in the works of the British, the German, the French, etc. Their works do not represent a national style or an 'ism' but the national filtered through the individual perception of the artist. The Italians are still Italian, the Germans are still German, etc. And never could a Russian be Chinese.

III. The Creative Approach

12. The latter point concerns the methods and devices implemented in the process of preplanning, programming and controlling the work of art.

There have been many dictators and "great" rulers who have tried to immortalize themselves by demanding a specific style in the design of their quarters. As is well known, a masterpiece cannot be designed according to a recipe or dictates. It cannot be forcefully prescribed.

Why is this?

Because the mechanism of the creation of a work of art is a complex one. There is an old saying: "Art is a gift given by the talent and the inspiration". What does that mean? It means that the process of creation is specific and multilateral. The discovery of the human genome on 15th February, 2001 and a number of other discoveries in the field of modern neuro-physiology give us the opportunity to understand more profoundly the process of creation and to find an answer to the contentious issue – is art a creation of the mind or of intuition?

What are the scientific facts?

The most important factor is that the human organism is a miniature model of the whole Universe and that it is in a constant organic relation with it. The growth of the human embryo not only follows the model of the evolution of life but all stages of development are stored in the human brain as well.

The first one – the fish brain
The second one – the reptile complex
The third one – the mammal cortex
The fourth one – the neocortex or the brain of Homo Sapiens

All these parts, along with the nervous and some other systems and above all with the genetic code, is the soul of the human being or the basic essence of his/her character and identity. These are the roots of the universality as well as of the identity of his/her works of art.

It is worth mentioning that the neocortex (reason, logic, consciousness) is a computer, actually, and it supplies and systemizes information which is assessed and processed by the limbs (the subconsciousness) where all the decisions are made. The intuition, the reflexes and instincts or inherited experience (wisdom) are in that sublayer, too.

The character, the emotions, desires, stimuli, impulses come from that centre.

13. The process of the art of architecture is on the one hand a rational synthesis of functions, structures and architectural spatial tectonics (the rational basis), while on the other hand – the attainment of supreme harmony – transformation of the matter into intellectual value, creation of harmony and happiness (the intuitive basis).

Thus the art of architecture is a joint program of the consciousness and subconsciousness, of logic and intuition.

The talent and the inspiration – that is the capability of the individual to create unity and harmony between the two spheres of the material ("earth") and the spiritual ("divine") values. That is the core of the creative approach. Inspiration is actually the capability to contact the harmonious energy of the Universe.The person who has never been in the mental state when his/her hand moves as if by itself on the white sheet of paper and before his/her astounded eyes the contours of his/her unborn work appear, will never be able to realize what intuition and inspiration mean.

And this is the process of the transition of consciousness from the level of routine life to the level of the everlasting values of the Universe.

What can be "learned" (rationally) and what does God give?

This is a subjective judgement, but it is said that the Truth, Beauty and Happiness are attainable for each innocent soul. They develop the character and this is the clue to the Eternal Universal values and identity.

As a conclusion we would like to repeat once again: We reject globalization and aesthetic imitation or "stylization" in art and architecture, and we do recognize the universal values and identity as the most important quality of great art.

The Americanization of the New Berlin

Janet Ward Boulder

There is a Las Vegas metaphor of postmodern city-creation – that is to say, the Strip's reputation of architectural bricolage, of a structural hyperreal so strong that it has become substance. Because it stands for Americanization in its most blatant sense, the Baudrillardian Vegas metaphor has long inspired a "fear and loathing" among the cultural-intellectual elite, particularly those whose raison d'être is provided by cities established in modern industrial times. Increasingly, however, we can identify the pervasive reach of the Vegas metaphor of image-based commodification into the root of all architectural practice.

Given the fact that Berlin was precariously positioned as the former outpost of Western democracy in the middle of Eastern Europe, it is understandable that the early intoxicated rush toward architectural reunification included an impetus to make Berlin look more "western" as quickly as possible, as if to provide three-dimensional evidence that the West had won.[1] This (American) "West" often appears as an alien, or at best uncomfortable, source of influence for Europe's latest capital. Nonetheless, two key transformational aspects can be identified in the Americanization of the post-Wall city: first, commercial architecture, including office space, is exerting a "considerable influence on the new face of Berlin" – perhaps more than any other building type.[2] Second, former public properties are being sold to privately owned investor groups, that is to say: increasingly, as in US urban practice, the market is saying what goes up where.[3]

A significant aspect of Berlin's ongoing recreation of itself as post-Wall capital of Germany is the desired recapturing of its glory years during the Wilhelmine and Weimar eras when it grew to become, by 1920, the third largest city in the world. Many conservative city boosters and liberal urban critics alike prefer to think that it is still possible to make the New Berlin into a quintessentially modern "European city". By way of contrast, UCLA architecture professor Dagmar Richter offers an extension of the underlying consumerist ethos of the Weimar years onto today's US-based practices of postmodern urbanism: "Berlin has learned from Las Vegas without knowing it... Berlin will become the first state-organized media city of surface. It will surpass our expectations of Las Vegas, Disney, and City Walk."[4]

With the current lack of confidence in corporate growth for Berlin, which is languishing in a bankrupt condition in the wake of the real estate boom of the early 1990s, all of Mitte will at this rate succumb to the commodification model, namely of history as theme park. This is epitomized by the trendily refashioned courtyards of Hackesche Höfe - themselves evidence of "final commercialization," states critic Frank Roost, and based on the hyping of the "myth" of what the area used to be in the days before World War II.[5] In the impulse to make Berlin into its previous images, we can find a too-neat elision of all the disruptions to the city caused by Nazism, wartime bombing, and Cold War division. The application of Critical Reconstructionist tenets – namely, re-hashing the old street plan, lot sizes, façade materials, and building height limits (22 meters for cornice lines, and 30 meters for rooftops) – has led to an instant-history approach. In this sense, commercial and non-commercial sites alike have been or are being reconstructed to fit the requisite iconology of heritage tourism ("Berlin must look like Berlin," as per Stimman's pronouncement [6]) that alone today seems to guarantee economic growth for urban centers. The many examples of this include Pariser Platz, the Adlon Hotel, the interior design of restaurants in Mitte, as well as plans to rebuild the Tacheles into a New Urbanist Johannesviertel, or to remake the Kommandatur, Bauakademie, and Stadtschloss into the images of themselves.

Paul Virilio agrees, observing that Berlin's rebuilt structures serve as "stage-sets" for the fledgling capital's boosterist productions. The most spectacular of these is the newly rebuilt Potsdamer Platz through which the Wall used to run. If we recall the unfulfilled Friedrichstrasse skyscraper competition of 1921, at which Mies van der Rohe first proposed his glass-skinned skycraper, we can see how this dream has now been belatedly achieved by the substitution of site and era. But in realizing its modernist wish to build a mini-Manhattan, Berlin inadvertently produced more of a mini-Vegas: a citational package with its own range of simulacra.

Chief architectural philosopher of Berlin's voids, Daniel Libeskind, brings attention to what he believes to be the underlying problem raised by Potsdamer Platz, despite its relative success. Libeskind remains highly skeptical of the ability of Potsdamer Platz to transform the city's image for itself and abroad, and he denigrates the use of this site for the promotion of millennial architectural commodity fetishism: "I believe that Berlin, like any city which deserves to be called that, is no commodity. It is not something which can be bought in a store like a refrigerator, washing machine, or computer. A city is something which you really need, like the air we breathe. That quality is something which you can't always say of commodities."[7] Perhaps Libeskind is stretching a few historical features out of shape here: for the pre-WW II Potsdamer Platz embraced Americanism as a modernizing and internationalizing business ethic with links to Taylorism, Fordism, and New Objectivity – the net effect of which was to salvage national reputation in the wake of the Versailles Treaty.

Nonetheless, many of us might well prefer to imagine what would have been had Libeskind been permitted to build his own design at Potsdamer Platz – a fantasy land of zigzagged structures celebrating the site's Cold War status of voids and divisions, and leaving untouched the destroyed street plan of old, with a "one kilometer long wilderness...in which everything," states Libeskind, "can remain as it is".[8] But perhaps this "as it is" is rather suspect: what about how Potsdamer Platz as it used to be before it was cut in two by the No-Man's-Land stage? It seems that all post-Wall imaginings concerning Potsdamer Platz, whether aiming at its refashioning as a commercial city or as a broken monument, have arisen from an imagined original condition of its urban fullness and completeness. This imaginary is what is most missed in the new Potsdamer Platz, even if it was never there: as Gerwin Zohlen states, the busy traffic junction of Potsdamer Platz was rather the very opposite of presence: "the non-form [Un-Form]...the non-square [Un-Platz] par excellence" careening full of tempo, traffic, and Einsteinian relativity.[9]

The reconstitution-process of Berlin's Mitte as a stage-set is leading toward a condition of over-encoding. By over-encoding I refer to a socio-geographical mapping through which not enough spaces are left unplanned for the sake of a slower, more manageable urban transformation; the result is a rigid network of fixed urban interfaces that remain bounded, or separate one from the other. Such is the offense that the Vegas Strip gives the critical visitor who insists on a frustrated search for habitat-connections that would have grown through time; and the same offense exists as a possible danger for the New Berlin. When the Wall fell, Berlin went through an initial phase of promoting both boosterism (or space-filling) as well as the "fragile and fragmentary network"[10] of the urban voids of its WW II and Cold War heritage. Now these spaces are being too rapidly developed, sanitized, bought out, discovered, and made over, as Berlin searches in vain for its lost modern city centers.

The resurrection, then, not just of Potsdamer Platz out of a lost No-Man's-Land but also of Berlin-Mitte as a focal point of reunification has led not so much to the wished-for re-centering of the city, but rather to the New Berlin's most "American sector," this time according to postmodern (rather than Weimar modern) dictates of cultural capital. Nor is this phenomenon limited to the former East Berlin: the electronically enhanced cinematic billboard on the new C&A store on the Ku'Damm is as effective as those in Singapore or Vegas. The city that once received the moniker of Frederick the Great's "Spree-Athen" – and then again, in industrial modernity, that of the forward-looking "Spree-Chicago" – is now on the way to becoming "Spree-Vegas." The catchy title for the recent German edition of a collection of Mike Davis' essays makes the case against Casino Zombies. Is the fate for a commercialized New Berlin in general, or Potsdamer Platz in particular, that of casino-cultural zombification? Or can we approach the condition of today's urban planning in a less over-defensive, atrophied way? A balanced view needs to be found that does not demonize economic and infrastructural renovation and the concomitant commercial development that has to accompany the city's long path to growth.

Let us turn to Richard Sennett, who, in The Spaces of Democracy, rejects the "all-at-once, massive development of urban centers" like Berlin's (Potsdamer or) Alexanderplatz and wants a pluricentral "collage city" à la Colin Rowe as the space of democracy, in the manner of Aristotelian "synoikismos – a coming together of differences." This is Sennett's understanding of the Athenian agora as the origin of fragmented, mixed social space for (albeit originally male) democratic interaction "The result of visual, decentralized democracy should be," states Sennett, "to shatter those images which attempt to represent the city as a whole."[11] We can reflect on the degree to which Potsdamer Platz offends against Sennett's concept of democratic urban space. It tries to be both the agora and a recentralization of urban commercial power according to global city standards. The question remains: can it (and can Berlin) be both?

---

1 Uwe Rada (editor of taz) is not so sure. The Polish bazaar that sprang up on the site of Potsdamer Platz during the Wende was just the beginning, as Rada satirically warns Western sensibilities. See Uwe Rada, Berliner Barbaren. Wie der Osten in den Westen kommt (Berlin: BasisDruck, 2001), 14.

2 Annegret Burg, "The Commercial Building: An Urban Component", in Annegret Burg, ed., Berlin Mitte. Die Entstehung einer urbanen Architektur (Basel: Birkhäuser, 1995), 25. Burg speaks of the belated, "rapid, stormy development" of office buildings for post-Wall Berlin Mitte (25).

3 As political scientist Elizabeth A. Strom emphasizes: "capitalist urban development doesn't require the mobilization of political support." Strom, Building the New Berlin: The Politics of Urban Development in Germany's Capital City (Lanham, MD: Lexington Books, 2001), 235 (emphasis original); and 117–34. See also urban sociologist Hartmut Häussermann's essay on the encroaching American style of urban growth on European cities, in "U-Bahn und Urban", Süddeutsche Zeitung (June 30, 2000): 11.

4 Dagmar Richter, "Spazieren in Berlin", Assemblage 29 (1996): 75.

5 Frank Roost, "Themenpark am Tacheles. 'New Urbanism' in Berlin", in Architektur in Berlin. Jahrbuch 2001, ed. Architektenkammer Berlin (Hamburg: Junius Verlag, 2001), 91, 89.

6 Quote by Hans Stimmann, in Paul Goldberger, "Reimagining Berlin", The New York Times Magazine (February 5, 1995).

7 Daniel Libeskind, "Potsdamer Platz", radix-matrix. Architecture and Writings (New York: Prestel, 1997), 161.

8 Libeskind, "Potsdamer Platz", 162.

9 Gerwin Zohlen, "Erblast des Mythos. Das Verfahren Potsdamer/Leipziger Platz. Rückblick nach vier Jahren", in Ein Stück Grossstadt als Experiment. Planungen am Potsdamer Platz in Berlin, ed. Vittorio Magnago Lampugnani & Romana Schneider (Stuttgart: Verlag Gerd Hatje, 1994), 17. See also Thomas Friedrich's citation of Einstein's reflections in 1920 on Potsdamer Platz in Der Potsdamer Platz. Ernst Ludwig Kirchner und der Untergang Preussens, ed. Katharina Henkel & Roland März (Berlin: Nationalgalerie Berlin, 2001), 63.

10 Giovanni La Varra, "Post-It City: The Other European Public Spaces", Mutations, ed. Rem Koolhaas, Stefano Boeri, & Sanford Kwinter (Bordeaux: ACTAR, 2000), 428.

11 Richard Sennett, The Spaces of Democracy (Ann Arbor: College of Architecture + Urban Planning, The University of Michigan, 1998), 41, 19, 41.

POP
ART

An Event City – A Pandora's Box

Hou Hanru Paris

Gwangju Biennale has been one of the most important global art events taking place in the Asia-Pacific region since it's creation in 1995. The 2002 edition (the 4th one), entitled "Pause", opened from 29 March 2002 till 29 June 2002. I was invited to co-curate the main exhibition "Project 1" with Charles Esche and Sung Wan-Kyung. This project, instead of repeating the aim of most international biennales to present the most fashionable trends and art stars, is conceived as a context-specific event, evoking new ways of critical reflections on the question of global art and it's relation with the reality of global-local negotiation.

To articulate this engagement, we also decided to turn the exhibition spaces into a kind of laboratory for new urban-architectural intervention in order to provide a specific context for the event – questioning the issue of urbanization as a new context for artistic activity today. On the other hand, such a process certainly also inspires art and architectural activities to confront each other and collaborate together. The merging of different disciplines is inevitable today; we are facing the challenge of reinventing new identities for different practices, including art and architectural urbanism. It's through this new negotiation that we can start imagining and defining new territories and strategies to cope with the tension between globally increasing urbanization and the necessity of local interests. Our ultimate intention was to create a biennale which functions as a kind of on-going laboratory of reinvention of urban conditions and new contexts for artistic practice.

1. From the beginning, the biennale has encouraged the understanding of the general theme "Pause" as a break from established modes of biennale, which limits art creation to the "correct" presentation of products. The idea was to start with a new direction: emphasize creativity, productivity instead of products. To open up space for "slow" works which represent resisting voices against the highly utilitarian evaluation system dictated by the cultural logic of globalizing, late Capitalism. To the opposite of the culture of the spectacle, projects which deal with slowness, emptiness and openness are conceived to provide such spaces of critique. Pause, is therefore a dynamic and critical process of reflection and differentiation.

2. Logically, the roles of the artist and the art work are deeply challenged: the subjectivity of the artist has to be opened to the dialogue of the other, while an active and interactive relationship between the artist and the public is established through the realization process of the works which are open, flexible, evolving in time and physically inviting the participation of the audience. Together, they create magic moments in which both the artist and the public can critically contemplate and reflect on the issues of art and society, globalization and locality, etc. Therefore, the biennale has been directly and consistently connected to the interests of the local public. This relationship is unique and irreplaceable.

3. Exploring further the question of "global" biennale and locality, one should naturally question the established forms of institutional structure and discourses. Striving against the white cube, the commonly taken for granted space for contemporary art presentation, or the "typical" form of biennale structure, we have encouraged projects and actions which go beyond such a confinement. However, more fundamentally, we have understood that the truly innovative aspect of the negotiation between artistic innovation and institutional framework happens in the efforts of self-organization on the part of the artists, who not only endeavor to create differences in art language, but also struggle to open independent spaces to provide new freedom for their creation. This is particularly significant in the Asia Pacific region where Western style of infrastructure has not yet existed.

Many artists, to make their radically experimental work possible, have organized their own alternative spaces out of everyday life contexts. They live and work in highly diverse and different cultural realities. The forms of their organizations are accordingly different and therefore strongly enriching. In the meantime, they have also started dialogues across the region and an important trans-regional network of self-organizations and alternative spaces have been built up to provide the most important conditions for artistic creation. This proves to be one of the most innovative efforts in the Asia Pacific region's contribution to the making of a new global art scene. They have shown examples of the necessity of decentralization of power and resistance to the homogenization caused by the accelerated globalization of communication and cultural "exchange".

Instead of simply denying the necessity of globalization, or global circulation and hybridization of different cultures, they have creatively proposed constructive solutions to make sure our future can remain rich and diverse, while opening to the other is an inevitable central task. Certainly, this kind of initiative has not been limited to the Asia Pacific region.

As a matter of fact, in Europe, Latin America and other parts of the world, similar self-organizational structures are becoming increasingly important. In the biennale project, we have attempted to bring those organizations to Asia, so they can meet up with their colleagues on the other side of the globe. A first step towards a global network of independent, self-organizational and resistant structures for creation is hence made in the realization of the biennale. The biennale itself is an ongoing event of such encounters and dialogues. It is not a once-for-all event, instead, it's like Pandora's box. Once it's opened, it can never be closed. On the contrary, it will be continued, developed, multiplied and spread in infinite time and space. It can also be seen as an echo to the current movements of critique and resistance of economic globalization and its geopolitical consequences. Once again, it proves that significant artistic activities are always engaged in reality; it is ultimately political.

4. The biennale, with a great number of site specific installations, and especially the adapted versions of "reproductions" of the alternative spaces, is by no means a simple presentation of objects. It's a permanent workshop, a lively urban space with real life events happening all the time. It's dynamic, complex and mutating system of creation and exchange. It's an event city. To put forward this point, a certain urban organizational intervention becomes inevitable. We have invited Chang Yung Ho and Kim Young-Joon to act as the exhibition architects. What is remarkable is that Yung Ho Chang (principle of Atelier Fei Chang Jian Zhu, Beijing) was the space designer if "Cities On the Move 1 – 4" (1997 – 1999). He has been closely collaborating in the visual art world since then. In fact, exhibition design has become a significant part of his activities, a condensed terrain for his urbanist experiments.

   Obviously, the roles of the architects in the Gwangju Biennale project, like in Cities On The Move, are far beyond designing formal structures to hold the works. Their main achievement is actually an ingenious urban planning project which, in spite of the lack of resource and time, has successfully injected powerful energy and complexity into the biennale. Their extremely intelligent and intense organization of the spaces between installations/pavilions and alternative spaces, between density and emptiness, between congestion and fluidity, between interior and exterior, etc. have brought a veritable new life necessary to the complex system of the project itself. This result recalls clearly the urban reality of Asian cities driven by the explosive urban expansion, density and dynamic economic, cultural and political modernization. Visual art practices are now re-contextualised in a completely new environment. Organically binding together, art and architecture are now generating a new knowledge and practice category.

Citytainment – Staging the City and Public Space

Dieter Hassenpflug Weimar

One of the birthplaces of the European City is the market place, the other is the cult place. The market place is the public space sui generis. It has been from the very beginning a narrative, theatrical space. Just think of the stage-like fronts of the houses surrounding the squares, the loud and varied activities of the tradesmen, shoppers, suppliers, foreigners and passers-by, the microcosmic theatrical process of buy and sell between shopper and tradesman, the chatting groups of people and children playing. When sociologists attempt to describe the present situation, they are inevitably confronted with the phenomenon of the disappearance of spatial qualities. Places are destroyed, eliminated, abandoned, ignored or reshaped. What is this thing being handled with so much disrespect? What is this thing we call a place?

Places are spaces that give the people who inhabit, visit and use them an identity. They are perceived by these people as a part of themselves. Places are spaces that keep our memories of history alive, making the present visible and thereby open for the future. Places are spaces with atmosphere. They touch us, we enter into an emotional relationship with them, we even identify with them. They can evoke approval or criticism, bring joy or pain. But they never leave us indifferent. Places can provide a stage for the great, the sublime, the dignified, the spectacular, but they can also be the scene of the small, the transparent, the simple, the every-day, in other words, spaces of the 'middle-range.' They are, however, always characterized by a stable stock of signs and symbols which make each place unique. Places accompany us people through shorter or longer spans of our lives and contribute in part to the meaning of our lives. In this respect, places are spaces of closeness, security and belonging.

It's possible to distinguish three spatial features, a social, a functional and an aesthetic one. If a city has these features in a high quality, I call it "urban" (a city that denotes urban qualities). The first feature refers to the ideal type of the tolerant, cosmopolitan and intellectual city dweller. The second refers to the utility value of the city, to its centrality, variety and mix of functions. And the third feature refers to the aesthetic dimension of the city, to its experience value, its atmosphere, aura, ambience and flair. Cities which provide these urban features have character, individuality and significance. They are unmistakable, coherent and legible places. In Europe this legibility can have its basis in what is called the "old European city".

However, places today are a highly endangered species. With their mazes of narrow spaces and alleys, their variety and intractability, their sensibility and unmistakability, their corporate and bourgeois self-portrayal, the old cities are only a nuisance in today's turbo-society. They are simply too slow. Thus they are accelerated, straightened, standardized, cleared out. Their structures are freed from one another and separated according to their function. The previously integrated uses such as living, working, relaxing and consuming, are separated from one another and each assigned to their own territory. Under this fordist treatment, streets are transformed into mono-functional traffic routes, squares into parking lots, the lively organized house fronts into smooth, sterile and cold facades. Cities become machines. The public places once populated by the young and old, the poor and rich, the bohemian and conventional, the locals and strangers, mutate into efficient functional spaces. The city falls apart and pours into the surrounding countryside. Business parks, single-family housing developments, traffic routes, shopping centres and recreational facilities push their way into the environs. Extenuated, confused, torn borders arise, chaotic structures of space, collage-like landscapes spoiled by development. A semiotically empty, indecipherable city-country-continuum is formed – a space without qualities.

However, the loss of places describes only one side of fordist modernization. The other is the overcoming of space (and time) itself. Through high-speed traffic systems, distances shrink and the availability of physical space increases. New media technologies open placeless, virtual communication spaces, rendering the existing, built arenas of public life superfluous. Under the pressure of economic globalization and mechanization, of growing mobility and permanent acceleration, places transform themselves – to borrow a term from Marc Augé – into non-places. What is meant here are efficient, fast spaces with a high turnover of people and goods, lacking any sense of character and meaning, places of power and efficiency that become more and more similar, until they can no longer be distinguished from one another. As a consequence, the ties that bond people to places dissolve until they do not feel responsible for their environment any more.

Amazingly the disappearance of places is contrasted by their reappearance! Places return in an altered form and function. Apparently, the loss of the qualities of place is experienced as a lack of something. A need for places develops. This need is translated into a demand, which is answered with a corresponding supply. A market for places – and with this an accompanying 'place industry' comes into being.

What type of places does this industry offer? In the attempt to answer this question, one comes across a particular type of place, which in and through itself represents the suggested contradiction between place and non-place. Places are produced which deny themselves as places – placeless places. But what is a placeless place? Well, it is a fiction of a place, a space that visually reproduces, anywhere in the world, qualities from arbitrary imaginary or real worlds.

The most important raw material for the manufacture of these placeless places is ideal nature: it is mined from those deposits of the imagination which we know as fairy tales, novels, comics, paintings, photographs, films and television programs. But 'place industry' also makes use of images and pictures from real places, prominent buildings, streets and squares. Even entire city ensembles are prospected for this raw material. Think about Disney's 'Main Street' concept or the fake of the inner city of Amsterdam near Tokyo or the reproduction of an old Toscanian village at the border of Johannesburg or the numerous fakes of cities in Las Vegas.

A type of cultural mining industry develops, which, instead of iron and coal, exploits pictures of spectacular nature scenes and attractive cultural details. The products of this branch are non-places which clothe themselves in the garments of place. I refer to these places as atopias, in order to highlight their close but complicated relationship to utopias.

The distinction between atopias and utopias refers to the difference between reality and possibility. Utopias are places which exist nowhere (only in the imagination). They are pure possibility and therefore a force for changing reality. They represent the eros of the political. Atopias, in contrast, are realized non-places with fictional qualities. They are simultaneously real and placeless. They are everywhere and nowhere at the same time. Whereas utopias, the stuff of which dreams are made, produce images of a better social life in an imaginary world, atopias unfold themselves in the here and now. They are available and materially present and at the same time without any connection to place, location and region. They are the topological expression of the fact that in today's world, everything is everywhere at the same time. They are the socio-spatial harbingers of a very powerful but superficial experience- or leisure-society – also called 'Erlebnisgesellschaft'.

With this term, sociology is reacting to a transition that is already in its advanced stages – the transition from the 'reason dominated' to the 'emotion dominated' modern age. According to this theory, people's actions today are no longer determined predominantly by reason, rather more decidedly by feeling. The rationalistic, fordist consumer society has become a mood-governed affluent society. Whereas instrumental-technical aspects such as efficiency, productivity, functionality, objectivity, etc. have dominated until now, emotional factors like atmosphere, ambience, aura, flair and other dimensions aimed at the senses and sensibility are gaining influence. An object must not only function, it must also be appealing to the senses, even exciting or fascinating them.

If one asks how the Erlebnisgesellschaft expresses itself in a spatial sense, then one comes up with the so called theme park. Theme parks are staged environments which should generate wellness, agreeable moods and feelings. The category of thematization refers to the origin of the ratio of experience (Erlebnisrationalität) in the medial "logic" of film and television. The medial habits of reception improve a kind of narrative space production that is oriented at the sequence of pictures and cuts of films. So each theme park reminds us that the time of functional space consumption and space organization has come to an end and that the time of story telling space improvement has already begun.

How can we Picture it?

Each Disney Themepark – in the USA, Japan and France – has it's own "Main Street". This street fakes the fake of the good old American small town main street as used in the film studios of Hollywood. Facades divided into small sections are staging a variety and mix of functions and some pleasant block border architecture. But these houses are nothing else but a set. Covering the walls of big stores they cut the informational connection between what is shown in the street und what is offered behind the facades. By inventing the mainstreet-concept, Disney has created the prototype of an inverse city-space. It is a private space, that pretends to be public. Here the customer is at the same time an observer and an actor of a game called "public life". By means of a materialized "secret instruction" Disney succeeded in producing something that is missed more and more in everyday life – the urban city.

With Celebration (near Orlando, Florida), the Disney corporation has just put the first private city on the market. This city is entirely a product of citytainment. Whoever buys here acquires not only a house, but also a lifestyle. It arises from the old American middle-class dream of a clean and tidy house in the country with a garden, dog and family. In the town hall of this atopian city are working or better say: acting employees of the Disney Corporation. They impose taxes, enforce design regulations and keep watch over the cutting of the lawns and hedges of the inhabitants and the condition of the house facades. Even the "mayor" is an employee staging a mayor.

Celebration is the ideal type of the new, inverse city. Here the public is private and the private is public. The classic relationship of regulating state and private economy has swung to the other extreme. It has changed into relations of economic protagonists. The inhabitant can and should not accept any responsibility beyond the contractual. Who buys a house in the "new city" buys at the same time a complete lifestyle, urban flair, security and simulative public order. However, he doesn't get one thing – participation in the public power over public space.

The situation is strange: At a time, where public order does not work properly the pseudo-public order in Celebration is guaranteed by private services! In the world that is still familiar to us, only the democratically legitimized state is entitled to lay down limits to private action and thus to intervene according to regulations. In Celebration the relations became reversed. Disney cares for the public order. While the state fails in guaranteeing it, the private enterprise is successful in creating a fictional public order. So the staging of public space is to be taken as the post-modern way of its declining.

Nevertheless, it makes no sense to condemn the development manifested in cities like Celebration as merely undemocratic. Its success demonstrates a widespread need for urban qualities. These qualities are however, as to their nature, public goods. Nobody can be excluded from their utilization. So, in order to avoid the social excluding capitalization of urban atmospheres and services, state engagement is needed for reconstructing our fordist treated – and in the end – executed cities.

Citytainment should not only be seen as an attack on public culture. It also embraces chances for the improvement of civil society. Postmodern citytainment can deliver useful ideas and experiences in the fields of management and aesthetics. In my opinion it should be possible to generate a kind of public "outdoor-imagineering" out of private "indoor-imagineering".

This consideration brings me back to the old European city, which, as already mentioned, does not fit into our fast times. Influenced by the postmodern condition we discover in our old cities a natural potential for "Erlebniswelten". We see hundreds of public "Mainstreets" and "Celebrations". Their experience-value is so formidable that the world champions of theme parks, the Japanese and North Americans, admire and occasionally reconstruct them with a lot of energy and money. We should not dismiss the postfordist trend towards the production of spectacular places as kitsch. We should rather comprehend it as an invitation to acknowledge the urban and rural world, in which we are living, as projects that have to stand up to emotions. So to take the postmodern challenge of citytainment for serious, means in the end to give back the urban space its proper values.

Citytainment admittedly should not only be left to the "bourgeois". It is also a concern of the "citoyen", of public culture. If not, postmodern cities will become the splendid graves of civil society.

A New Naturalism (6 Micromanifests)

Iñaki Ábalos Juan Herreros Madrid

1. Latent Garden

Only as a puritain rest of modernity can it be understood that even today the word beauty is proscribed from professional arguments. The simple statement of the desire to construct a new notion of beauty, away from the tricks that we meticulously tend ourselves, is exactly what gives sense to our work and gives way to imagine a convincing action plan. If we accept this proposal the own work will transform into a singular trip notebook with which one can build a mental atlas, an imaginary geography able to illuminate each place from this projection thereby inverting traditional contextualism and giving shape, step by step, to a latent garden. This vision might act as a reminder to us of the surviving pintoresque vein underground, engagening names such as Uvedale Price, Frederick Law Olmsted, Le Corbusier, Bruno Taut, Roberto Burle-Marx, Lina Bo Bardi, Robert Smithson, Archigram, Cedric Price and others... The fusion between nature and artifice, the dissolution of the disciplinary limits of architecture, art, landscaping and thinking, the organization of experiences as narrative sequences, the primacy of the visual and the movement, also in the immaterial and the invisible, the construction of a public space and architecture which reflects new sensibilities are ingredients that lead to a new garden that is seen or not seen, but without which there can be no new architecture, only variations.

Whoever imagines this, must not only feed himself/ herself from abstract announcements and last names but must build a parallel cartography that illuminates his projects by conquering new formal references, new symbols, new words and techniques that allow him to abandon the old ones which lack the necessary attraction. The protagonist must rebuild his audience and the world to which he aspires. The notebook is composed with sketches (works and projects), notes, borrowed and original ideas, names, images, conversations, multiple materials that reflect upon themselves and that generate our own gaze. The architect´s work is assimilated into the work of a gardener: choosing species and sowing them in an organized way, scoping that time, will make a good work.

2. Descountrylized/Areas of Impunity

The dissolution of the natural-artificial opposition we can see on every scale requires a works programme, which is nothing but the rediscovery through architecture of the contemporary human position in the world. 'Areas of impunity' are precisely the zones where, as an exception, we find this ambiguous condition with an imprecise definition as public or natural spaces. Formerly negative zones, endowed with a new urbanity by the gaze of the new social subjects and their practises. Look at the wasteland beyond the outer suburbs, look at the way that almost all the emerging forms of socialization have been constructed in these wastelands although (or precisely because) they are deregulated territories. We are tempted to ask whether they might contain a metaphoric model, a near-model, or whether it is possible to think of their complement, the de-edification, given that the term 'wasteland' embodies a fascinating concept: land that has lost its attributes as the city approaches, sterilized before proceeding with the occupation, but also given a transcendental role in its new context. We ask ourselves whether architecture could be constructed in the same way....

3. Ecomonumentalism

We have got used to thinking of architecture in terms of place, believing that it holds the keys to our ability to tackle the project. Many forms of anchorage to the location have been developed in recent decades, from those involving the phenomenological root ('Anchoring' is the title of an important text by Steven Holl), via the Bersonian influence in Moneo's work and the structuralist effect of the 'genius loci' in Aldo Rossi, to attitudes that arose at the Frankfurt school (Frampton and his contextualism). In recent years we have witnessed an important shift: every location has started to be regarded as a landscape, either natural or artificial, and it has ceased to be a neutral background on which more or less decidedly sculptural, artificial, architectural objects stand out. Instead, it has become the primary interest, the focal point of the architect. In this way, by changing the point of view, the landscape loses its inertia and becomes the subject of possible transformations; it is the landscape that can be designed and become artificial.

At the same time, architecture has begun initially diffuse processes involving the loss of traditional definition in which there is an obviously growing interest in including a naturalistic type of condition in the geometric and compositive as well as the constructive aspects, all part of a search for environmental sensitivity and formal complexity which responds to the new values of our society. The project is validated insofar as it constructs a complete redescription of the place; as it primarily proposes the invention of a topography. This dual movement from nature to the project and from the project to nature thus resuscitates an 'ecomonumental' condition which is beginning to push inexorably beyond any argument of opportunity, in a way that others would not hesitate to call 'the spirit of the times' or 'the desire of a period'.

4. World Material

This artefact-city we are talking about appears to us as an amalgam, a hitherto unknown material, a conglomeration of natural, artificial and immaterial elements or lows which at the same time is porous and fibrous, with dense stable areas laden with memory, and vast unread areas lacking qualities, almost liquid, made up of antithetical elements that have broken the link with the precision of traditional limits between natural and artificial. If we were modernist architects, we would think of this city in moral terms and would produce reformist policies. However, we think it is much more necessary and, if you wish, more closely linked to architectural practice, to find poetic a substratum in this magma, to understand it as something that is inviting us to experiment in a new thing, other than this material; the dissolution of the natural-artificial opposition we can see on every scale, involves a works programme which is nothing other than the discovery through architecture of the contemporary human position in the world...

5. Hybrid Technique, Mestizo Style

Sensitivity towards nature-oriented policies has influenced the technical paradigms, with interest shifting from high tech experiments, no doubt a residue of the modern spirit, towards hybrid models in which the accent has begun to be placed on the interaction between natural materials – massive and energetically inert – and highly sophisticated artificial materials – light and energetically active –, which respond sensitively to environmental variation, giving rise to composite systems in which the former are responsible for accumulating and reducing exchanges while the latter act as generators, capturing energy resources. This new technological model implies a shift from the aspects of material organization – mass production, simplified assembly, time and cost optimization etc. – towards the rational organization of the energy consumed during both the production and upkeep of the building. This shift now enables us to conceive 'systems', not from the perspective of congruence and unity in the materials, but rather in terms of their environmental congruence, thus opening the way to experiments in which the congruous mixture of heterogeneous materials becomes a new visual feature. This hybrid materialism implies a profound transformation of aesthetic ideas in harmony with the crossbreeding of our human landscapes.

6. A New Naturalism

Paradoxical conclusion: a new naturalism with no natural references. How did the colloquial use of the term 'natural' arise? The word 'natural' or 'naturally' probably could not be used in its present sense in so many languages until nature was domesticated, understood, subjected to taxonomic organizations which gave a reasonable explanation to something which was previously construed as an unapprehensible, threatening mystery; until it could be contemplated as worthy of representation, and a picturesque concept was superimposed on a certain degree of cosmogenic organization as the effect of multiple journeys that provided the necessary distance and capacity for observation. This is a plausible hypothesis which needs no ratification. Its mere mention allows us to imagine a new naturalism arising from the profound ambiguity in which nature is presented as the subject of knowledge and aesthetic experience, a hybrid, crossbred, entropic, humanized conglomerate that is confused with its former enemy, the artifice, tightly wound into the political space, a carbon copy of what was once public space, a turbulent, flowing, random magma. Perhaps the key to the illumination of this expanding naturalistic gaze is to be found in the journeys yet to be taken, the dark zones of the atlas of the picturesque, those constant continents unconnected to the set of routes that hold it together. A new naturalism should begin by integrating these zones, bringing them to life and giving them a voice, demanding architectures that can be equally meaningful in Lagos and Quito as in New York or Dusseldorf, capable of articulating an immediate, unified sense of beauty, something which could never be considered insulting or arrogant. Who can claim to have achieved such a thing today? But perhaps this fruitful journey can only be taken in the opposite direction from tradition, from those dark holes towards us, now reincarnated in the new indigenous peoples of a different form of wild nature. These journeys may well have begun already and we are still incapable of understanding them, now the objects and not the subjects of a turbulent beauty to come.

The Virtual-Physical Nexus: Future City Scenarios for the Arab World

Mohamed Ayman Daef Assiut

Focus

The electronic movement of information in almost real time transcends individual physical places and provides an arena for a new set of economic rules. The contemporary influences of digital technology are pervasive, and it is hard to escape the economic, social and environmental footprint of this profound technological change. The new digital revolution makes a truly borderless society possible. Distinctions are disappearing between town and country, private and public, here and there, and perhaps even between reality and fiction. At last, physical construction can be reduced, and long distances can be eliminated, whilst presence and social interaction can still be maintained, if not enhanced. Location and physical space requirements of major urban activities are being radically changed. Even the long-lasting phenomenon of rural-to-urban migration might also be reversed. From the "Arabianranta"[1] in Helsinki to "Netville"[2] in Toronto, empirical evidence is accumulating, and major examples are emerging both in Europe and North America. Yet, the complexity and magnitude of urban problems in the Arab countries such as Egypt would justify a similar concern.

Problem Area

As in many other Arab countries, Egypt has witnessed the phenomenon of accelerated urban sprawl and the resultant informal housing development. Historically, most city nodes in Egypt are separated from the desert hinterlands by agricultural barriers. Therefore, most uncontrolled urban sprawl in the country has occurred at the expense of fertile agricultural lands, and that represents a major socio-economic challenge that faces urban planning and development strategies in Egypt (Figure 1). Similarly, in many other Arab countries, cities are separated from hinterlands by either harsh desert or mountainous landscapes. The situation as such has led to an increased urban spread of existing city boundaries, while smaller towns and villages in the hinterlands are witnessing a socio-economic decline and an increased loss of their well-educated and well-trained inhabitants. In spite of the recognition of the problem by most governments in Arab countries, rural-to-urban migration continued, existing major urban centres kept growing, and the urban environment has become more threatened. Regional imbalances have become the norm in most situations, and conventional urban and regional planning methods appear to have failed in addressing the pace and magnitude of the problem efficiently and sufficiently.

A second dimension of the problem is represented in the gender aspect of development planning in most Arab countries. Regardless of the ever increasing rates of women's education in Arab countries, a considerable percentage of them still prefer not to join the work force after education. For cultural reasons, this tendency is due to their preference to stay at home after marriage.[3] Obstacles to combining family responsibilities with employment still persist in most Arab countries. This problem becomes even more noticeable in areas far away from major urban centres.

Questions & Hypotheses

This paper will seek answers to certain questions:

- How does the invention of almost light-speed ultra-fast information media influence our perception of space and our spatial behaviours?
- What does this mean for future attempts of city planning in the Arab World?
- What are the benefits of placing the physical-virtual nexus on the urban planning agenda in the Arab World?

Tentative answers to these questions would lead to the postulation of research hypotheses as follows:

- In the electronically restructured cities of the twenty-first century, our perception of space and our spatial behaviour will be deeply influenced by the framework of a new economy of presence.

- Teleworking made possible by the present digital revolution could be utilised as a solution to many urban planning problems in the Arab countries. This can be achieved distinctively through the inducement of a migration reversal from rural-to-urban to urban-to-rural, and the opening-up of smart opportunities for women's participation in the labour force.
- Planning Agenda and areas of policy intervention in the Arab World could be adapted to make use of the new possibilities opened-up by the digital revolution.

Economy of Presence

The technological advancements in the past 10 years or so have had a profound impact on the organisation of space and time. The new technologies make a truly borderless society possible. In his E-topia, Mitchell rightfully asserts that in the electronically restructured cities of the twenty-first century, perception of space and spatial behaviour will be deeply influenced by the framework of a new "economy of presence".[4] The digital revolution has recently opened up a new dimension of communication which is both remote and asynchronous.

	Synchronous	Asynchronous
Local	Requires Transportation Requires Coordination Intense & Personal Very High Cost	Requires Transportation Eliminates Coordination Displaces In Time Reduces Cost
Remote	Eliminates Transportation Requires Coordination Displaces In Space Reduces Cost	Eliminates Transportation Eliminates Coordination Displaces In Time & Space Very Low Cost

Table 1 Advantages, Disadvantages and Costs of Various Interaction Modes

Source: Mitchell, William J (2000: 138).

This mode of interaction goes to the extreme of separating participants in both space and time. Advantages of this newly introduced mode of interaction in comparison to other previous conventional modes can be summarised as illustrated in Table (1). With the development of digital networks, a rapid and massive shift has occurred towards the very low cost, remote-asynchronous quadrant of the table. This has been the most fundamental effect of the digital revolution.

Confronted with the new developments of the digital revolution, conventional theories and instruments of both architects and urban planners do not seem to be sufficient – if not valid – anymore. The architect does not design objects anymore, but rather relations, and although the space as an object is diminishing, there is still a possibility of planning. The multitude of both physical places and virtual ones – as represented in the structure of sites on the World Wide Web – have now become intricately intertwined. Ties between places are no longer confined to transportation links, as electronic hyperlinks are also used in this very present age to link virtual places. Mitchell depicts the interrelationship amongst all of these components as illustrated in Figure (2).[5] Thus, in different contexts, very different patterns and mixes may make sense. Architects, urban designers and planners will need to consider the trade-offs among the many emerging possibilities.

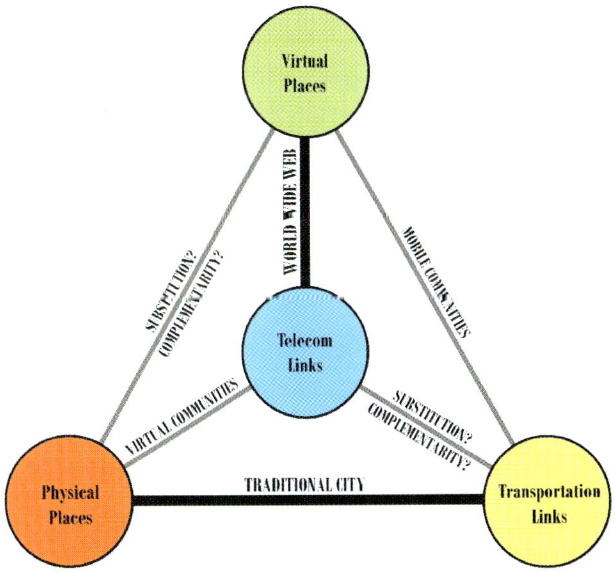

Fig. 1 Physical and Virtual Structures and their Interrelationships

Source: Mitchell, William J (1999: 126).

With the combination of both physical and virtual structures, numerous hybrids are becoming possible, where both physical and virtual places play significant roles, and where linkages are formed both by transportation and electronic connections. Such hybrids are possible to include – amongst other things – established cities on which networks have been overlaid, and scattered rural communities tied together and to the existing cities themselves by networks.

The emergence of powerful and convenient electronic alternatives to traditional means has opened up many new possibilities for the arrangement and distribution of office work space, shopping areas, educational facilities, home banking, entertainment facilities, housing neighbourhoods and many other services and facilities. Thus the economy of presence has already started – and will continue – to produce new patterns of urban structures based on the physical-virtual nexus; e.g. teleworking, online shopping, distance education, home banking and online entertainment. Home-based working could possibly introduce new criteria for deciding upon the best location for housing neighbourhoods, as locating near to workplaces is no longer a preference that is solely measured in a physical sense.

Teleworking

The economic future of Arab cities will be defined by their capacity to generate, process, and distribute information. With the emergence of a new economy of presence, businesses can reassess where jobs and work are to be located. Powerful economic factors, involving pressures to increase productivity and reduce costs, are influencing the way that businesses take these deci-sions. When information is stored in digital form, the data required can be extracted, worked on and eventually refiled from a computer located hundreds or thousands of kilometres away. In theory, the workers can be based anywhere in the world. Accordingly, work is being relocated to other geographical areas within a country's boundary; e.g. to back offices in rural areas where overheads and labour costs may be cheaper, or from conventional offices into the homes of workers.

On the one hand, teleworking may be a tool for employers to move work to geographical areas where working conditions, salaries and collective bargaining rights are the poorest. But on the other hand, teleworking may be an interesting alternative for employees in certain phases of their lives; e.g. as an attractive alternative to married women with household responsibilities.

Teleworking offers many advantages for the employee. It widens the choice for the individual, and enhances flexibility in working time and methods. It also gives more people the opportunity to work, and enables combinations of work, personal life and caring responsibilities. Teleworking also removes the strain of traveling to the place of work. Above all, teleworking can thus foster a reversal of migration from rural-to-urban areas, and help retain workers who want to spend more time caring for their families – especially women.[6]

Fig. 2 Uncontrolled Urban Sprawl on Agricultural Land – Alwalidiya – Assiut – Egypt
Photographed by the Author on 4 August 2001

Conclusion: Areas of Policy Intervention

Within the expansion of information and communication technologies (ICTs), the poor and the rural populations in the Arab countries can be marginalised as the market interest in establishing costly access to internet for these communities is low. Nevertheless, teleworking opens the whole array of possibilities for rural socio-economic development. New employment opportunities can be generated through telework in rural areas. ICTs would enable a greater number of people an easier access to educational opportunities irrespective of their geographical location. Increased levels of education and knowledge combined with an increased scope of social participation might lead to empowerment of rural and remote populations, and greater grassroots participation in community affairs.

Teleworking may also help curb rural migration in Arab countries, as the cities lose their exclusivity as centres of employment due to the growing opportunities for income-generation locally through remote channels. In addition to the retention of rural population, there is a likelihood of migration of previously urban teleworkers to rural areas. There is also a possibility of migration from areas of high livelihood costs in industrialized countries to cheaper and environmentally more attractive regions in Arab countries. Teleworking can be harnessed in Arab countries to empower women and assist them to gain better employment opportunities and improve their livelihoods while preserving and distributing the gender-specific knowledge; e.g. in agriculture, environmental resource management and health. Home-based teleworking often appeals to women as it allows them to retain their careers while taking care of children. Yet, more ought to be done in Arab countries to encourage the capacity mobilization of the female population through ICTs.

Reduction in the cost of telecommunications services as well as the provision of suitable ICTs infrastructure is a precondition to the successful utilization of the ICTs by Arab countries. Replacement of bricks with bits, and mortar with clicks, can have a profound impact on the organization of space and time and, hence, future city scenarios in the Arab World in the third millennium.

References

American Sociological Association. 2002. ASA News. Available at: http://www.asanet.org/media/neville.html, accessed on 13 April 2002.

Di Martino, V and Wirth, L. 1990. "Telework: A New Way of Working and Living" In: International Labour Review. Vol. 129, No. 5.

Helsinki Virtual Village. 2002. The Arabianranta Virtual Community.

Available at: http://www.teamware.com/top_story/smartcities/ virtual_village.htm, accessed on 13 April 2002.

Mitchell, William J. 2000. E-topia: Urban Life, Jim – But Not As We Know It. Cambridge, MA: The MIT Press.

Mitchell, William J. 1999. "The City of Bits Hypothesis" In: D. A. Schön, B. Sanyal and W. J. Mitchell (Eds.), High Technology and Low-Income Communities: Prospects for the Positive Use of Advanced Information Technology. Cambridge, MA: The MIT Press.

Nilles, J. 1985. "Teleworking from Home" In: Forester T (Ed.), The Information Technology Revolution. UK: Blackwell.

Shaw, L. 1996. Telecommute! Go to Work Without Leaving Home. NY: John Wiley & Sons.

United Nations Statistics Division (UNSD). 2001. Social Indicators. Available at: www.un.org/Depts/unsd/social/unempl.htm, accessed on 26 December 2001.

1 Helsinki Virtual Village is a public-private initiative that was established in order to develop the new Arabianranta area, into a virtual community providing the companies and residents of the area with the services and technology that the information society has to offer. The initiative for the Arabianranta development was launched in Finland in 1996. (Source: Helsinki Virtual Village – Website).

2 "Netville" (a pseudonym) is located in suburban Toronto. It was one of the world's first residential developments to be equipped with a broadband local network. The neighbourhood was built from the ground up with a 10Mbs high-speed computer network supplied and operated free of charge by a consortium of private and public companies. (Source: American Sociological Association – Website).

3 For example, statistics compiled from the International Labour Office's (ILO) database have shown that in 1998, 19.9% of women available for work in Egypt were unemployed, as against only 5.1% of men. Similarly, in 1991, 14% of women available for work in Syria were unemployed, as against only 5.2% of men. (Source: United Nations Statistics Division – Social Indicators Website).

4 The term is coined by William J. Mitchell. See Ch. 9, in Mitchell, William J. (2000).

5 Mitchell, William J. (1999: 126).

6 This part is based on surveying existing literature on teleworking with special emphasis on its relevance to Arab countries; e.g. Shaw, L. (1996), Di Martino, V. & Wirth, L. (1990), and Nilles, J. (1985).

The Multi-Layered City – An Alternative Urban Design Paradigm

Michael Burt **Lior Datz** Haifa

Introduction

The urban habitat is claiming it's place as the principal living environment for most of humanity. Toward the end of the 21st century more than 90% of the world population will live in cities. We are going to witness larger 'urban fields', stretching to the horizon, and the development of megalopolises, with frequent collapse of infrastructures and rising conflict between the built and the natural environment and subsequently, with a deteriorating environmental quality. This forecast, propelled by demographic, economic, ecological and social-cultural processes, which are now in motion, contributed also to the emergence of the 'sustainable development' doctrine. It emphasizes the need and produces the pressure to contain our urban sprawl and slow down (significantly) the conquest of new 'green lands', which in it's turn will compel us to adopt high density solutions to historically inexperienced levels and induce a way of thinking and perceiving, which could not rest on familiar historical precedents. The structure of our cities and their organization, with all their conflicting physical manifestations, are destined for far-reaching evolutionary transformation already seeking a conceptual framework and it's new theoretical and applicative definitions. With the passing last decades it becomes clear that the emerging urban reality finds itself in a rising conflict with some of the most fundamental concepts and presuppositions of the prevailing urban design paradigm.

The Prevailing Urban Design Paradigm

After some heart-searching it seems to rest on the following, deeply rooted presuppositions and convictions, permeating the urban planning approach and practice, some of which go back to immemorial times, and some of a more recent (last centuries) origin.

1. The belief that in spite of the prevailing anthropocentric conception of man's uncontested supremacy in and over nature, it's supervision, management and maintenance is still the responsibility of God, or, in more secular terms 'nature will take care of itself.'

2. The comprehension that the compelling needs of human socializing, security and interaction are the (still valid) propelling power and the rational belong to the creation and maintenance of the urban habitat.

3. The conviction that man is not in a position to threaten and endanger nature and ecology and jeopardize long-term environmental quality, and that nature will always embrace and envelope benevolently the urban habitat with an abundant, friendly, life-sustaining blanket.

4. The conviction that the immensity of our planet earth and the seeming abundance of nature's resources preclude the possibility that we will ever be short of means and buildable land resources for our urban expansion.

5. The belief that urban population growth, its rate and intensity, is a process which could be met and resolved within the prevailing design and planning traditions and long tested urban design models, with their standing disposition toward land and space utilization, and their historically proven organizing principles and ethical imperatives.

6. The accepted urban design doctrine that a strong and binding relationship exists between urban gross densities and normative levels of servicing infrastructures, to preserve and ensure normative urban welfare standards for the greater majority of it's population.

7. The conviction that increasing effectiveness of the transportation and communicationsystems may balance and compensate for the excessive sprawl of a city.

8. The understanding that the principal key to urban development, it's timing and effectiveness is in controlling the appropriate land resources.

9. The acceptance of the moral-cultural conception that land is a commercial commodity, to be parceled, manipulated and exploited for (private) profit or power and control gain, to a point of interfering with, and obstructing of the interests of public development.

10. The understanding that urban structure carries stark resemblance to an evolving organism, it's morphology and hierarchical nature, with physical growth and an expansion process resolved to least of effort and energy consumption. It imposes a design and evolutionary development mode, adjusted to incremental growth and staged completion.

11. The conviction that we can rely on our accumulated historical heritage of proven urban precedents, to solve any present and future developing malfunctions of urban systems or infrastructures.

12. And if not, 'time is on our side' and we will have enough time to engage the existing 'checks and balances' of the prevailing corrective measures and mechanisms, to restore order and equilibrium, and to correct for our design shortsightedness and miscalculations.

In Search of an Alternative Design Paradigm

As it turns out, most of the beliefs, convictions and assumptions on which the prevailing urban design paradigm rests, have lost their solidity and assurance, under the assault of changing demographic and urban conditions, needs and new technologies, some of a revolutionary character.

The 'sustainable development doctrine' is already making a profound impact on our design policies, with it's imperative of containment of urban sprawl and the consequent insistence on urban densification, and induces us to resolve the conflicting issues of high gross urban density and preservation of appropriable levels of proper performance of urban infrastructures and its ecological environmental quality.

It should be stated categorically, that very little of our historical urban heritage and the prevailing design paradigm equips us to deal with this complex multiple task, although the issue and some proposed general or partial solutions were raised by urban design thinkers over the last century, some of which offered wise, local solutions in the field.

What is urgently required is an alternative urban design approach and a comprehensive framework of principles which conform with the sustainable development doctrine and amount to no less than an alternative urban design paradigm.

The Multi-Layered-City

This new conceptual approach, which the authors suggest naming. The multi-Layered City, is in its early evolutionary stages and its framework of principles may run as follows:

· In conformity with the sustainable development doctrine, it is imperative to take responsibility over the planet Earth, its natural resource management and maintenance, with a long-term vision and true consideration for the welfare of future generations.

· It is imperative to slow down, dramatically, the conquest of new green lands and the associated nature and landscape resources, by limiting urban expansion, as far as possible, to the existing city bounds.

· It implies a far-reaching urban densification process, utilizing the 3-D space, air and subterranean rights, to the limit, to generate buildable land resources for realization of private and public built volumes and all the servicing and the biotic infrastructures, which are required for a full and proper urban functional performance.

In order to facilitate the aspired density rates and urban (environmental) quality levels, a new code of priorities, for the utilization of the ground level, should be adopted.

· The design morphology of the Multi-Layered-City will be characterized by layered distribution of the urban mass and fabric and spatial hierarchical structure which leaves most of the ground area for public pedestrian use and biotic and social-educational Infrastructures.

This design policy requires a drastic revision of prevailing land ownership concepts and practices, trying to substitute them with ownership rights of volumes in space, thus leaving the public in control of the urban development process and its priorities.

· New definitions of the relation between the built and the natural environments should be promoted, giving rise to 'hanging gardens', vertical-spatial parts, rooftop play grounds and sky promenades, far in excess of the total developed ground area, above (and below) the ground level.

· Social-cultural-economic interaction spaces should be allowed to develop on few layer-levels, as climatically controlled malls, bridge avenues, spatial public squares and spatial pedestrian traffic grids, assisted by advanced mechanical systems, joining the subterranean, the ground level and the hovering public spaces, into a continuous urban whole.

· The resulting compactness of the urban fabric will reduce the dependence on privately owned motorized transportation, make public transportation more profitable, with many more positive by-products.

· The core of the required applied morphology and design proficiency lies in successful generation of built masses, characterized by high sponge-like permeability to climatic environmental and accessibility parameters, while protecting privacy and resolving all other density induced conflicts.

· The Multi-Layered-City, although encouraging high intensity development and large scale projects, still should be solved for incremental growth and staged completion, mostly for economical reasons especially of the public support systems of bridge avenues and other elevated built facilities.

The Multi-Layered-City concept could be applied to an existing urban situation (with the required design compromises), or to a new urban development.

The Multi-Layered City may present a viable alternative to currently practiced urban development policies, in connection with the following issues:

· Clearance (by demolition) and construction, as part of the urban renewal development strategy.

· Contribution to 'urban continuance' in problematic, stricken urban situations.

· Usage of space over sizable public functions within the existing urban fabric.

· Urban development in the (topographically) mountainous urban expances.

· Urban development in the marine (artificial islands) environment.

In reference, a few names should be mentioned, of people who contributed over the (recent) past to the evolution of the ideas and concepts forming the basis of this presentation.
In particular:
Sanct Elias, Le Corbusier, Jona Friedman, Buckminster Fuller, Paul Rudolph, the Japanese Metabolists, Peter Cook and the Archigramm, The WCED (1987) publication ('Our Common Future', Oxford Univ.Press), the Rio declaration (1992) and the UIA-Chicago declaration (1993), concerning the sustainable development doctrine and it's urban ramifications.

Power Station, Berlin Architects/Artist: Jourdan & Müller.PAS/Kasper König

Reflections on Architecture Identity of Space as a Resource

Krzysztof Chwalibóg Warsaw

Identity of the built environment is now becoming a resource as precious and scarce as clean air and water. Resources of identity represented by the buildings and urban landscape, characteristic for a given area, are shrinking.

In the past architecture, like landscape, has been mostly defined by local factors such as: topography, climate, building techniques and cultural traditions. At present the importance of them all is progressively receding under the pressure of technological uniformity on the global scale. The local and regional specificity is constantly decreasing.

Easy to distinguish identity of space is an important feature for the general quality of life of its inhabitants and visitors, as well as from the psychological, social and cultural point of view. Therefore it should be considered as vital for human beings as ecological factors. However this fact doesn't seem to be reflected, neither in so-called "commercial buildings" nor in some "fashionable" tendencies observed in contemporary architecture and based on the extensive use of futuristic technology and overemphasized scale.

In consequence, natural and local cultural factors are giving way to the prevailing technological impact. As a result the environment is becoming more technological and less human. Taking into consideration the above reasons, legible identity has to be recognized as an important resource and a basic factor of the quality of architecture and of the urban landscape.

In order to see how the above problem is reflected in urban design and architectural policy, short observations from a few European cities are going to be presented. Chosen cases concern at first cities destroyed by the war and then reconstructed (Warsaw, Rotterdam and Berlin), and also Amsterdam and Marseilles. As far as it concerns an attempt to comprehend how the identity is treated in the works of some of the most renowned top architects, more significant examples will be shown.

Cities

Warsaw

More than 70% of Warsaw was destroyed during the second world war of 1939–45, and its urban identity almost totally disappeared.

When reconstruction works started in 1945 a violent quarrel burst out on the subject of the character that should be given to the recreated capital. The modernists opposed the idea of rebuilding the Old Town with medieval structure, arguing that it would be far more rational and less costly to build the modern city instead. Fortunately (as we can say now) the historical part of Warsaw has been carefully recreated with the help of prewar surveys and the eighteenth century paintings by Canaletto. Thanks to this process during the fifties and sixties, traditional scale and the variety of design of old churches, palaces and houses has been brought to life again.

And then modern construction started. As an outcome of an urban design competition for a part of the central area of the City, a cliché of Cité Radieuse appeared. In the place of the dense, continuous pattern of street development seventeen identical, fifteen-story high slab blocks have grown. The other parts of Warsaw became a mixture of remaining prewar buildings and new office towers looking like any other city in the world. And peripheral housing areas became endless repetitions of the archetype created by Le Corbusier (as in all Communist countries).

Till now the market square and the surrounding narrow streets of the Old Town are the most popular place to go, both for the inhabitants of Warsaw as well as the visitors, while a business area in the very center is unable to provide attractive public space among its overgrown glass towers.

Rotterdam

Rotterdam was bombed and the problem of recovering its identity appeared extremely difficult. Reconstructing the Old Town was not an issue here. Instead modern a commercial center called Lijbaan was constructed by Van dem Broek@Bakema in combination with blocks of flats. Let me quote now what Lucien Kroll said about this project in an interview: "And Lijnbaan in Rotterdam, isn't it Dutch? O yes, there are some traces, however rather shy, but most of all it is pretending to the International Style."

And the whole Veena area is also modern and nice looking, but a newcomer could hardly say in which part of the world is it situated. The same, but on an even larger scale, can be seen in the development area of the future new center called, by the way, "Manhattan on the Maas". This complex seems from its very beginning doomed to be a poor cousin of the real Manhattan and no office tower, no matter how huge, is going to change this obvious fact. You can put as many skyscrapers designed by famous international and local architects there as you wish but this will not give any quality to the urban landscape, apart from the skyline. What counts is a lively public space defined by well-scaled, interesting architecture and, above all, with its own character instead of a borrowed "nowhere look".

Berlin

Reconstruction of Berlin is a process of knitting together the city divided by the iron curtain and split into two separated entities, one of which was heavily marked by the destruction of all remaining traces of the urban tradition. The challenge of rekindling the spirit of this city has been such an enormous task, and the scale of the whole process so important, that the possibility of eliminating mistakes was hardly imaginable. And the debates among the professionals and the general public reflect this. However from the point of view of a visitor who is able to compare what has been achieved compared with some other cities, there is no doubt that a new urban quality has been provided.

At first that should be pointed out is the strong feeling of necessity to rediscover the identity of place by recreating the continuity of development, and traditional city scale is a basic value of the idea of "critical reconstruction" formulated by Professor Josef Paul Kleihues and implemented by the Senate of Berlin.

It is normal for there to be a difference between the point of view of an artist and an urban planner. But a city has to be a lively environment for the people and not a frozen monument to a past catastrophe. This is why Daniel Libeskind is wrong in opposing the continuity of street facades in favor of loose, open spaces. His other mistake is to fight against stone and regulated building height in the name of imagination and freedom of expression. It would be completely stupid to create a world full of chaotically monotonous, glass-covered, unrelated structures. This would not be an expression of democracy but rather of a frightening anarchy. When one looks at thousands of chaotic megalopolises growing all around the world, one has to consider any regulated city as a miracle and an extremely valuable resource. A city worthy of this name cannot be a bunch of individualistic buildings, even if they are of very high artistic quality, but a harmoniously organized entity. And Berlin is trying to make it.

Amsterdam

Amsterdam – one of the most beautiful cities of Europe – is undergoing spectacular transformations of its docks, similar to Rotterdam. After the evacuation of port facilities the city has taken over the old docks areas called: Java, KNSM, Sporenburg and Borneo.

However the way of development adopted here is displaying a totally different approach from that of Rotterdam. One can only guess that, thanks to the long tradition of planned, harmonious urban growth, which started in early Middle Ages, Amsterdam is simply continuing its own principles instead of trying to import its new face from outside. Let me refer again to the words of Lucien Kroll: "It is again an issue of culture proper to the Dutch, and International Style could not invade this culture, however Calvinist, deeply anchored to the polders."

And probably because of this, Amsterdam refused to let high-rise buildings grow close to the central part of the city. Some of them are only visible on the outskirts. The recently built housing areas are, as far as possible, retaining certain features of the historical development. Small canals cut across the shape of Jawa island, and the traditional form of the single family four-floor house is appearing. The new urban tissue is providing a variety of shapes, colors and materials. Amsterdam is not trying to imitate Manhattan, but sticking to its own way.

Marseilles – Between Identity and Global Design

1. Transatlantic and the Old Town

When, in 1952, Le Corbusier built his famous "Unité d'habitation", looking like a huge oceanic steamer, he believed he was creating a new valuable model of urban life. And he was partly right, because during the next forty years more or less similar copies appeared in most European cities and, above all, in all urban areas of the Communist empire.

Now those buildings have been destroyed or totally reshaped in the countries able to afford it. In the others (like in Central and Eastern Europe) they will remain as the ghettos of the poorer social groups unable to leave for a better environment.

Today Unité d'habitation stands alone among low developments of the southern district of Marseilles. The local people call it "la maison du fada" which means "the madman's house". Several architects living in it are satisfied, and this seems to confirm the theory about "architecture for the architects" being different from the "architecture for normal people".

Conforming to the ideas of Corbu, the old, unrational and unfunctional cities have been doomed to be destroyed. However the old town of Marseilles with its narrow streets and squares and houses with shops, bars, and small gardens in the backyards is fortunately still there.

2. "Cinque Avenue" against the "Grand Bleu" Spaceship

Although this seems like a title of another science-fiction thriller, it is only a striking comparison of the two urban design projects separated by more than a century in time and two kilometers in space.

When the works connecting the new channel bringing the fresh water to the city have been accomplished, an important complex of monumental buildings and landscape composition with an artificial waterfall and park will have been realized. The composition creates a focal point in the middle of a housing area where five avenues converge. The relationship between the pattern of the whole housing area and the public buildings is perfect, and the beaux arts architecture is wonderfully connected to the hill and interplay of the water and greenery. Important views from this "belvedere" are presented along the avenues. The whole composition is not something very local but represents a Mediterranean identity, as its archetype is the Roman "mostras" (being built on the end of any aqueduct).

Now when we move about two kilometers from this point, succeeding to find our way from the subway's corridors, we will emerge in totally different surroundings. Somewhere in between an important road rather than an avenue, multifamily housing blocks and scattered little houses remaining from the old village of Saint Just, you will finally see "Grand Bleu" (local people call it the blue "cosmic vehicle") which by some accident landed in this casual place to become the headquarters of the regional government. It was designed by one of the top British architects as a result of a won competition and is without any doubt extremely well studied, climatically sensible, and interesting from the point of view of interior space relations. Unfortunately it is completely unrelated to the surrounding space.

This situation seems to be typical for the way many urban areas are now growing. More and more futuristic shapes are "landing" in many cities, with all the emphasis being made on the "fashionable form". The notion of classical "genius loci" has so far been eradicated by the modernistic idea of machines as models for buildings. Be it a steamer or a spaceship. And those do not care for the context.

Identity and Star Architects

After trying to see the identity problem in the existing urban environment of a few European cities, it would be interesting to change the observation field and learn something about the textures of the world's top architects. Any architect from this group is easily recognized because of the specific way of design. The manner of organizing spaces, composing forms, materials and colors, playing with scale and meaning, all make up the architectural expression.

All this is so personal, and it would be impossible to confuse the architecture of Norman Foster with that of Frank O. Gehry or Jean Nouvel. The intensity of this individual style is not tuned to any local context but is somewhat like producing the effect of a stamp being applied to a given location in a given country. It is easier to recognize "this is Gehry" than in which part of the world it has been built. And this always occurs when the author is so fascinated by his own idea to the point of not being able to care about the context. Keeping in mind a very different scale of the phenomenon, it is however possible to see a parallel in the violence of invasion of commercial investments (such as a supermarket or Mac Donald restaurant) dropped somewhere without any regard to the environment in any point of the globe.

Fortunately this approach is characteristic only of some star architects. Several of them have quite a different understanding of the genius loci and identity issues. Their buildings, designed for the specific area, reflect not only their personality but also the context of the place and culture. The dividing line lies between those who still believe in the dogmas of the modern movement, defining how the building should look and relating it not to the place but to the archetype of a machine or a vehicle. In order to get a closer look at the reality of this problem, let us take the case of the international invited competition for the directional center of the region of Galizia in Santiago de Compostela.

Santiago de Compostela Competition

The result of this competition provides a spectacular comparison of two opposite design philosophies. The one is modernistic and universal, and the other is local-identity oriented. In this confrontation of ideas among other conventional proposals, two radical schemes have been competing .

One – the idea of the horizontal megastructure invented by Le Corbusier to be put in any hilly city (Algiers, Rio de Janeiro etc.) and "recycled" for Santiago de Compostela by Jean Nouvel.

Second – (by Peter Eisenman) a form based on the archetype of the shell – a symbol of Santiago from the early Middle Ages, but also a straight reference to the form of a medieval town and to the organic shapes of the hill on whose slopes it is situated.

While the scheme of Nouvel clearly expresses the modernistic idea of domination over nature by levelling the hilly country and introducing outscaled technological structure, Eisenman shows extreme sensitivity as well to the culture of the place and then to its topography. His proposal clearly shows the right way of treating the landscape of the earth with respect and building the future on the roots of the past. Looking at the verdict of the competition's jury, one cannot but agree. Eisenman's scheme won the competition. This is a very instructive example of ideological confrontation which we are witnessing around us. Unfortunately not always the respect of nature and culture prevails over technological brutality. For this reason such values should be most appreciated in the creative process of some of the renowned architects. This kind of approach can be seen in the works of Charles Correa reflecting deeply the Indian culture, and you can see this when Alvaro Siza is constructing in Berlin and expressing the local character, or when Raphael Moneo is designing a "Scandinavian" building for the museum in Stockholm.

Conclusions

Conclusions of this short trip across a few cities and the observations on some phenomena of the present architectural scene lead to the statement about the utmost importance of the identity of the place and culture – resources which are very often in danger of disappearance and were only preserved thanks to the high quality of architecture and very careful and consistent architectural and urban policies. Local/regional identity is a basic resource which architecture should provide.

Towards Zero-Energy Urban Infrastructures

Thomas Spiegelhalter Pittsburgh

We all know that we are far away from achieving sustainability. But how far? A good place to begin the discussion starts with one of the most resource hungry regions in the world: the USA. The U.S. with only 6% of the world population consumes over 30% of the world energy resources (Resource: National Geography Institute). American cities and towns account for over 85% of national energy use. Land use planning and urban design affect negatively about 70% of that, or 56% of the nation's total energy use (Anderson, 1993), excluding resource infrastructures, national and worldwide facilities.

The Current Status?

The second half of the 20th century saw a tremendous infrastructure expansion into the American landscape, especially right after WW II. A surprising amount of this expansion had failed by century's end, and what were left were the remnants of thousands of post-industrial relics of infrastructures of every type. Many left over structures are former bypassed highway urban relics, their economies devastated by the construction of the Interstate highway system several miles from their original "Mother" highways.

All kinds of resources are funneled through these industrial and urban systems without any concern about their origin, and about the destination of waste; the linear metabolism of these inputs and outputs are treated as largely unrelated.

Mobilization to Where?

Today we have additional increased exurban growth, which is reshaping American landscapes and especially the coastlines. A new American amenities migration, one that rivals the exodus from the Frost belt to the Sun belt a generation ago, is transforming the Atlantic and Gulf shorelines. An estimated 41 million people – more than one in seven Americans – now try to reside in a county that abuts the western, eastern or southern sea-board. That number swells by several million when inland residents with second homes near the shore are included. An unprecedented influx of car-orientated residents is converting aid-back seasonal resort towns into sprawling year-round communities with burgeoning non-sustainable economies.

These coastal migrants are transforming seasonal resort towns that used to bustle for just a few summer months into sprawling, air-conditioned year-round communities. Up and down the coast, more car-oriented development is spreading for miles inland attracting new businesses to serve them. The average ecological footprint of an American uses 25 acres to support his or her current lifestyle. This corresponds to the size of 25 football fields put together. In comparison, the average Italian lives on a footprint 60 per cent less, not to mention people living in developing countries!

As a summary, in these fast-growing "new urbanism economies" of the U.S., more and more car-oriented urban sprawl is being built at an uncontrolled phenomenal rate and density with little thought for future social or worldwide environmental impact.

Power Market

Today, the production of building markets is mostly conquered and democracy is being stripped of its power by transnational and decentralized business decisions. These kinds of structures depend always on a vast system of external supply lines to rural areas and manufacturing centers, facilitated by global transport and communications infrastructures. And global supply lines are still required for securing oil, coal and natural gas from the limited number of highly productive fields in production: the mining, shipping and processing of the raw material and its world-wide distribution has necessitated a vast network of logistics, aggressive military management, nowadays high-alert homeland security arrangements and strategic diplomatic agendas – as well as specialized economic systems.

A majority of current and recent wars, regional and local armed conflicts are resource wars, in part or entirety. For perspective, consider global military budgets, which run at $ 700-800 billion per year globally, around $ 2 billion per day (According to the Stockholm International Peace Research Institute, a figure combining internal and external expenditures on arms and the military). Consider a few comparative examples: With a billion dollars the solar industry would be able to built two to ten PV manufacturing plants, depending on technology type, producing products able to generate electricity cost compitively even on today's skewed energy playing fields. Such plants, once built, would be quickly replicated around the world, unlike war option – where the returns will be measured in new generations of terrorists with horrifying weapons more effective than civilian airliners (Jeremy Leggett. In: The Solar Revolution: Micro power and Security in the Changing World, REFOCUS Jan. – Febr. 02, page 17).

Urban Analysis and Policy Development

Unfortunately, it is often the same environmental tragedy: A lack of regional government planning fosters sprawl. For example in metros in the US Northeast and Midwest, dozens of local governments regulate development. The result is fragmented planning. Efforts by local and state governments to restrict sprawl are growing but have had limited impact. For example, Detroit Metro has 280 local governments; metropolitan New York has more than 560. Each of them can adopt their separate zoning and housing and land-use codes. These disparate interests work against regional planning and promote sprawl. One municipality can restrict growth, but if a neighboring town welcomes it, sprawl continues. Furthermore, in all these remnants of postwar industrial, urban and suburban areas, the poor fall out of the consumer society and are abandoned and isolated in the inner-city ghettos. In the last decades, thousands of manufacturers have moved away from the inner cities, taking valuable job opportunities and leaving behind environmentally hazardous sites. These sites are now expensive to clean up when aban-doned, and the lenders may be held liable as a last resort.

This situation has created a lack of incentive for remediation of these sites, preventing poor inner-city communities from realizing meaningful economic opportunities. Now a large number of urban analysis based on GIS System statistics have challenged this as "environmental racism" and have correlated health, crime and suicide effects with lower economic status. If all this is left unchecked, the ecological and social problems of these dubious areas will increasingly dominate the human scene. It's unbearable, that the rich few can continue to turn their backs on the pollution and poverty of these infrastructures and operate in comfortable isolation from these seats of desolation in single-minded, gated-residential suburbs.

Measuring Progress towards Sustainability?

Right now, the main pollutant of American urban infrastructures is absolutely not based on a Input-Output-Quantification using international Live-Cycle Analysis Methods:

New holistic urban planning concepts, regional scenario tools and eco-audit measures on an international competitive level (the big picture) are needed that integrate social and ecological responsibilities for open-minded, multifunctional urban spaces and urban growth and consumption management.

Measures should be readdressed to strengthen the participation of individuals and their sense of belonging to the city, to the neighborhood in which they work and live. It is the individual commitment of everybody's city, which is absolutely central to achieving real sustainability.

A more precise transparent assessment is thus needed, with the aim to reduce the contribution of polluting infrastructures to various environmental problems, from a global scale (e.g. greenhouse effect etc.) to a local one (waste production etc.). Statistics are the basic ingredients in the "what if" and "if this", then that design and planning speculations with which the spatial problems of environmental structures are tackled. Unfortunately many current institutional approaches to sustainability are propaganda based and not achievement based. It is not what ought to be done that counts but rather what is effective that is done. Every city and every suburb and their various supporting infrastructures must be comprehended and managed in their evolutionary entirety as a circular self-contained, ecological long-living organism.

Their various elements and structures in different districts and neighborhoods must be interpreted and constantly updated as a complex system, that is characterized by flows as a continuous process of change, transformation and development. It must be possible to overview, monitor and manage the constant changes in its use and appearance, as well as in technology, in order to ensure a minimum of disturbance and a maximum conservation of resources.

It is a global goal to reach sustainability, from a renewable energy technology, industrial innovation, total services and personal lifestyle point of view. They also should foster a better understanding of urban systems through regional modeling, traffic and transport impact simulation, visualization and dissemination of alternative scenarios, to build democratic networks of experts and expertise, of international, national and regional companies and a multitude of committed citizens.

In most European cities transit investments complement, and are coordinated with, major land use decisions. For American Cities, only 5% of home and work trips are made on public transit, and only 2% of all urban travel (Warren, 1998). The tough reality is that American cities must begin to offer a radically better way of getting from "A to B" to reduce a significant amount of greenhouse gases. For example, the Ecological Footprint method is particularly useful for this urban analysis task because it starts to quantify the biologically productive area necessary to provide the supplies of a given human population and infrastructure and to absorb its wastes.

The U.S.A. has the world's largest research and development need addressing urban, energy, and climate change issues, the output of which greatly influences the availability and application of new urban infrastructure and building typologies, policies and master plan concepts towards future zero-energy-emission urban infrastructures. Both conventional and renewable energy resources for living and working environments, transportation and community infrastructures have an uneven geographical distribution in the U.S. Similarly, the local social and environmental impacts of energy consumption and urban living vary greatly between locations in a wide array of climatic, cultural, geographic and political self-governed regions.

But behind these efforts of improvements within global networks is certainly an irony: Parallel to these sustainability projects on earth the NASA and consortia are developing thousands of advances for living and working technology on the idea that ordinary people would someday live and work in space communities powered by solar fuel cells.

The NASA's Space Shuttle, Life and Micro Gravity Sciences and Applications Program is one of these numerous steps in turning this scenario into reality. Private and state funds are much higher dedicated to these goals (including necessary military facilities), than the budget to minimize resource consumption and to make efficient long-term zero-emission living and working infrastructures on earth.

Looking at the big picture of our spaceship earth, we know that, more than ever, we need biophysical accounting systems and a faster progress in the application of modern scientific climatology to architectural and urban design in the 21st century.

Resources

IPCC: "Climate Change 2001: Impacts, Adaptation and Vulnerability", http://www.unep.ch/ipcc/
Renewable Resource Data Center (U.S.) http://rredc.nrel.gov/

Sony Center Berlin Architects: Murphy/Jahn

Architects and Architecture for a New World

Tay Kheng Soon Singapore

Preamble

The XXI UIA Congress in Berlin is framed by two earth shaking events, which has shaped its themes. The first is 9/11 and the other is the global economic downturn. In a deep sense, these are tidal waves of the 20th Century. They will continue to wash over the 21st Century before it comes into its own. Both events have immense implications in society, nature and architecture. Their far-reaching implications lead us into many unfamiliar areas of thought. So we must let it. Architecture thinking has to go beyond the materiality of the object.

"The end of history" was greeted in triumph. But the 'post modern age' awoke to a rude shock. "The primacy of the subject" was demonstrated in a most horrific and shocking way. Post-modernists have cause to ponder. Predictably, there is a swing to the political right accompanied by clamours for law and order, sabres have been drawn, and aggressive rhetoric against a presumed "axis of evil" has been uttered. Architecture journals continue to dish out the old diet of ever more exuberant architecture seemingly unaffected by the momentous events that grip the rest of the world. Rearranging deck chairs on the Titanic. Is it paralysis or conservativeness in this time of crisis? What do the disciplines have by way of bold environmental, cultural and intellectual initiatives about their own milieu and bridges between civilisations?

Modernisation, Alienation and Homogenisation

Disparity, disorientation and dis-equilibrium result from rapid change. Yet the world accepts that rapidity is part of an equation wherein northern markets justify increased direct foreign investment into southern growth centres so that the trickle down can address the plight of the poor while making an affluent but culturally timid new middle class. The result is the wholesale transformation of the entire human and natural landscapes. Thus, the non-west is witness to the rapid erosion of sacred notions, loss of humane traditions, despoliation of beautiful sites, history muffled and tranquillity dispelled. All this, with the complicity of the new middle classes but accompanied by a sense of loss and the disorientation. Calls by developed societies for environmental concern and human rights must be backed by remediation if they are to have moral clout. Otherwise, bad conscience goads cynical support for extremists posing as champions of the oppressed or critique is dismissed as hypocritical sermonising by everybody else. Despite the stark prognostications unpalatable to the academic elite of powerful states, the Huntington hypothesis is taken seriously precisely by those who feel most violated. Outward shows of solidarity can conceal private reservations and regrets. Still, on the surface, life goes on and the making of a new life gathers pace.

Cath-Up Architecture

Asia, having passed through the post-1945 phase of nation building, gives way to fashionable temptations emanating from the West. The architecture from the spirit of self-reliance up till the 70s is inexorably displaced by a stylistic prostration before the altars of Western Star Architecture, even if strenuously denied by its perpetrators. The drive to catch-up gathers momentum in every field. Even the nation-builder-generation of architects is stampeded by younger ones. A new generation of managers and designers are 'sold' on the 'universality' of Western taste in architecture and design. Their idea of being up-to-date is to subscribe to western styles and tastes. They all assert that independent creative expression is confined to the expression of the nuances of place alone. The stage is set for the total homogenisation of the world according to the West. This is "the triumph of the West". If it is fashionable among the star architects of the West to ignore or pay lip service to the environment in their quest for hyper-subjective design, then non-west architects will follow suit too. They, of course, claim that any similarity is purely coincidental. Since style is referential, creative consideration of local realities is not a priority in the mind. The possibilities of an environmental aesthetic are thus obviated.

Fast track, cut-throat conditions of architectural practice in Asia do not favour any deep design considerations for program, function or form. Given this, eco-design considerations simply do not enter into the mental horizon. Ready-made design concepts selected on reputation and brand name obviate any need for critical discrimination or localisation. Under such conditions of practice, any breakthrough is a miracle.

Human Economy in a "Full" World

In an 'empty' world the eco-system can cope with the debris of the human economy. In a 'full' world, the eco-system is stressed to its limits. It will be hard put to supply the commodities and process the waste generated by an economy of six billion people. And this is not helped by human instincts of use and throw. Human judgements, instincts and cultures were formed in a relatively 'empty world'. Thus being local does not mean being global unless there is a sea change of attitudes. This is the dilemma of Asia's entry into the world arena. If Asia tries to succeed with the outmoded technology and consumption pattern, any hope for global environmental resuscitation would be foreclosed for everyone. It will help if there is a global environmental barometer to measure comparatively cumulative damages. Deficiencies in current accounting methodology also need review. "Externalities" in conventional economics must be accounted for in the "full" world otherwise a false picture is formed.

Noia and Paranoia and the Possibility for Substantive Dialogue

Methodology apart, human appetite is the major contributing cause of environmental degradation. It is thus necessary to understand the nature of the human appetite beyond needs. The starting point has to be an understanding of the dynamics of the human condition.

From its original state of undifferentiated consciousness i.e. bliss or 'noia', human consciousness increases through categorical differentiation we call knowledge. This increasing differentiation is experienced spatially and psychologically as separation and alienation. This is thus, an existential condition of anxiety we literally described as being beside oneself, i.e., 'para-noia'. The paradox accompanying this condition is a simultaneous longing for noia and a reaching out for ever greater differentiation. They pull in opposite directions. To assuage anxiety an artificial self is constructed at each stage of life to represent self. It is not self. It is persona. In other words it is identity. Maturation is achieved when paradoxically there is a certain release or abandonment of the constructed self. The matured person needs no other externally derived identity other than self-identity. If indeed this description is accurate of the maturation process, indeed this is the motor of history. It is that which drives economy and culture.

Sharing such a perception of human nature is deeply humanising if not humbling, necessary pre-conditions for proper dialogue untainted by presumed purity or superiority across peoples, cultures and faiths. Perceived this way, human dis-equilibrium and its impacts on the environment can be understood as phasic with each stage acted out characteristically but clothed differently depending on cultural and environmental contexts. But they are all essentially the same.

Identity

Culture, as a basis of identity, is a divisive form of politics when used as a barrier. The truly modern individual has the identity of a minority of one. Group identity, if exclusivistic, is an invitation for political exploitation. Architecture based on such identity is thus a tool of politics, not something to be exonerated in the "New World". The modern culture project must be emancipative. To do so it must be premised on a process of raising human capacity. This is the nexus between architecture and human development. The modern individual is an autonomous person whose communitarian instincts and affections are real and authentic. His or her compassion and empathy for others does not have to be obligated to any exclusivistic claim of any group. It is derived from trust gained through successful sharing. The aesthetics of identity should thus consist of the particularistic realities of each and every place and with no chauvinism about it. There are natural differences to be recognised and expressed as distinctive qualities but belonging to a shared universe of civil values.

Kitsch, the Narcotic of Change

Kitsch is a widespread phenomenon in all urban societies undergoing rapid change. Because kitsch is wasteful and callous of nature, it needs to be understood. The explanation that kitsch is linked to education or the lack thereof is insufficient. Kitsch results from the workings of existential anxiety. Kitsch is thus not to be interpreted as failures of culture, faith or civilisation.

Kitsch is pleasurable anaesthesia. It is thus a narcotic. Thus the typology of Asian metropolises are predictably similar given the trauma of rapid change they all experience. All city centres consist of steel and glass towers à la New York or Houston; the outer city is made up of kitsch condos and pseudo Mediterranean villas. While kitsch is a narcotic against the pain of change, terror is the other response. Its expiation is however different, it is through inflicting pain on the presumed perpetrators of the torment they endure. Terror is thus the counterpart of kitsch. The two are collaterally inverse.

Big or Small: How Much is Enough?

We cannot halt large-scale developments and only encourage small-scale interventions when pent-up needs are great. Social justice demands resolution in quick time. Politically, how will it then be possible to legislate incentives and disincentives to prevent the worse aspects of environmental degradation which affect bottom-line profits in countries eager for direct foreign investment to fuel the needed change? Equally, how can appetites be moderated on the demand side of the equation in a media space dominated by advertising? These are difficult questions given the dependency on global capitalism and the growth economy as the means to solve outstanding developmental problems. An ageing population in the developed world with dropping birth rates is a major factor in the next decade. Lowered western import will prompt a shift to local and regional markets. Export substitution may become a necessity. Weaning away from fossil fuel dependency is going to be traumatic. All these will have serious effects on globalised economies. Looked at positively, this will lead to greater local environmental consciousness and possibly stimulate a different eco-regional economy.

While some have advocated the "small is beautiful" strategy, and this is attractive for societies where the basic infrastructure has already been built, the reality is that the deprived masses of the non-west cannot wait for the promise of slow incremental gain. The proposition is in any case academic. China and South East Asia have forged ahead with their market-driven growth premised on global capitalism. Universally, the impact of science and technology cannot also be denied. There is no going back. Closing in is no option either. It is an invitation for old tyrannies to gain ascendance. The growth model has to be moderated. Presently it holds undisputed sway. To introduce needed modifications, civilisations and cultures must bring up to the contemporary surface all the reverence for life and nature they can find in their traditions and heritages. To do this, they must be willing to re-open the books that have been closed from critical review due to politico-institutional contingencies of the past or due to the grip of orthodoxy. The stewardship of nature enjoined upon all humans in sacred Hindu, Buddhist, Jewish, Christian and Islamic texts can together form the common moral basis for a new dialogue aimed at sustainability. The disappointment is that thus far such a dialogue has not taken place to any appreciable extent. This is an important project in the context of the environmental ethics of a "full world". The resolution of ethical issues cannot be bombed or terrorised into existence.

Re-Reading History

An analysis of western history since 1500 shows the rise of materialism with the corresponding decline of non-material values up through the industrial revolution to the present. Architecture is a faithful representation of the trends. Sometimes direct sometimes paradoxically. In my reading, consciousness of the contention between non-material and material values in the West was sparked by the crisis of the military-industrial complex experienced culminating in the first oil crisis of 1973. With it arose the anti-war movement, banning of the bomb, eco-feminism, and the hippy movement in the 'making of the counter culture'. If such a historiography is to be applied to the non-west, the suggestion is that this scenario will repeat itself. The pattern is already clear. The non-west is predicated on catching up with the West. The inference to be drawn is that a level of material satiation must first be attained, as it did in the West, before a value shift will occur. What the level of material satisfaction might be in the non-west is a security of livelihood question. But the definition is not in isolation. It will be also be defined by good local and Western lifestyle exemplars. Traditional Asian examples in the divestment of self have limited appeal for those climbing the ladder of success and they are the majority. Thus, the West has a great responsibility to set a good example of living well on less. It must also transfer the best non-polluting technologies to the non-west in their rapid growth phase to limit the damage to the eco-space of everyone. There must also be a global consumer movement to establish desirable standards for durable, low in embedded and operating energy, environmentally responsible products. Architecture must do likewise.

Towards a New Vision

Who is responsible in the big picture is less important. What to do is. We need to make a "New World" based on a better appreciation of the human condition. Examples of successful regional eco-repair work, eco-responsible buildings and settlements, co-housing, eco-villages and eco-urbanisation demonstration projects need to be commissioned and built in both the West and in the non-west. Architecture must go beyond eco-gesture. International sponsorship for good research and community embedded new model projects to counter the fast-track mode in all non-west building projects. What is clear is that thinking architects must lead the profession in a dialogue across the disciplines, with developers, city officials and financiers etc. to set a new agenda at least for eco-responsible demonstrative projects. Sadly, so many awarded 'green' buildings do not inspire the public or the profession being unattractive. The contribution of architects to the new environmental dialogue is our ability to visualise beautiful and fully adapted architecture.

This is a task for the best minds and the most sensitive of souls be they civic, artistic, philosophical or religious. Thinking on architecture must return from its hyper subjectivist detour to a spirit of accountable critical inquiry premised on deep moral commitment to all life. When design extravagance will be seen as anxious adolescent experimentation, and no more than this, it is a sign that health has returned and a public can respect the architect again. This is the professional dimension of the new environmental ethic: an invitation for professions to dialogue across civilisations, cultures, disciplines and the public. It is a new professional vision.

Specifically, better ways must be found for big to work with small. Big capital with small capital, big infrastructure with small infills. Many kinds of social, financial and administrative structures to accommodate many small kinds of small initiative. Architects, especially those in the context of locations undergoing rapid transformation, should conceive of extensive yet inclusive spatial geometries, which are capable of integrating large-scale eco-systems with many small incremental interventions in the creation of habitat. This is architecture not conceived as buildings but as infrastructure.

New Architects for a New World

Architects need to see themselves as artists whose egos are not separated from the entire living community of peoples, animals and plants. The exoneration of infantile excesses in design currently portrayed as creativity and high achievement is to be criticised if wasteful and polluting. Architects could become the 'Udangi' (traditional Balinese architect and master of sacred ceremonies) again. Architects should not only be building designers but social agenda facilitators and choreographers in the celebration of life, nature and the profound. Architects have to be among the 'wise persons' in society not the wild exhibitionists of wanton imagination. Every wanton Western architect breeds many more wanton non-western architects. Such is the asymmetrical nature of current culture-politics in an unequal world. The design and consumerist media presented as 'lifestyle' thrives on this.

Megacities: Aggregation and Transformation

Given the development equals growth paradigm, the dynamics of the rapid transformation of the non-west is characterised by Mega cities. Since cities are regarded as the most efficient engines of economic growth they attract capital and migrants. Within the historiography cited earlier, it is posited that a point in time on the trajectory of growth, when a level of material saturation is attained, is a critical moment. It is at that moment that a value change occurs. Non-material values come into contention with the dominant material values. Two critical issues need clarification. The first is what constitutes a definition of saturation. This is essentially cultural in nature and also linked to a perception of long-term security. Ideas can be shaped and livelihood and income distribution made more equitable. The second issue that is apparent is that the faster the arrival at that point the better, so that the painful part in infrastructure building is got over quickly. The caveat to this, of course, is that eco-efficient technologies be fully employed in the duration to control and limit the damage.

Conclusion

The task ahead is a mighty challenge to architects and thinkers everywhere, to imagine a new, quality-life-modern-human-culture with spatial and technological features that are compatible with nature and human nature. It is not a utopia but an entopia, one that is achievable. If the kind of energy architects are capable of in pursuing imaginative and exuberant designs should be applied to eco-responsible design thinking, much headway can then be made with clients and governments.

The purpose of development must finally be about building human capacity. A participatory system of building habitat has to replace the technocratic regime of 'robocrats' and technocrats who presume to do it for people. This requires architects to assume a new role as facilitator besides formulating designs. To build big structures they conceive, they must allow for many small interventions by disparate groups and individuals. The old totalised solution is not in sync with 21st Century Information Culture everywhere. Buildings must be designed for changeability and adaptability so as to last long. Architects have also to be masters of ceremony in the building of homes, civic places, industries and communities. In other words, they must help shape a new development and spatial economics that is premised on the sustaining of all life forms. In anticipating such a move, architects must become active dialogue partners with thinkers in other disciplines and active in enlightened community formation.

Architects must articulate a new spatial system that facilitates human life but which co-exists with nature in an economy that has close-to-zero impact on the eco-system starting from now.

Konrad Adenauer Foundation, Berlin Architects: Van den Valentyn Architektur

Holism and the Sacredness of Shelter A Southern African Experience

Llewellyn van Wyk Philadelphia

If ever there was any doubt, the events of the past 11 months have confirmed that the defining issue of this century is going to be social sustainability. However, of the triple bottom-line accounting approach to sustainability, this is the one aspect that is the most difficult to define. It is also the one aspect that rests most properly with the architectural profession with regard to the built environment. In line with the UIA Accord, it is architects alone among the built environment professions who have to satisfy the cultural aspirations of communities in human settlements.

Architecture has historically been a fundamental component of the human need to find meaning to existence: it follows therefore, that a socially responsible architecture will be able to assist its users to find and add meaning to their lives. There rests therefore a very real obligation and duty on contemporary architects to review their design methodologies against the need for social relevance. In my view, parachuting – where the architect drops a predetermined form onto a site – cannot satisfy the requirement for social justice.

The urge to find meaning for one's existence goes back to the first peoples as is evidenced in all the wall paintings of early man. Their paintings transformed the caves from natural 'spaces' to cultural 'places'[1]. Contemporary Man is living in an era in which more than ever people are asking philosophical questions about identity and reality. How do I experience it? As Slavo Zizek[2] puts it "Today everyone can imagine the end of the world, but nobody can imagine a different social order. Although we live in an era of allegedly free choices, they are often vain ones – like the choice between Pepsi or Coke."

One of the most obvious consequences of globalization has been its impact on local building traditions. Joe Aranha[3], a Fulbright Professor from Texas Tech University, has noted that the major forces of change during the 20th century, such as the universal spread of the modern movement together with the more recent phenomenon of globalization, have affected traditional built environments and have tended to crush regional identity throughout the world.

Our own experience in southern Africa has confirmed this. More recently, these losses of traditional building culture and environment together with the placeless character of contemporary "global" architecture have initiated local and foreign architects working in southern Africa to search for ways to re-establish regional and local architectural identity. As Laurens van der Post has put it "without history we have no meaning."[4] For most societies the notion of heritage and the continuity of tradition in architecture is a complex one that requires careful consideration when viewed in terms of the forces of modernization, globalization and urbanization. The argument focuses on regional identity and architectural authenticity in the context of imitative versus derivative approaches to the carrying forward of tradition in the built environment. This approach measures an architect by his ability to translate the social values of the community into the building fabric. As Ivor Prinsloo[5] puts it "In the southern African context, an architecture of experience would make reference to the ultimate users themselves and focus on their emerging experience as a totality."

In our contemporary world, and given the population numbers of our cities, it is not possible for each person to undertake the construction of his shelters. So it is no wonder that the rapid expansion of cities and villages has lead to a breakdown of social structure and health standards. The current forces of globalization has the power to further aggravate the separation of the individual from an 'architecture of experience'.

The South African Institute of Architects has just completed the review of submissions for their Award of Merit and Award of Excellence Programmes and it has been both rewarding and refreshing to note that, unlike many of our contemporaries overseas, our meritorious work is not reflective of the new Internationalism, or what Hans Ibelings[7] has labelled 'supermodernism'. The key notion of that philosophy is to deliberately disconnect architecture from the cultural and physical landscape: in this way, it becomes global and a-contextual.

Professor Errol Haarhoff[8] has noted that one of the characteristics of the architecture of the southern hemisphere is its continuing validation of the significance of cultural and physical landscapes in defining architecture, irrespective of any other design agendas. The physical qualities of Africa have much to do with this ability: no one in this country can be untouched by the vast spaces, the quality of the landscape, or the brilliance of the light. 'Students of human psycho-dynamics demonstrate that all of humankind's actions and interactions, both conscious and habitual, are based on the way in which information is processed from the environment. They acknowledge that whilst the foundation of this processing is intrinsic to human nature, it is substantially subject to conditioning. It follows that our information processing is learned'.[9]

One of the negative impacts of Globalization and Colonialism – the forerunner to Globalization – is that the indigenous pattern of information processing learned from one's predecessors over generations, is sublimated to a new processing order that is hardly understood and a-contextual. Many of the indigenous people of southern Africa have consequently forgotten key components of their cultural traditions and are having to relearn those concepts from the few Elders who remember them. However, and here is the good news, the process of re-learning has begun.

Holism

It is important that this re-learning extends beyond the borders of our regional experience. According to Laurens van der Post[10], "the Bushman was a walking pilot scheme of how the European man could find his way back to values he had lost and he needed for his own renewal". This renewal is dependent upon us recovering reverence for all forms of created life.

General Jan Smuts, one of South Africa's great sons, described his philosophy of Holism as the tendency in nature for disparate elements to coalesce in greater wholes – the result of which is more than the sum of the parts. This functions at the basis of chemistry, biology, all living organisms including the individual being and his relationship to society, the elements of his character and social character, nations, the planetary system and culminating in supra-mundane reality – the essence of which underpins the divine character of man. Holism is a function of the viability of relationship (in itself requiring the capacity to hold the greatest picture view) as an evolutionary potential and is an ever-growing progression of creativity into more sustainable wholes. It is process orientated rather than systemic.

Many other contemporary thinkers shared this notion of interdependency and integration. According to van der Post[11], "…to me, every insect and animal somehow is part of the Word made life, made flesh, made real. It is the word in action. Everyone and everything carries this charge. It carries, as it were, a signature of creation. Nothing is unsigned in life: a great seal is stamped on great and small. The Bushman felt known wherever he went. The stars knew him. He felt known. Wherever he went he felt he belonged. But modern man has lost this feeling of being known". Their experience is also a warning to us. "If we do not recover this sense, nature will turn on us one day, and we will be eliminated as the Bushmen were eliminated."[12]

You cannot but be aware that in Africa it is not Man that is in charge, but Nature. If you come to a river in flood, you have to just wait for the water to subside before you can cross. Out of his experiences in Africa, Jung – a close personal friend of van der Post – developed his theory of 'collective unconscious' in man. In all human beings, he argued, there is such an area in which the whole life participates mystically. People in Africa, he felt, were in touch with it and realized that the meaning of their lives depended on their not losing touch with the collective unconscious. It is from this collective unconscious that all our greatest energies come, all the patterns that give meaning. The Africans had ceremonies and dances, music and stories to express this, and ritual to contain it.

The impact of Colonialism on this 'collective unconscious' can be measured in the response Jung received from an Elgonyi man when he asked him what great dream he had, to which the man replied; "I'm afraid we don't have great dreams anymore now – the District Commissioner has them for us instead.[13]" That is the curse of modern life – it is so easy to let others dream for you. And how easy it is for architects to dream for others.

The Lessons Learnt

Applying the philosophy of Holism to the act of building would imply that structures that are holistic would include respect, even reverence, for the multi-dimensional evolving potential of the human experience way beyond the mere material and poetic experience. It responds to all the physical, social, emotional and spiritual needs of the inhabitant. In giving consideration to implementing the principles of Holism into shelter, the following design principles emerge as worthy of consideration:

- Ergonomics of the shelter including respect for privacy of individuals and facilitation of gathering.
- Ergonomics of social interaction within the community facilitating multi-cultural meeting and expression and social enrichment.
- Expression of own culture in art and all forms of creativity and openness to enrichment.
- Expression of own culture in worship taking note of symbolic aspirations and having regard to the encouragement of openness to novelty.
- Participation in construction, even though this may not be in the physical act itself but through consultation during the design process.
- Opportunities to exercise human interaction with the natural environment, through, for example, gardening, so that one may be constantly recreating and developing one's environment.
- Economic considerations, viabilities and the capacity for growth and innovative development.
- Ecological considerations with inherent flexibility and offering a natural synthesis with the environment.

The Lessons Applied

Viewed from this perspective, the architect has an awesome role in facilitating this thrust towards wholeness. It begins by an understanding that the structure is more than just a defence mechanism from the elements. The notion of building has a deeper significance than the mere creation of use: the 'act' of building together with the 'meaning' of shelter has a spiritual value that is embedded in both the structure and the environment. In African philosophy, any act that sustains life is part of the divine: hence the mother's breast is considered the first god that a newborn experiences. A building as shelter forms an essential part of sustaining the life of the individual, the family and the community. As such, it is recognised as an integral part of the sacred whole. In essence, the structure creates divine imminence.

Peter Rich, recipient of numerous architectural awards in South Africa, has done much research over many years into the rural dwellings of the Southern Ndebele ('mapogga') and the Bantwane. His work embodies the following design notions learned through his reinterpretation of the architecture of the Mapogga

· Formality and Open-ended-ness. The formal layout has an open-ended-ness, which deals with asymmetry where the need arises.
· Social organization. The planning of the structure is expressive of the social order. It takes cognisance of the needs and demands of the occupants, recognising that this will change over time.
· The Village concept. Ndebele homesteads are at once both house and village. The grouping of the huts is the result of incremental growth, but each addition makes the most of creating new spaces of equal quality. Rich puts it thus: "A house is not a home unless it is a village."
· The Outdoor Room. The buildings are placed to form enclosing loose fitting outdoor rooms and courtyards. The internal and external spaces act as one. The outdoor spaces also attenuate climatic conditions either trapping heat in winter or cooling breezes in summer.
· The Entrance Threshold. The transition from public to private space is clearly defined wherever it occurs. Entrance thresholds are extended in many ways. This includes designing the formal entrance route – the garden, the entrance court, and the front door – differently from the daily routes. The Molatedi Community, for example, has 39 different words to describe a threshold
· The Implied Diagonal. In Mapogga Homesteads the unit grows to the back or on an implied diagonal and therefore forward. The sculptured embellishments to the courtyard's corners serves to enhance this visual diagonal. The Mapogga are able to achieve an implied diagonal within an orthogonal grid without the awkwardness of the rigid 45 geometry of much contemporary architecture.
· Form. In Mapogga architecture the house form is kept simple: any adornment is restricted to the walls and the garden.
· Landscaping. The dwelling becomes a stage set to be viewed from outside. Similarly, an awareness of the outside is created from inside by framed views.
· Raw Architecture. The structure of the dwelling reveals how it was made and of what it was made.

Sustainability ultimately must imply more than the judicious use of resources: for our world to survive its component parts need to be re-united into a cohesive and interdependent whole. Social sustainability has to do with developing those components that sustain life and provide meaning. We will achieve social relevance when we create architecture of experience that goes beyond our experience.

1 Ouzman, Sven. An Account of Three Bushman Rock Paintings Sites in the Eastern Free State of South Africa. Rock Art Department, National Museum

2 Zizek, Slavo. Senior Researcher Institute for Social Studies, Ljubljana Slovenia: Time Magazine Special Issue Summer 2000/1

3 Aranha, Joe. Uses and Abuses of Tradition. University of Science and Technology Zimbabwe

4 van der Post, Laurens. (1980) A Walk with a White Bushman

5 Prinsloo, Ivor Architecture SA Autumn 1981

7 Ibiling, Hans. Supermodernism: Architecture in the Age of Globalization. Rotterdam. NAI 1998

8 Prof. Haarhoff, Errol. Kz-Nia Journal, Durban. 2001

9 van Wyk, Claudius. Holism and Holistic Settlements in Africa. IMSA Journal, South Africa

10 van der Post, Laurens. A Walk with a White Bushman

11 Ibid

12 Ibid

13 Ibid

Reichstag, The new German Parliament, Berlin Architects: Foster and Partners

The Innovation of Innovations: Works of Architecture in 2030

Richard Kroeker Halifax

What works of architecture will most change the shape of our cities in the next decades? The year 2030 isn't very far distant. It is an easy assumption that it will bring us more of what we currently face. It is more interesting to consider the particular milieu in which innovation will have the most profound effects, and to look for the quarter from which innovation is most likely to come. Leapfrogging technological development might crowd the headlines and publication lists in the next few decades, but the fundamental needs will change not a bit for the majority of the earth's human inhabitants. Innovation, as discussed in the context of this kind of gathering, must resist the impulsive search for novelty, rather looking for design intelligence to emerge where it matters most. Habitation, food, water, and cultural survival will become ever more acute concerns for much of the world's population. Architects who choose to be involved in communities where the needs are most basic and most acute will be confronting new issues that will profoundly shape the physical and political context. Designers will be called upon to consider issues beyond the narrow context where Architecture is an activity for the economic elites driven by excess capital. The most significant and interesting innovation will emerge where it is a necessity, not a luxury.

The greatest concentration of design activity and critical reflection is currently engaged in those sectors where the issues are the least critical? The manipulation of desire for the purpose of directing and controlling consumption currently occupies far too many of our best minds. This milieu values novelty is over innovation, and the two become easily confused. The next few years will require that our brightest, most innovative designers be involved with issues directly related to the daily lives of communities whose fundamental preoccupation is with survival. It is our task as educators and design leaders to ensure that an emerging generation of innovative designers has a proportional sense of the issues and their potential to address them. They should not be lulled into accepting the role of tertiary service providers for the elite, competing with each other for stakes that are far too low; where the concentration of intelligence and expenditure of human energy is entirely misplaced. Innovative work, by definition, will emerge from the margins. The current state of affairs puts the majority of the world's population on the margins. How much more vital for talented designers to be involved in addressing fundamental issues of collective existence in situations where new ideas are urgently needed, to immediate and profound impact.

The next decades of this century could increasingly mean a shift from looking at global influences to looking at the tribe and the family. The focus will need to be with children, with women, and with the very aged; in any culture with a long view, these are always the most powerful forces. They remain powerful forces, even when excluded from the institutions of political control. Corporate headquarters and national capitals will be increasingly irrelevant as architectural condensers or architectural symbols, lacking the meaning and means to provide the needed communication links. The village theatre, in whatever form it is variously defined, will increasingly replace these. The inherent risks of focusing at this scale have to do with communication. The last century should have taught us that international movements are no substitute for effective cultural exchange. They may have a neutralizing, even unifying, effect on cultural forms, but not on tensions between societies and interest groups. Innovation will come from those places where it is most needed, where the stakes are highest. Those engaged, as architects, in the making of culture will be needed most in the context where the issues are most acutely experienced. The context has long existed, but will be in sharper focus in the next few decades. Communication between cultures will be most productive in terms of innovation where the conversations are not one sided.

Changes are inherent in any living culture. It is innovation that will ensure that they are not catastrophic. Marginalized cultures are prone to being treated as irrelevant: completed, finished, consigned to history. In this context, innovation is sometimes seen as inappropriate distortion of a received artifact. Innovation is the acknowledgement of the ongoing cultural viability of a society. Innovation is emergent, not imposed. Innovation addresses fundamental issues; it is radical. In other words, it is rooted.

Architectural innovation in the next decades will continue to look to our place in the natural world. The most progressive buildings, cities, and communities will be those which are responsive to their immediate impacts on the natural context, and increasingly draw their support from this context. The precedents for doing this have always been there in natural history and in cultural history.

Negative inertia will continue to come from the vested interests. Vested interests by their nature resist fundamental change as long as possible. When they do change, it is often only in the face of a crisis, and they bring huge resources and intelligence to bear in making those changes. As a result, the shifts are often convulsive. We have brought ourselves to the brink of our own destruction as a species at the beginning of this century. Hopefully we have put the darkest aspects of what human activity can offer behind us. The world has been through a century of sporadic and convulsive changes at the hands of concentrated vested interest groups.

The spectacular increase of available information over the past few decades has been mirrored by a similarly rapid information loss. Languages, scientific knowledge, mythology have disappeared at an unprecedented rate from indigenous cultures and marginalized societies. It is cross-cultural influences that usually lead to innovation in any given group. If continuing developments in communication technology can create reciprocal relationships rather than hegemonic ones, then innovation will ensure cultural continuity. A culture defines itself and its history by what it dreams as much as by what it remembers. Cultural retrieval in the decades ahead will be as much a matter of dreaming as it will be a matter of documentation and remembering.

If Architecture can be innovative in addressing basic aspects of existence for communities who need it most, it will assume its proper role as the most fundamental of all cultural activities. It is part of a survival impulse and a poetic impulse that addresses and expresses all aspects of our biological and spiritual life. Myths of the Mi'kmaq people of Atlantic Canada tell us that we learned the art of building from the birds. We will continue to draw technique and inspiration from our changing understandings of nature and our place within it. The creation of Architecture is the result of a fundamentally optimistic impulse. This optimism shouldn't be misdirected to an acceptance of a mediocrity or the inevitability of the current condition. We will make increasingly critical choices as societies over the next two decades, and it is part of the architect's role to provide the visions that will help direct those choices.

Designing and Building in the Tropics

Bruno Stagno San José

Resource Architecture

The general theme of this congress, Resource Architecture, includes energy and culture as design resources. Resource Architecture calls for a response as to how architecture, urban design, and landscaping, can contribute to a global development strategy that highlights cultural values and respects diversity in its quest for sustainable constructions and habitable environments. The application of such standards should strengthen the so-called marginal cultures and initiate a universal dialogue that leads to an accentuation of identity and increased comprehension and tolerance.

I am an architect who works in the tropics, in a developing country, and I will answer the questions posed by the Congress focusing on the urgent need to strengthen the architectures of the different latitudes as a step towards a global cultural equilibrium of respect. I will concentrate on two design resources that are determinant to tropical architecture. The first is cultural, as it refers to the need to renovate the conception of space in our contemporary "dish-culture" architecture. The second is more architectural, in that it regards shade as the great qualifier of tropical architecture and as its best resource of passive energy.

Contemporary Space as a Design Resource

The four-dimensional conception of time and space that characterized Modern Architecture has changed. Nowadays, in our global culture of communications ("dish-culture"), the sense of space is different, for through the windows of television, e-mail, internet, and other instruments we can navigate in virtuality, transport ourselves to other latitudes, and thereby enrich our own immediate reality. When we are before these screens we communicate with a different reality that can be effectively added to our own. This juxtaposition of realities obviously has a direct effect on our understanding of architectural space.

This entirely novel possibility has an enormous potential as a design resource. Not only has it come to characterize our present period and actualize our sense of space, it has in practice superseded the time-space conception inherent to Modern Architecture. Today, we are accustomed to the juxtaposition and cohabitation of parallel realities, creating in us a special and novel sense of association. Just as the early 20th century modern concept of space-time generated new architectural forms that remained valid for many decades, for some years now, technological advances and cultural changes have increasingly demanded a new definition of space. In the history of architecture, it is once again necessary to consider what we understand by space. A cultural redefinition of space is more effective and representative for the introduction of the present time in architecture than the pretended generalized introduction of vanguard construction technologies in the planet, for quite simply, not everyone has access to them.

If we try to find a meaning to architectural space in the 21st century, it would be interesting to determine what role generalized access to virtual worlds has played in its characterization. Some decades ago, reality was statically defined as a mere reflection of ourselves (mirror) and later on dynamically, as a projection towards other horizons (window). Today, I perceive it as a mirror that is simultaneously a window, that is, as a mirror with an opening in the center through which one can see the distance. In such a mirror we are at once reflected in our reality while we can see other horizons through the window, that is, we are synthesizing simultaneous sensations with ambiguous spatial situations.

With the mirror-window metaphor, I have wanted to express the existence of a virtual timeframe that coexists with the actual timeframe and the relativity of the spatial dimensions, as the proximate space is redimensioned and juxtaposed with the distant space. Time ceases to exist when we are reflected in the mirror and simultaneously seeing through the window, we simultaneously exist in two worlds without moving. That is to say, we are immersed in an ambiguous spatial situation in which there is continuity of the space before the mirror with the space behind us and with the distant space framed by the window. This continuous space makes belonging relative while bestowing it with a connotation of plurality.

We can conclude that our situation is governed by the concepts of simultaneity and ambiguity. If we translate this reflection to architecture, we can conclude that this spatial continuum has limits, but these limits are wide and fuse the different contexts. Ambiguity arises from experience as a result of the assimilation of other identities and simultaneity from the perception of different images at the same time. In architecture, the ambiguity of space and the simultaneity of time combine into a new experience that has the potential to create an atmosphere of enormous wealth in which man can feel immersed and his spirit stimulated with ineffable sensations. Today more than ever architectural space becomes "the space of all dimensions".

Reflections, difficult to capture in photography, are images of a building or a surrounding that are projected onto shiny surfaces and that enrich the experience of space, saturating it with unusual images. This astonishing experience of multiple visions fills the space with subtleties and incorporates the ambiguous and the simultaneous, constituting an interesting design resource in architecture. Conceived as such, architectural space allows for the opportunity to create buildings without solemnity because its areas are not static nor symmetric, and do not have a single reading because the limits are at times virtual planes of light or shade. These spaces shelter, are bright yet lie in the shade, focus the landscape, and juxtapose a reflection onto other images. In synthesis, these are spaces that summon alternatives and sum up possibilities.

In the Light of Shade

The sun is the most abundant resource in the tropics. Diverse technologies have successfully harnessed its energy potential. However, as a design resource, I believe it is shade more than light that truly defines architecture in the tropical latitudes.

Light has been a fundamental element in architecture. Although it is true that light can shape space, it is unquestionable that shade does too, particularly in the tropics, where the shade summons and generally conditions behavior. Just as light is a prized source of warmth in certain latitudes, in the tropics it is the coolness inherent to shade that allows for action.

In the dry hot latitudes, such as in North Africa, man creates closed spaces that are embraced by walls to conserve the cool interior. The opposition of light and shade is dramatically expressed in these architectures of walls. The spatial relation between the interior and the exterior is limited to a few apertures in the thick walls through which the light penetrates, revealing the thickness of the walls. In such cases, the border between the interior and the exterior has depth. On the other hand, in the architecture of Mies van der Rohe, for example, in which the horizontal planes of the roof and floor stand out, the space slides towards the exterior and towards the light. In these architectures, the limit is the difference in clarity.

Just as architecture in dry climates is of walls, in the tropical latitude it is of roofs. The tropical roof is a parasol that fulfills a particular function. The overhang extends far to provide protection from the sun and the rain. In the tropics, to achieve well-being it is necessary to ventilate the shade to lower the temperature and reduce the humidity. The heat inherent to light combined with the humidity produces such lethargy that only ventilation and shade can neutralize. For this reason, walls are eliminated and the openings increased to allow the breeze to move without hindrance. Because it allows for well-being, shade thus becomes the principal source of passive energy.

Open spaces, covered by large roofs, are interesting due to the mix of diffuse light and shade that are an essential aesthetic factor in these buildings. In them inhabit the depth, the reflections, the veils, the chiaroscuros, and the attenuated clarities that are specificities of the shade.

The transition from the intense and crude light of the exterior and the shade of the interior via the eaves, pergolas, and other elements produce an intermediation of chiaroscuros and create an interior shade that is perceived as a shade surrounded by half shades. These half shades are a limit that has depth and width, and of considerable dimensions with its universe of lights, shades, and experiences, becoming an important theme of design to resolve the transition between inside and outside. Furthermore, this transition is an opportunity to characterize tropical architecture beyond the traditional umbrella and convert it into a truly architectural entity. It is in this space where the building is refreshed, where the light is subdued, where the shade invades the apertures and where the breeze improves ventilation.

These limits are spaces in which a baroque game of half shades in transparency predominates. These ambiguous and simultaneous architectural entities interpose between the lighted exterior and the shaded interior. The shade as an ever-present need, with the ambiguous and simultaneous space as an update, are design resources that in unison with ventilation, roofs, eaves, and vegetation, grant tropical architecture its identity and character.

UIA Berlin 2002

XXI World Congress of Architecture

Thank you to all our sponsors and patrons for their kind support

Main Sponsors

GRAPHISOFT® · autodesk® · Deutsche Post World Net – MAIL EXPRESS LOGISTICS FINANCE · SCHÜCO INTERNATIONAL

Co-Sponsors

HOCHTIEF · Schindler · bluepool AG · Audi

Media Partner

BertelsmannSpringer bauverlage

Patrons

CAPAROL · Die Bahn DB Immobilien · Vivico REAL ESTATE · se'lux Semperlux Aktiengesellschaft · media spree · GROHE WATER TECHNOLOGY

vitra. · a-matter architecture and related · hebel Porenbeton

Official Airline

Lufthansa

INDEX
CHAIRPERSONS AND SPEAKERS OF THE XXI WORLD CONGRESS OF ARCHITECTURE

VERZEICHNIS
MODERATOREN UND REDNER DES XXI. ARCHITEKTUR-WELTKONGRESSES

Chairpersons

Gunnel Elisabeth Adlercreutz
Dr. h.c., Architect SAFA
Finnish Building Information Foundation
RTS, General Director
Lives in Helsinki, Finnland

Johannes Busmann
Dr., Publisher
Edition Müller and Busmann,
Managing Director
Lives in Wuppertal, Germany

Bruno Flierl
Dr., Architectural Critic and Historian
Lives in Berlin, Germany

Donald Hackl
Architect, FAIA
Loebl Schlossman & Hackl
Lives in Chicago, USA

Ikmal Hisham Albakri
Dr., Architect APAM, MSIA, RIBA
Kumpulan Akitek SDN BHD,
Executive Director
Lives in Kuala Lumpur, Malaysia

Manuel Cuadra
Dr., Architect and Publicist
Hesse Branch of Bund Deutscher
Architekten BDA, Secretary
Lives in Frankfurt/Main, Germany

Kay Friedrichs
Dr., Architect
Lives in Aachen, Germany

Joseph Hanimann
Journalist
Frankfurter Allgemeine Zeitung,
Cultural Correspondent
Lives in Paris, France

Franz Alt
Dr., Author and Journalist
Lives in Baden-Baden, Germany

Cynthia C. Davidson
Architectural Critic
Anyone Corporation, Director and Editor
Lives in New York, USA

Burkhard Fröhlich
Architect and Journalist
Deutsche Bauzeitschrift - DBZ, Chief Editor
Lives in Gütersloh, Germany

Falk Jaeger
Prof. Dr., Architectural Critic
Bauzeitung, Chief Editor
Lives in Berlin, Germany

Alexandru Beldiman
Architect
Union Internationale des Architectes (UIA)
Vice President
Lives in Bucharest, Romania

Taoufik Ben Hedi El-Euch
Architect
Union Internationale des Architectes (UIA)
Second Vice President
Lives in Tunis, Tunisia

Joaquim Guedes
Prof. Dr., Architect and Urban Planner
University of São Paulo
Lives in São Paulo, Brasil

Gert Kähler
Prof. Dr., Architectural Critic and Publicist
Lives in Hamburg, Germany

Heinz Kull

Dr., Urban and Regional Planner,

Senior Expert

Lives in Berlin, Germany

Thies Schröder

Landscape Architect, Author and Journalist

ts redaktion, Director

Lives in Berlin, Germany

Heinrich Wefing

Dr., Architectural Critic and Art Historian

Frankfurter Allgemeine Zeitung - FAZ,

San Francisco

Lives in Belvedere, CA, USA

Philipp Meuser

Architect BDA and Journalist

Stadtforum Berlin, Coordinator

Lives in Berlin, Germany

Petra Schwarz

Journalist

Lives in Berlin, Germany

Frank Werner

Prof., Architectural Critic and Architect

Bergische Universität Gesamthochschule

Wuppertal

Lives in Wuppertal, Germany

Yannis M. Michail

Dr., Architect, Urban Planner

Deputy Ombudsman for the Quality of Life

Lives in Athens, Greece

Alejandro Sokoloff Moreno

Architect

Union Internationale des Architectes (UIA)

Vice President

Lives in Bogotá, Columbia

Gerwin Zohlen

Architectural Critic and Publicist

Lives in Berlin, Germany

Jean-Claude Riguet

Architect

Union Internationale des Architectes (UIA)

General Secretary

Lives in Paris, France

Carola Wedel

Journalist

Zweites Deutsches Fernsehen - ZDF

Cultural Correspondent

Lives in Berlin, Germany

Speakers

Iñaki Ábalos and Juan Herreros
Architects
Ábalos & Herreros
Live in Madrid, Spain
www.abalos-herreros.com

Arrouf Abdelmalek
Architect
University of Biskra
*1967 in Khenchela;
lives in Biskra, Algeria

Dirk Althaus
Prof. Dr., Architect
I:KEB Institute for Cyclical Economies
and Energy Efficiency in Building
*1940 in Hannover, Germany
and lives there
althaus@ikeb.de

Silvia Arango
Prof., Architect and Historian
National University of Columbia
*1948 in Bogotá, Columbia
and lives there

Kaoko Aromaa
United Nations Center for European Crime
Prevention and Control (HEUNI), Director
Lives in Helsinki, Finland
kaoko.aromaa@om.fi

Rasem J. Badran
Architect
DAR AL OMRAN –
Architects and Consulting Engineers
*1945 in Jerusalem, Israel;
lives in Amman, Jordan

Achraf Bahri-Meddeb
Prof., Architect and Urban Planner
National School of Architecture
and Urbanism
*1955; lives in Tunis, Tunisia
rm@comete.com.tn

Milica Bajic Brkovic
Prof. Dr., Architect
International Society of City and Regional
Planners - ISoCaRP, Secretary General
Lives in Belgrade, Yugoslavia
www.isocarp.org

Shigeru Ban
Architect
Shigeru Ban Architects
*1957 in Tokyo, Japan
and lives there

Jairo Aldemar Bastidas Gustin
Architect
University of Brasilia
*1971; lives in Brasilia, Brazil

Stefan Behling
Prof., Architect
University of Stuttgart
*1962 in Madison, USA;
lives in Stuttgart/London, Germany/UK
www.uni-stuttgart.de

Daniel Biau
Dr., Civil Engineer
United Nations Center for Human
Settlements (Habitat),
Deputy Executive Director;
lives in Nairobi, Kenya
www.unhabitat.org

Harald Bodenschatz
Prof. Dr., Urban Sociologist
and Urban Planner
Technical University of Berlin
*1946; lives in Berlin, Germany
harald.bodenschatz@tu-berlin.de

Kurt Bodewig
Federal Minister of Transport,
Building and Housing
*1955 in Rheinberg;
lives in Berlin, Germany
www.bmvbw.de

Gabor Bojár
Physicist
Graphisoft R&D Rt., Chief Executive Officer
*1949; lives in Budapest, Hungary
www.graphisoft.com

Konstantin Nikolov Boyadjiev
Dr., Architect, UAB
Bulgarian Academy of Sciences,
CAS Director
*1942 in Vraca;
lives in Sofia, Bulgaria
konstb@dir.bg

Ruzica Bozovic-Stamenovic
Dr., Architect
National University of Singapore
Born in Belgrade, Yugoslavia;
lives in Singapore
www.arch.nus.edu.sg

David Chipperfield
Architect
David Chipperfield Architects
Lives in London, UK
www.siulon.com/cindex.html
to be confirmed

Mohamed-Ayman Daef
Dr., Urban Planner
University of Assiut
*1959 in Assiut, Egypt
and lives there
daef@runbox.com

Wei Dong
Prof. Dr., Architect
Southeast University
*1957; lives in Nanjing, China
dongwei@seu.edu.cn

Gerhard Bremmer
Prof., Architect BDA
Hesse Chamber of Architects, President
*1936; lives in Wiesbaden, Germany
www.akh.de

Krzysztof Chwalibóg
Architect
Architectural Magazine "Architekt",
Chief Editor
Lives in Warsaw, Poland

Phillipp Daffara
Architect and Urban Designer
Maroochy Shire Council,
Principal Urban Design Coordinator
Lives in Mountain Creek Qld, Australia
daffarap@maroochy.qld.gov.au

Peter Droege
Prof., Architect and Urban Designer
University of Sydney
Born in Germany;
lives in Sydney, Australia
droege@solarcity.org

Michael Burt
Prof. Dr., Architect
Technion, Israel Institute of Technology
Lives in Haifa, Israel
architecture.technion.ac.il

Jean-Louis Cohen
Dr., Architect and Journalist
Institut Français d'Architecture, Director
*1949 in Paris, France
and lives there

Klaus Daniels
Prof., Civil Engineer
HL-Technik AG (Consulting Engineers),
President of the Board of Directors
*1939; lives in Munich, Germany
www.hl-technik.de

Werner Durth
Prof. Dr., Architect and Historian
Technical University of Darmstadt,
Faculty of Architecture, Dean
*1949 in Mengeringhausen;
lives in Darmstadt, Germany
gta-tu-darmstadt@web.de

Chris Butters
Architect
UIA Work Programme "The Road from Rio",
Director
*1950 in South Africa;
lives in Oslo, Norway

Peter Conradi
Architect
Federal Chamber of Architects, President
*1932 in Schwelm;
lives in Stuttgart/Berlin, Germany
www.bak.de

Vakhtang Davitaia
Prof., Architect
Academy of Architecture, Georgia,
Chairman
*1934 in Dzveli Senaki;
lives in Tbilisi, Georgia

Dietmar Eberle
Prof., Architect
B & E Baumschlager-Eberle GmbH
*1952 in Hittisau/Vorarlberg;
lives in Lochau, Austria
office@be-g.com

Peter Eisenman
Prof. Dr., Architect
The Cooper Union, School of Architecture
*1932 in Newark;
lives in New York, USA

Lord Norman Foster
Architect RIBA
Foster and Partners
Lives in London, UK
www.fosterandpartners.com

Georg Gewers
Architect BDA
Gewers Kühn & Kühn Architekten BDA
*1962 in Bevergern;
lives in Berlin, Germany
www.gewers-kuehn-kuehn.de

Michael Gordon
Urban Planner
City of Vancouver, Planning Department
Senior Planner
Lives in Vancouver/B.C., Canada
Michael_Gordon@city.vancouver.bc.ca

Barbara Engel
Architect
Brandenburg University of Technology
Cottbus
*1969 in Offenbach/Main;
lives in Cottbus, Germany
engel@tu-cottbus.de

Cettina Gallo
Prof., Architect
Italien National Agency for New Technology
(ENEA), Energy and Environment
Lives in Rome, Italy
gallo@sede.enea.it

Yury P. Gnedovski
Dr., Architect
Union of Architects of Russia (UAR),
President
*1930 in Sverdlovsk;
lives in Moscow, Russia
www.uar.ru

Donald Morrison Grant
International Federation of Surveyors (FIG)
Born in Bairnsdale;
lives in Sydney, Australia
dongrant@ozemail.com.au

Ugo Farinelli
Prof.
Italian Agency for Energy and Environment
(ENEA), Former Director for Energy
Lives in Lund, Sweden
www.lu.se

Karl Ganser
Prof. Dr. rer. nat.,
Former Director of IBA Emscher Park
Deutsches Architektur Zentrum DAZ
Chairman
*1937 in Mindelheim/Schwaben;
lives in Breitenthal, Germany
foerderverein@daz.de

Reinhard Goethert
Dr., Architect and Urban Planner
Massachusetts Institute of Technology (MIT)
Lives in Cambridge/MA, USA
architecture.mit.edu

Hou Hanru
Critic and Curator
*1963 in Guangzhou, China;
lives in Paris, France

Gilbert Felli
Architect
International Olympic Committee (IOC)
Sports Director
Lives in Lausanne, Switzerland

Meinhard von Gerkan
Prof., Architect BDA
von Gerkan, Marg and Partners gmp
*1935 in Riga/Latvia;
lives in Hamburg, Germany
www.gmp-architekten.de

Bart Goldhoorn
Project Russia, Publisher and Chief Editor
*1963 in Deventer, Netherlands;
lives in Moscow/Amsterdam,
Russia/Netherlands
www.prorus.ru

Newton Harrison
Prof. Emeritus, Architect
The Harrison Studio, Principal
Born in New York;
lives in Del Mar, USA
harrstudio@aol.com

Dieter Hassenpflug
Prof. Dr. phil. habil., Sociologist
Bauhaus University Weimar
*1946 in Harare, Simbabwe;
lives in Weimar, Germany
hassenpf@archit.uni-weimar.de

Peter Herrle
Prof. Dr.
Technical University Berlin
The Habitat Unit, Head
Lives in Berlin, Germany
www.habitat-unit.de
to be confirmed

Louisa Hutton
Architect, BA (Hons), AA Dipl, RIBA
sauerbruch hutton architekten
*1957 in Norwich;
lives in Berlin/London, Germany/UK
www.sauerbruchhutton.de

Natasa Kaspari
Civil Engineer and Urban Planner
*1973 in Nicosia;
lives in Nicosia, Cyprus

Manfred Hegger
Prof., Architect BDA
HHS Planer + Architekten BDA
*1946 in Korschenbroich;
lives in Kassel, Germany
www.hhs-architekten.de

Thomas Herzog
Prof. Dr., Architect BDA
Technical University of Munich,
Faculty of Architecture, Dean
*1941 in Munich;
lives in Munich, Germany
www.herzog-und-partner.de

Christoph Ingenhoven
Architect BDA
Ingenhoven Overdiek und Partner
Lives in Düsseldorf, Germany
www.ingenhoven-overdiek.de

Harald Kegler
Dr., Urban Planner
Laboratory for Regional Planning
*1957; lives in Lutherstadt Wittenberg,
Germany
www.dr-kegler.de

Andreas Gottlieb Hempel
Architect BDA
XXI World Congress of Architecture UIA
Berlin 2002, President
*1941 in Dresden;
lives in Berlin, Germany
www.agh-berlin.de

Brigitte Holz
Architect BDA, Urban Planner
Freischlad + Holz Architekten BDA
*1954 in Straelen;
lives in Darmstadt/Berlin, Germany

Helmut Jahn
Prof., Architect
Murphy/Jahn Inc. Architects
*1940 in Nuremberg, Germany;
lives in Chicago/IL, USA
www.murphyjahn.com

Derrick de Kerckhove
Prof. Dr.
University of Toronto,
McLuhan Programme, Director
Lives in Toronto, Canada
www.mcluhan.utoronto.ca/derrick.html
to be confirmed

Gunter Henn
Prof. Dr., Architect BDA
Henn Architekten Ingenieure
*1947 in Dresden;
lives in Munich, Germany
www.henn.com

Bernd Hunger
Dr. Dr., Sociologist and Urban Planner
Federal Union of German Housing
Cooperatives and Associations
*1953 in Apolda;
lives in Berlin, Germany
www.gdw.de

Andrei V. Kaftanov
Architect, Historian
The Union of Moscow Architects,
Member of the Board
Lives in Moscow, Russia
s.volkova@relcom.ru

Peter Kern
Prof. Dr.
Fraunhofer Institut für Arbeitswirtschaft
und Organisation (IAT), Managing Director
*1941 in Heilbronn;
lives in Stuttgart, Germany
www.iao.fraunhofer.de

Gottfried Kiesow
Prof. Dr., Historian
Deutsche Stiftung Denkmalschutz,
Chief Executive Officer
*1931 in Altgennin;
lives in Bonn, Germany
www.denkmalschutz.de

Mathias Klotz
Prof., Architect
Diego Portales University, Director
*1965 in Viña del Mar;
lives in Santiago, Chile
www.udp.cl

Rem Koolhaas
Prof., Architect and Urban Planner
O.M.A. Office for Metropolitan Architecture
Lives in Rotterdam, The Netherlands
to be confirmed

Lars Krückeberg
Wolfram Putz
Thomas Willemeit
GRAFT
Live in Los Angeles/Berlin, USA/Germany
www.graftlab.com

Unmesh Kirtikar
Architect
Kunjbihari and Sen Architects, Principal
Lives in Calcutta, India
unmeshkirtikar@caltiger.com

Christian Koch
Architect BDA
*1953; lives in Berlin, Germany
www.christian-koch-plus.de

Kaspar Kraemer
Architect BDA
Bund Deutscher Architekten BDA, President
*1949 in Braunschweig;
lives in Cologne, Germany
www.bda-architekten.de

Dogan Kuban
Prof., Architect
XXII UIA World Congress Istanbul 2005,
President
*1962 in Paris, France;
lives in Ankara, Turkey

Åse Kleveland
Former Minister of Culture
Swedish Film Institute, Managing Director
*1949 in Stockholm, Sweden
and lives there
www.sfi.si

Hans Kollhoff
Prof., Architect BDA
Swiss Federal Institute of Technology Zurich
*1946 in Lobenstein/Thuringia;
lives in Berlin, Germany
www.arch.ethz.ch

Michael Krautzberger
Prof. Dr., Ministerialdirektor
Department of Architecture and Urban
Planning, Federal Ministry of Transport,
Building and Housing, Head
*1943 in Reichenberg;
lives in Berlin, Germany

Friedrich Kuhlmann
Landscape Architect
Technical University of Berlin
*1967 in Giessen/Lahn;
lives in Berlin, Germany
friedger@t-online.de

Arnold Klotz
Prof. Dr.
City of Vienna, Planning Director
Lives in Vienna, Austria
www.wien.gv.at

Ulrich Königs
Architect
königs architekten
*1964 in Cologne, Germany and lives there
koenigs@baunetz.de

Richard Kroeker
Architect
Technical University of Nova Scotia
Lives in Halifax, Canada
www.tuns.ca

Manfred Kühn
Dr. rer. pol.
Institute for Regional Development and
Structural Planning, Erkner, Project Leader
*1960; lives in Erkner, Germany
Kuehnm@irs-net.de

Kisho Kurokawa
Prof., Architect, Theoretician and Lecturer
Kisho Kurokawa Architect and Associates, President
*1934 in Nagoya;
lives in Tokyo, Japan
www.kisho.co.jp

Luc Lévesque
Architect
University of Montreal,
School of Landscape Architecture
Lives in Montréal, Canada

Romuald Loegler
Dr., Architect
Atelier Loegler i Partnerzy
*1940; lives in Kraków, Poland
office@loegler.com.pl

Engelbert Lütke Daldrup
Dr., Urban Planner
City of Leipzig, Department of Planning and Construction, Head
*1956 in Kranenburg;
lives in Leipzig, Germany

Pablo Miguel La Roche López
Prof.
University of California Los Angeles
*1961; lives in Los Angeles, USA
plaroche@ucla.edu

Richard S. Levine
Prof., Architect
University of Kentucky,
Center for Sustainable Cities, Co-Director
Lives in Kentucky, USA
rlevine@uky.edu

Peter Lorenz
Prof. Dr., Architect
Peter Lorenz Architect and Partner
*1950 in Innsbruck, Austria and lives there
www.peterlorenz.at

Greg Lynn
Prof., Architect and Architectural Theoretician
Greg Lynn FORM
Lives in Venice, CA, USA
to be confirmed

Fernando Lara
Prof. Dr., Architect IAB
PUC-Minas University, Brazil
Lives in Belo Horizonte, Minas Gerais, Brazil
ferlara@pucminas.br

Diane Lewis
Prof., Architect, FAAR
The Irwin Chanin School of Architecture of the Cooper Union
Lives in New York, USA
www.cooper.edu/architecture

Maria Lubocka-Hoffmann
City of Elblag, Department of Urban Planning, Chief Executive Officer
Lives in Elblag, Poland
umelblag@elblag.com.pl
to be confirmed

Hang Ma
Architect
Shenzhen University
*1974 in Changchun;
lives in Shenzhen, China
chinamahang@eyou.com

Lykke Leonardsen
City of Copenhagen
Lives in Copenhagen, Denmark
lykke@keb.kk.dk

Jorge Francisco Liernur
Prof., Architect UBA
University of Torcuato Di Tella
Lives in Buenos Aires, Argenitina
www.utdt.edu

Regula Lüscher Gmür
Architect
City of Zurich, Department for Urban Planning, Deputy Director
*1961 in Basel; lives in Zurich, Switzerland
www.stadt-zuerich.ch

Winy Maas
Architect
MVRDV
Lives in Rotterdam, The Netherlands
www.mvrdv.archined.nl
to be confirmed

Christoph Mäckler
Prof., Architect BDA
Prof. Christoph Mäckler Architekten
*1951 in Frankfurt/Main, Germany
and lives there

Horst Matthäus
Dr.
Deutsche Gesellschaft für Technische
Zusammenarbeit (GTZ), Brazil
Lives in Recife, Brazil

Peter Mörtenböck
Prof. Dr., Architect SBA
Vienna University of Technology
*1966 in Ansfelden;
lives in Vienna, Austria
pmoerten@email.archlab.tuwien.ac.at

Peter Niehaus
Architect BDA
Siemens Real Estate GmbH & Co. OHG,
Executive Director
*1941 in Hanover;
lives in Munich, Germany
www.siemens.com

Stamatina G. Malikouti
Dr., Architect
Technological Educational Institute of
Piraeus (TEI)
Lives in Athens, Greece
mbedimarhio@ath.forthnet.gr

R. Larry Medlin
Prof., Architect
University of Arizona
Lives in Tucson, USA
rlmedlin@u.arizona.edu

Eric Owen Moss
Prof., Architect FAIA
Southern California Institute
of Architecture, Director
Lives in Los Angeles, USA
www.sciarc.edu

Peter Noé
Business Economist
Hochtief AG,
Member of the Executive Board
*1957 and lives in Essen, Germany
www.hochtief.de

Garrik Andreevic Maloian
Prof. Dr., Architect
Russian Academy for Architecture
and Construction
Lives in Moscow, Russia

Rahul Mehrotra
Architect
Rahul Mehrotra Associates
Lives in Mumbai, India
rahul@bom3.vsnl.net.in

Walter Nägeli
Prof., Architect BDA
Nägeliarchitekten
*1953 in Mainz;
lives in Berlin, Germany
buero@naegeliarchitekten.de

Crispino C. Ochieng
Dr., Architect and Urban Planner
Jomo Kenyatta University of Agriculture
and Technology
*1958; lives in Nairobi, Kenya
crispino@jkuat.ac.ke

Peter Marcuse
Urban Planner
Columbia University New York
Lives in New York, USA
pm35@columbia.edu

Steffen Mildner
Deutsche Gesellschaft für Technische
Zusammenarbeit (GTZ), Project Co-ordinator
Lives in Herrmannstadt, Romania
www.gtz.de

Julian Nida-Rümelin
Dr.
Federal Government Commissioner
for Cultural and Media Affairs
*1954 in Munich;
lives in Berlin, Germany
www.bundeskanzler.de

Philipp Oswalt
Prof., Architect and Publicist
Brandenburg University of Technology
Cottbus
*1964 in Frankfurt/Main;
lives in Berlin, Germany
www.oswalt.de

Frei Otto
Prof. Dr., Architect BDA
*1925 in Siegmar/Saxony;
lives in Leonberg, Germany
www.freiotto.com

Roberto Pirzio-Biroli
Architect
Studio for Architecture and Urban Planning
Lives in S. Margherita, Italy
r.pirzio.biroli@libero.it

Elena Rodina
Architect and Urban Planner
Lives in New York, USA
elenades@hotmail.com

Jurij Sadar
Architect
Sadar in Vuga Architekti
*1963 in Celje;
lives in Ljubljana, Slovenia
biro@sadarvuga.com

Dominique Perrault
Architect D.P.L.G., Urban Planner S.F.U.
Dominique Perrault Architect
*1953; lives in Paris, France
www.perraultarchitecte.com

Valerie Portefaix
Architect
map.office gutierrez + portefaix
*1969 in Saint Etienne, France;
lives in Hong Kong, China
www.map-office.com

Frank Roost
Urban Planner
Technical University of Berlin
Lives in Berlin, Germany
www.tu-berlin.de

Frank Samol
Architect
B.U.S. Urban Management and
Environmental Planning, Principal
*1953 in Heide;
lives in Dresden, Germany

Evgeni Pertsic
Prof. Dr.
Moscow State University
Lives in Moscow, Russia

Eckhart Ribbeck
Prof. Dr., Architect and Urban Planner
University of Stuttgart
*1942 in Oranienburg;
lives in Stuttgart, Germany
ribbeck@si.uni-stuttgart.de

Peter Ruge
Architect
Pysall. Ruge Architekten
Lives in Berlin, Germany
www.pysall-ruge.de

Matthias Sauerbruch
Prof., Architect BDA, AA Dipl, ARB
sauerbruch hutton architekten
*1955 in Constance;
lives in Berlin/London, Germany/UK
www.sauerbruchhutton.de

Günter Pfeifer
Architect BDA, DWB
pfeifer roser kuhn freie architekten
*1943 in Schopfheim;
lives in Freiburg, Germany
pfeifer.roser.kuhn@baunetz.de

Sonia Ricon Baldessarini
Dr., Architect
*1956 in Rio de Janeiro;
lives in Berlin, Germany

Alexander Runser
Architect
Runser.Prantl.Architekten
*1955 in Vienna, Austria and lives there.
a.runser@tuwien.ac.at

Wolfgang Schäche
Prof. Dr., Historian and Architect
BDA, DWB, DASL
University for Applied Science Berlin
*1948 in Berlin, Germany and lives there.

Herrmann Scheer
Dr., MdB, Economist and Social Scientist
EUROSOLAR e.V., President
Lives in Berlin/Bonn, Germany
www.hermann-scheer.de

Gerhard Schröder
Chancellor of the
Federal Republic of Germany
*1944 in Mossenberg;
lives in Berlin, Germany
www.bundeskanzler.de

Allan Shulman
Prof., Architect
University of Miami
Lives in Miami, USA
allan@shulmanarchitect.com

Werner Sobek
Prof. Dr., Architect and Consultant Engineer
Werner Sobek Ingenieure
*1953 in Aalen;
lives in Stuttgart, Germany
www.wsi-stuttgart.com

Jörg Schlaich
Dr.-Ing., Civil Engineer
Schlaich Bergermann und Partner GmbH
*1934 in Stetten i. R.;
lives in Stuttgart, Germany
www.sbp.de

Monica Schümer-Strucksberg
Berlin Senate for City Development
Lives in Berlin, Germany
www.stadtentwicklung.berlin.de

Elizabeth Sikiaridi
Prof., Architect
University of Essen
*in London, UK;
lives in Essen, Germany
sikiaridi@uni-essen.de

Tay Kheng Soon
Architect
Akitek Tenggara Singapore/Malaysia
*1940 and lives in Singapore
www.akitektenggara.com

Arno S. Schmid
Prof., Garden and Landscape Architect BDLA
International Federation of Landscape
Architects (IFLA), Past President
*1937 in Munich;
lives in Leonberg, Germany
www.iflaonline.de

Martha Schwartz
Landscape Architect
Martha Schwartz, Inc.
Lives in Cambridge/Ma, USA
www.marthaschwartz.com

Gerd Sippel
Deutsche Gesellschaft für Technische
Zusammenarbeit (GTZ), Jakarta
Lives in Jakarta, Indonesia

Albert Speer
Prof. Dr., Architect BDA and Urban Planner
AS&P Albert Speer & Partner GmbH,
Managing Director
*1934 in Berlin;
lives in Frankfurt/Main, Germany
www.as-p.de

Elmar Schossig
Architect BDA
Gatermann + Schossig und Partner
Architekten BDA
*1950 in Chemnitz;
lives in Cologne, Germany
www.Gatermann-Schossig.de

Vassilis Sgoutas
Architect
Union Internationale des Architectes (UIA),
President
*1934 in Athens, Greece and lives there
www.uia-architectes.org

Alexander Dimitrov Slaev
Dr., Architect
University of National and World Economy,
Sofia
*1960; lives in Plovdiv, Bulgaria
a.slaev@plovdiv.techno-link.com

Brian A. Spencer
Prof., Architect, AIA/IAA
Lives in Madison/Wisconsin, USA
basarch@aol.com

Thomas Spiegelhalter
Prof., Architect and Urban Planner
Carnegie Mellon University, Pittsburgh
Lives in Pittsburgh, USA
thomas@andrew.cmu.edu

Georgi Stoilov
Prof., Architect
International Academy of Architecture,
Sofia, President
*1929; lives in Sofia, Bulgaria

David Sutherland
Architect
Fender Katsalidis (Aust) Pty Ltd Architects,
Director of Planning
Lives in Melbourne, Australia

Christiane Thalgott
Architektin BDA, SRL
State Capital of Munich, Planning and
Building Department, Chief Executive Officer
Born 1942 in Wroclaw, Poland;
lives in Munich, Germany
plan.s@muenchen.de

William Henry Ssentoogo
Architect RIBA, Urban Planner RTPI
African Union of Architects (AUA),
President
*1936 and lives in Kampala, Uganda

Dörte Stollberg-Barkley
Dr., Architect
Southampton Institute
Lives in Southampton, UK
dorte.stollberg@solent.ac.uk

Leszek Stanislaw Swiatek
Dr.
Technical University of Szczecin
*1964; lives in Szczecin, Poland
akcent@akcent.com.pl

Sara Topelson de Grinberg
Architect
Union Internationale des Architectes (UIA),
Past President
Lives in Mexico City, Mexico
www.uia-architectes.org

Bruno Stagno
Architect
Institute for Tropical Architecture, Director
Lives in San José, Costa Rica
stagno@racsa.co.cr

Peter Strieder
Jurist
Senator for City Development Berlin
*1952 in Nuremberg;
lives in Berlin, Germany
www.peter-strieder.de

Ugur Tanyeli
Prof., Architect and Historian
Yldiz Technical University
*1952; lives in Istanbul, Turkey

Klaus Töpfer
Prof. Dr., Former Federal Minister
United Nations Environment Programme
(UNEP), Executive Director
*1938 in Waldenburg/Silesia;
lives in Nairobi, Kenya

Anne Stenros
Dr., Architect SAFA
Design Forum Finland, Managing Director
*1954; lives in Helsinki, Finland
anne.stenros@designforum.fi

Franz Sumnitsch
Architect
BKK-3
Lives in Vienna, Austria
to be confirmed

Necdet Teymur
Prof. Dr., Architect
University of Dundee
*1945 in Gaziantep, Turkey;
lives in London, UK
n@teymur.co.uk

Sigurd Trommer
Architect
City of Bonn, Planning and Building
Department, Chief Executive Officer
*1944 in Weißenfels/Saale;
lives in Bonn, Germany
www.bonn.de

Klaus Unger
Architect
Office of Public Works,
Assistant Principal Architect
*1947 in Dublin, Ireland and lives there
www.opw.ie

Frans Vogelaar
Prof., Architect
Academy of Media Arts Cologne
Lives in Cologne, Germany
vogelhaar@khm.de

Janet Ward
Prof. Dr., Historian
University of Nevada Las Vegas
Lives in Las Vegas, USA
www.unlv.edu

Minja Yang
Sociologist
UNESCO World Heritage Centre,
Deputy-Director
*1952 in Tokyo, Japan;
lives in Paris, France
www.unesco.org

Andrés Escobar Uribe
Dr.
Metrovivienda, Public Company
of Land Banking, Manager
*1960 in Medellín;
lives in Bogotá, Colombia
gerencia@metrovivienda.gov.co

Bostjan Vuga
Architect
Sadar in Vuga Architekti
*1966 in Nova Gorica, Slovenia;
lives in Ljubljana, Slovenia
biro@sadarvuga.com

Irene Wiese-von Ofen
Dr., Architect
International Federation for Housing
and Planning (IFLA), President
Born in Berlin; lives in Essen, Germany
www.ifhp.org

Ken Yeang
Dr., Architect, AA Dipl, APAM, RIBA, AIA
T.R. Hamzah & Yeang Sdn.Bhd.
Lives in Ampang, Malaysia
www.trhamzahyeang.com

Donata Valentien
Prof., Landscape Architect
and Urban Planner
Valentien+Valentien+Partner
*1944 in Coburg;
lives in Weßling, Germany

Uwe-Jens Walther
Prof. Dr., Sociologist
Technical University of Berlin
Lives in Berlin, Germany

Leigh Woolley
Architect
Leigh Woolley Architect and Urban Design
Consultant, Principal
Lives in Hobart, Tasmania, Australia
leighwoolley@ozemail.com.au

Raymond Young
University of Glasgow
Lives in Glasgow, UK

Michael Vesper
Dr., Sociologist
Minister for Urban Development
and Housing, Culture and Sports
of North Rhine-Westphalia
*1952 in Cologne;
lives in Düsseldorf, Germany

Weijen Wang
Prof., Architect AIA
University of Hongkong
Lives in Hongkong, China
www.hku.hk

Llewellyn van Wyk
Architect MIA, CIA
The South African Institute of Architects,
SAIA, President
*1955 in Cape Town;
lives in Philadelphia, South Africa
www.saia.org.za

Anke Zalivako
Architect
*1963 in Stuttgart;
lives in Berlin, Germany

Moshe Zarhy
Architect
UIA Work Programme "Architecture for
Science and High-Tech Facilities", Director
*1923 in Jerusalem;
lives in Tel Aviv, Israel
zarhyar@zarhyarchitects.co.il

Bernd Zimmermann
Urban Planner
Office of the Bronx Borough,
Special Advisor on Policy and Planning
Born in Germany;
lives in New York, USA

ANNEX:
THE FUTURE OF ARCHITECTURE
PROPOSITIONS FOR THE BUILT ENVIRONMENT

UIA WHITE BOOK 2002

Editor

International Union of Architects UIA Paris

The Future of Architecture
Propositions for the Built Environment

Preface

What started as an innovative European manual, prepared by the Architects' Council of Europe, has now evolved into a document for the whole world.

The White Book has clearly won a bet.

It has proved that values are universal and that all the architects of the world share the same concerns about the present and, more importantly, nurture the same fundamental aspirations about the future.

The White Book highlights what needs to be done if we are to live in a more sustainable and equitable built environment, how our mission relates with the architectural and cultural heritage and how the natural environment must unfailingly be taken into consideration in all of our endeavours.

The White Book thus opens the door to a "preferred" future.

But it does more than that.

It reiterates that architecture is of public interest and that, as such, it serves society and the public at large.

This is becoming increasingly important in the light of consumer protection where we need to incessantly repeat that quality in architecture and quality in architectural services give tangible benefits to the public.

The White Book is to be deemed as a reference book of principles that will inspire and lead. It will be our duty to disseminate it to as wide an audience as possible – architects, public and decision makers.

The UIA will encourage a continuing debate on the White Book. Such a debate will vindicate the book. Regional specificities will only strengthen our conviction that the basic core principles of our mission are universal.

Our thanks to all who have contributed to make this book possible.

Vassilis Sgoutas
UIA President

July 2002

Introduction

People, Politics and Architecture.

Our circumstances differ widely in this fragmented world.

Most of us work hard for a living, but some have materially comfortable lives, while others lack even basic necessities. Some experience unprecedented contact with other peoples and cultures while most people still travel little beyond their own district. Some can freely express themselves in public while to others the lack of such freedom is of relatively little importance, given their daily struggle.

Within these circumstances we all share two fundamentals. The first, timeless, is our common humanity. For architects this implies our need for shelter and for surroundings worthy of our spirits and memories. The second is of today: the globalization of trade, communication. For the first time in history, most of us know what is happening to each other and to our globe as soon as it happens.

This White Book is about better places for people to live and about retaining the life-giving properties of our towns and cities in the face of today's challenges and opportunities.

This Book is about politics and architecture. Government, politics and policies affect us all. This is as true for buildings and the environment as for any other part of life. If we want more and better houses and schools, cleaner and more coherent cities, and if we want to retain our cultural identity while profiting from freedom of trade and movement, we need the laws and policies to obtain this. A good environment only comes about to the extent that politicians and all implicated in its procurement actually see that this happens. Good places to live are not a matter of chance.

This Book is about the market, the natural environment and architecture. Global movement of people, money, ideas, goods and services offer unprecedented personal and economic possibilities. But the impact of global trade, with the concentration of capital, the flight to the cities, urban sprawl, the loss of identity, and habitat destruction is unprecedented. It is not clear whether a "win-win" situation is possible, where we can all be more comfortable while yet ensuring our world survives until tomorrow. But if this does not happen, tomorrow will not be worth living for.

This Book is about people and architecture. Very few people desire to see so many of us living and dying in poverty and squalor. So what more can be done to "beat the swords into plough-shares"?

This Book is about the reflection of the world architectural profession on the circumstances in which we today find ourselves. It follows work by the UIA Professional Practice Commission which has as its agenda the adaptation of the architectural profession in the face of world trade.

These reflections are partial in the small facet of our lives – seen alongside food, health, literacy, freedom of belief and expression – which is the built environment. Architects are conscious that for many people the engineering of clean water, sanitary conditions and access to energy and to global information are the most important aspect of having better places to live and work.

But engineering is not enough. If we do not act, the built repository of memory which is our historic monuments, towns and cities will tomorrow be reduced to a tourist stage set. If we do not act, tomorrow everywhere will look even more the same than it does now. Every day without a vision for the urban periphery, it makes the attainment of full human dignity on the edges of our cities even more remote. In the continued absence of environmentally sustainable sources of electricity, every day where air conditioning is preferred to the natural cooling of buildings hastens the rise of oceans.

Therefore, this White Book is written primarily to inform and make recommendations on our built environment, to international organizations, and national and regional governments. It is addressed to those decision makers who care about the quality of our towns and cities as places to live, work and play; and to those who are prepared to address the difficult task of seeing that the freedom to buy and sell, and to improve one's lot, is balanced by a respect for the majority who are outside decision making structures.

Eoin O'Cofaigh
Author, "Europe and Architecture Tomorrow"
President of the Architects' Council of Europe, 2000

Preparation

This publication started with the 1995 White Book of the Architects' Council of Europe ACE: "Europe and Architecture Tomorrow – L'Europe et l'Architecture Demain". The UIA owes a great debt of gratitude to the ACE for its collegial co-operation. It is particularly indebted to Eoin O'Cofaigh, colleague, friend and visionary for the future of the profession and ACE President in 2000, who wrote the ACE White Book with the support of Hervé Nourissat and other active members of ACE.

The aim of this UIA publication is to present the view of the architects of this world organization on the future of architecture to politicians and to the public. The work follows contributions, between 1999 and 2002 from all UIA regions, on behalf of the UIA President, Vassilis Sgoutas.

For Region I (Western Europe) – Andreas Gottlieb Hempel, 1. Vice-President
For Region II (Eastern Europe and the Middle East) – Alexandru Beldiman, Vice-President
For Region III (The Americas) – Alejandro Sokoloff, Vice-President
For Region IV (Asia) – Hisham Albakri, Vice-President
For Region V (Africa) – Taoufik El-Euch, 2. Vice-President

The five Vice-Presidents owe a debt to the kind collaboration of Louise Cox, Councillor of Region IV, Krzysztof Chwalibog, Councillor of Region II and Jean-Claude Riguet, General Secretary of UIA.

This UIA version of the ACE White Book was prepared by Andreas Gottlieb Hempel.

Content

View over the Landwehrkanal towards the new part of the Deutsches Technikmuseum Photographer: Florian Profitlich

1
Prologue

1.1 Contrasting Visions

The city represents one of the cultural cradles for humanity. Millions visit cities and towns, living testament to millennia of history and cultural endeavour. Much of the world's best architecture and urban design, great or humble, shows how our forebears solved the problems of living, with fewer economic and technical means at their disposal than we have, a legacy we enjoy in our daily lives. Thanks to a free market and open economy, the world is today materially richer than ever. Our living standards were unknown to our ancestors. But whereas some countries became richer and richer, others – exploited by the mighty – became simultaneously poorer and powerless. And hence, life in many parts of the world is far from satisfactory.

Despite immense advances in the past hundred years, many live in cramped surroundings – often even in bitter poverty – or in urban peripheries which do not properly support day-to-day life, much less the variety of social and cultural interaction needed for creative potential to blossom. The contrast between rich and poor side by side is particularly intolerable. Many people – old and young, rich and poor – cannot satisfy their most simple needs: an adequate water supply and sanitation, fresh air and a quiet environment, adequate shops, convenient traffic connections to work and to cultural centres. Despite our immense wealth and knowledge, many contemporary physical surroundings are uglier than some we have had for centuries. The countryside and the few remaining unspoilt areas are constantly eroded by inadequately controlled construction. At the same time, studies show that the construction process is far from perfect, and social or political intervention is controversial in a fragmented world. Beside the urban regions, the remaining landscape and natural places have to be protected carefully to reinforce a breathing place for the urban spaces.

1.2 Shared Concerns, Shared Challenges

This White Book sets out the Architectural Profession's position on the built environment in a globalised society: that among the world's citizens' many concerns, the wish and the need for a humane, culturally coherent built environment and manageable cities are as strong today as ever. This is often unrecognised. Building is too often treated as an exclusively economic process, whereas it is primarily an ecological, social and cultural phenomenon, responding to people's needs and aspirations. As a matter of fundamental economic importance, the construction industry must give the building promoter better performance and value for money. However, over and above this economic necessity, and beyond the challenge of reconciling the market economy with ecological, social and cultural values, is the challenge to improve our shared built environment on behalf of all citizens and the whole of society. Our built environment is too important to be a matter of mere industrial and economic policy. Unless this is recognised, our built environment will continue to stagnate and deteriorate. Developing countries who are imitating the way of life of the rich countries without reflecting alternatives for their own particular development, will be the poorer – both socially and culturally.

To improve our built environment is a specific, everyday challenge in city, suburb, and countryside. We must improve standards of accommodation for living, working and meeting others. We must foster our regional identities. We must make, and remake, our built environment, so that to shelter, light, heat, and cool ourselves does not destroy our planet. We must safeguard and retain a meaningful role for our built heritage, which we in turn pass to those coming after us. We have to encourage our governments to provide appropriate shelter for the poorest – where architecture plays no role at present but where a roof over one's head is more important than anything else.

This challenge is urgent. Deepening global transport and media links between continents and regions result in far-reaching regulations for those involved. Such international contracts (GATT/GATS), intended to promote economic efficiency and establish harmonised and free trade in goods and services may in fact harm the common good. The negative consequences of deregulation when opening up international markets under the sole aspect of economic competition cannot be overlooked. By its force, contemporary economic thought dominates social and cultural concerns as well as the environment to an extent which would have been unthinkable in the past and which is now unacceptable. Recent economic studies, intended to chart the future, posit only a partial vision which, because of a pretended completeness, is even less acceptable than present reality.

Above all, however, this challenge is urgent to eradicate poverty and because many new phenomena – a mania for growth and change, long-term unemployment, exploding information technology, advanced manufacturing processes, global communications, an awareness of regional identity, social and cultural pluralism, consumer protection and a deepening and urgent concern for the environment – suggest the need for new ways of thinking: a new paradigm of thought.

These new thoughts have to lead towards a new policy of civil peace between all systems of economies, societies and culture and in relation to the precious resources of our natural environment.

1.3 Architectural Policies for Tomorrow

This White Book seeks to sketch and to influence that coming paradigm. It details the contribution architecture can make to life throughout the world. It maps the architectural profession's view on recent political and administrative measures, which promote economic competitiveness as an end in itself, at the expense of the public good in our buildings and cities. It reminds architects that their duty is to attend to the joint interests of building promoters, building users and the public, the consumer, to create architectural quality that can be considered our future heritage. In particular, it asks and challenges politicians and administrators, at local, regional, national and international levels, to again ponder the effect of public policy on the built environment, and to shape that policy so as to meet the world's urgent economic, ecological, social and cultural needs.

This White Book constitutes an invitation to citizens all over the world, public representatives, building promoters and our construction industry colleagues to enter into dialogue with us, the architects, to how best improve our shared built environment, and also the building design and construction process in view of economic, ecological, social and cultural aspects. It focuses on the quality of infrastructure and building design as the key determinant of built environmental quality for our societies. Only careful planning can bring about the highest construction standards which craftsmen, contractors and the building industry have the capacity and obligation to deliver.

1.4 Making the Future

Despite economic power, entrenched ways of thinking and human limitations, we can choose the shape of the future. We must aim for a public consensus that only new ways of thinking lead to the future. The objective must be to create infrastructure, urban design and architecture whose quality is equal to the values of the building culture of the past. This results as a duty of co-operation of all disciplines.

2
International Planning and Building

2.1 Globalisation

The general political context of our time

We are familiar with the various paradigms of political, economic, social and cultural thought which have shaped our continents, countries and regions.

Politically, these paradigms have provided us with the nation state. At the moment, we note that nations all over the world tend to abandon or at least limit a part of their sovereignty in favour of regional or continental co-operation (examples: MERCOSUR, EU). Former power blocs are being transformed into optional alliances of states (CIS). Intensive negotiations (such as GATT/GATS) aim to abolish customs barriers and other hindrances to the fair and international competition of products and services. Global market access is the goal. Life today is largely determined by the impact of this globalisation. Even if there are different opinions of the goals and speed of these developing processes, people and nations will have to prepare themselves for international co-operation; there can be neither peace nor prosperity without it – nor on the economic and social level or in the religious, ethnic and cultural spheres.

In the age of globalisation, the importance of nation states will not only diminish in the framework of international co-operation, but it will also shift in favour of regional links and identities. Ethnic and geographical particularities will become more visible.

International economic thought, too, focuses on ideals such as full employment, freedom of competition in an open market, transparency in public procurement, economic rationalisation, value for money, consumer protection and unlimited economic growth. We produce ever-more goods at ever-lower present day financial cost, always seeking improved efficiency. Investment capital abounds in the international money market; however, what is missing are jobs and a global investment strategy which structurally also favours poorer countries. We need to develop employment creation and investment schemes, avoid the over-exploitation of natural resources and limit or halt ecological damage.

Migration for different reasons (political or ethnic repression, pessimistic economic prospects etc.) has internationally led to complex structures of society. Multicultural aspects will need to determine metropolitan politics, above all, to achieve the necessary balance and ensure a large degree of freedom for the individual. There's no telling what will be the consequences of an exploding information and communications system on industry, trade and service, on the evolution of society and the creative powers of individuals. One cannot overlook the danger that a culture of virtual reality seen in the media art worlds will take the place of home-grown cultural traditions. This development goes hand in hand with the loss of identity, character and the sense of home. The same holds true for the expansion of "anything goes" architectural designs, which threaten to replace traditional use of forms, thus contributing to intellectual rootlessness especially in countries and regions of the "developing world".

2.2 Globalisation of Trade

Globalisation of trade in services is the context in which we construct.

Globalisation of trade in goods and services is no more than the logical continuum of the free market in goods, services, and the free movement of capital and workers.

Since the dawn of history and in different ways and to varying degrees there has always been free trade between different parts of the world. Today's globalisation of trade is new in its spread and extent. Global traffic is no longer confined to luxuries, now fuel, textiles, electronics, and steel, cement, building stone, are traded globally. The thrust is to total liberalisation but intellectual services need a certain regulation for quality and consumer protection. Money moves around the globe in the blink of an eyelid, only people are not always free.

For our built environment, globalisation of trade brings:

- more investment, not as an end in itself but as an adjunct to return on capital employed;
- cheaper materials, through greater competition, and thus, the possibility of more plentiful building, a boon where there is not enough accommodation;
- more exotic materials, through greater choice;
- where invited, cheaper workers;
- mutual understanding by knowledge and tolerance

On the other hand, globalisation of trade also brings:

- environmental stress through transport related consumption of fossil energy;
- neglect – or worse, destruction – of local identity where foreign promoters are indifferent to all except their own self-interest;
- loss of sense of local place in the new placeless warehouse, factory, or office;
- destruction of smaller locally owned industries through supplanting by outside materials and construction methods;
- emotional opposition against influences from abroad;

To discuss the loss of local identity, the loss of hundreds of languages, the advance of the English language across the globe, and the destruction of the natural environment is merely to restate what is already known.

So is it possible to accommodate the benefits, perhaps the inevitability of global trade in goods and services with some limit on the destruction of identity and of the environment? If not, we will have to decide which is the most important. Until now the answer has been clear: the capital and the return wins every time but the identity of regions is being lost to the detriment of local societies.

Several responses are possible.

In the rich and "developed" countries there is an increasing search for personal and social as opposed to economic values. Shorter working weeks and better public services (health, education, child care) are valued more than higher earnings and lower taxes. In like vein, there is a real awareness that a high quality Urban Environment is a worthwhile end in itself.

Most of us live in cities: if we want good lives, we need good spaces in which to live. This argument is almost over. However, it is by no means clear, that the argument or the rationale are accepted or even understood elsewhere. Religious fundamentalism or insistence on ethnic preferences excludes human potential.

Another response is to reject global movements and to promote local industries and culture. Think globally – act locally! Educated people become more politically aware and wish to shape their own destinies. But this argument will be a long time in taking root all over the world and by that time it will be too late for

· the hutongs of Chinese cities;
· the marble facades of the Taj Mahal;
· the Brazilian jungles;
· the villages of central France;
· the rural societies in "developing countries";
· and so on.

We should disentangle the picturesque qualities, the indigenous values and the climatic adaptation of vernacular architecture from the other realities: bad sanitation, inadequate heating and cooking facilities leading to shortened lives, especially for women in many countries, lack of privacy leading to familial stress, inhuman working conditions etc.

Authorities should recognise and accept that different commodities have different values. Hidden materials are of themselves culturally neutral, but what we see and touch has also an irreplaceable value. The environmental terms effect the manufacture of materials and transportation. Efficiencies in manufacture, even where they exist, and where cheapness does not derive from underpaid workers, are often offset by environmental lunacies of transport.

So what is to be done? Options exist:

a) re-tilt the playing field to favour local materials where these are of cultural importance. Stop transcontinental trade in rough and heavy materials for construction and in façade and paving components. This will affect relatively little construction trade by value and is a bottom line requirement. However, precisely because this will not touch on the bulk of trade, neither will it significantly reduce environmental damage from transport.

b) Factor environmental cost into all trade decisions. But how is the cost of carbon dioxide and other damaging emissions to be factored into an assessment of the acceptability of a given trade deal? The only device presently favoured in directing trade is the market: given the absence of taxes on fossil or nuclear energy, environmental cost accounting is in reality never applied. Unless there are to be trade restrictions, carbon taxes are inescapable for proper environmental cost accounting.

c) Require that the "outside" architect partner with a local architect. There is however an issue not only of materials but also of design. Notwithstanding every architects' understanding of architecture as a cultural commodity with a history and locale, only the exceptional foreign architect has the skill and time to learn the local climate, customs and building traditions

d) Encourage so-called "barefoot-architects" within the profession. They have to advocate the basic needs of the poorest and the homeless in the world, a contrast to the "star"-architect and be a catalyst in the local community.

e) Investors from abroad should be encouraged to engage carefully skilled local architects They should be preferred to "second class" commercial architects coming also from abroad with ready schemes for foreign investment projects not respecting the cultural and regional value of the local built environment.

However the advantage of a local association is not always clear. There are two problems with advancing culture as a defence against increased global competition:

· the need for local involvement can be advanced as a necessity to protect local identity, whereas in reality this can be an excuse for an unwillingness to compete and a shield for poor practice,

· the need to comply with regulated requirements for local involvement is used as blackmail against outsiders so that a fee is charged which is disproportionate to the value of the local input.

The need for local involvement might therefore better be seen as an ethical issue rather than a regulatory one. In other words: the professional outside architect will of their own volition seek a local partner, the better to protect local identity – and in so doing, also protect local employment. But is ethics an adequate power with which to oppose global capital? On the other hand we have to consider that architectural quality is not connected to the technical equipment of an office but to the intellectual and creative potential of its architect. In many countries with foreign investment in real estate the bad performance most often comes from abroad.

2.3 Life and Building

Our relationship with our surroundings – both personal and social – is of great importance in our everyday lives.

In our everyday lives, we interact with our environment at every scale and at all levels, from the smallest to the greatest. At its smallest scale, this environment may consist of a chair or window: personally manipulable, touched by hand, seen close up. We dwell in and use individual buildings, experiencing these in a prolonged relationship and at many levels. We inhabit towns, cities or landscape, experiencing these as inhabitant or onlooker, sometimes frequently, sometimes rarely. At every point, environmental quality affects our lives for better or worse. The fact that a sound and secure environment raises the quality of life is reason enough to promote the best possible environment. But there are other reasons to strive for the highest standards in our buildings and urban spaces.

Every building project impacts beyond the technical and economic aspirations of its realisation, entailing ecological, social, cultural, aesthetic, atmospheric and other changes that also influence the sense of identity. Most buildings last at least several generations. The way we now design our buildings, towns, and cities, will endure as future testimony of the social and cultural awareness of our age, as our ancestors' testimony has endured for us. What is built today helps shape not only our own lives but also those of our children.

2.4 Building and Economics

Current concerns with economic interest must not cloud the most important issue – the best environment for everybody. To function properly, and take the interests of all citizens into account, contemporary society in general and environmental management in particular must be based on law and regulation. It is not, however, easy to legislate in relation to quality and cultural dimension. These 'soft facts', which are often the most significant, cannot be reduced to a formula without diluting their real content.

At the same time, the built environment comes into being through the free market. The building promoter has the right to the best value for money. The construction industry, including architectural services, operates inside the 'hard reality' of society's legal and economic framework. However, professional rules increasingly seem to be replaced by pure economic competition. Agreements on the protection of intellectual services that take into account both the change in market conditions as well as the required quality of design services are necessary.

Contemporary trends in political thought, seen in the economic objectives of free competition and rationalisation, have developed to threaten consumers and the common good. In recent years, legislation e.g. at European Union and Member States levels, designed to promote harmonised and transparent procedures for the creation of our built environment, focuses on the building procurement process, and seeks to optimise this in terms of reduction of initial cost, defect minimisation, and quality assurance. The construction of dwellings, workplaces and neighbourhoods has been reduced to a matter of short-term and private interests. But these political and economic policies also have a human dimension: the resultant building always impacts on people other than those directly involved in the procurement process. We must not allow the legitimate and necessary optimisation of the design-and-construction process to impede the quest for the best results of that process: the best built environment. Today, given the dominance of economic thinking at institutional levels, this danger is great.

2.5 Contemporary Environmental Concerns

Environmental damage has reached an alarming extent. In the construction sector, too, we need to focus on the careful use of our resources.

In recent years, it has been realised that economic growth – in industrialisation, production and in consumption – is not consonant with the greatest good. It is generally agreed that unlimited growth is not sustainable. We no longer believe newer technology will inevitably be better than what it replaces.

We are coming to realise that to continue to organise our environment as we have done in the century just past will result in social, economic and ecological disaster. And the answer does not consist solely in increased environmental engineering, necessary though this is. Two trends can be discerned in the increasing attention we now pay to our physical surroundings: firstly, grow-ing concern about pollution, environmental impact, increasing traffic volumes, demolition of the human-scale urban fabric, erosion of the traditional landscape, both natural and constructed and exploitation of natural goods and resources mainly in developing and poor countries, and secondly, growing recognition of the social, cultural and economic significance and potential of our physical arrangements for living.

Our society has fundamentally changed. There are no longer any simple or necessarily correct solutions to the challenges to our way of organising society socially and culturally. To seek to use contemporary aesthetics or planning methods as a solution is to manifest a false consciousness, and ignores three quarters of the globe. Any answers must be found at a deeper level.

2.6 Today's Debate

Society can and must now state its comprehensive and sustainable objectives for the built environment. Any partial vision would be faulty.

We should study the relationship between the built environment, the natural environment, political and economic policy, building promoter, owner and user, the construction market economy, peoples' social and cultural aspirations, and the economic and social objectives of states and their fractions on the regional as well as the continental level. At this time of unprecedented wealth of the developed countries and dramatic poverty on the other hand, specific social, cultural, economic, environmental, aesthetic and political challenges relating to buildings and the built environment confront us and demand a response. We should now direct public policy at global, national, regional and local levels, to meet the following fundamental objectives:

- Ensure human welfare and shelter for all.
- Ensure that every person has the basic technical infrastructure: clean water, sewage, disposal, energy and communication.
- Ensure that every person has somewhere suitable to sleep each night.
- Ensure that every home permits quiet and reflection and allows family gathering and interaction.
- Use construction processes in a life-enhancing way for all, and not in a way which merely maximises private profit.
- Ensure that megalopolises, the exploding metropolises especially of developing countries, are provided with an infrastructure that enables humane urban coexistence for all social classes, and that they are well designed.
- Improve public transport infrastructure for sustainable commerce so that life in the city, suburb, and countryside does not depend only on private traffic means.
- Remake town, city and suburb so that our fragmented society regains social cohesion and achieves better social interaction.
- Remodel high-rise prefabricated housing areas – socially and functionally mono-structured – which are inhabited by up to 70% of the urban population in some parts of the world including the public spaces as well as the buildings themselves.
- Reconstruct the industrialised centres and peripheries of thousands of towns and cities to reflect human scale and values.
- Devise new concepts on the integration of nature and the city that render the housing sphere of urban dwellers more attractive and ensure that natural resources are protected and increased including comprehensive water management and clean air.
- Ensure that our historic centres resist mass tourism and continue to support ordinary life.
- Limit building infrastructure so as to properly conserve the few remaining unbuilt spaces as a legacy to the future.
- Make the construction process safer for those involved.
- Heat, cool and light our buildings without further destroying the planet.
- Make every building, where people live or work, admit sunlight, daylight and fresh air, contain the essential infrastructure.
- Ensure that our indoor environments do not injure our health.
- Ensure that all buildings are designed according to the corresponding economic, ecological, functional and aesthetic aspects.
- Retain the regional architectural character in our villages and towns and develop the local or regional character against global uniformity.
- Have contemporary architecture echo contemporary achievement in technology and knowledge and come up with innovative, compatible and intelligent systems by way of research.
- Make contemporary built environments of a standard of construction, architecture and ecological sustainability, which matches the best in our past.
- Better support the activities of the micro, small and medium

enterprises in our design and construction industry.
- Introduce specific regulations for the intellectual services of the independent professions such as planners and architects so as to ensure the material background for competent and qualified performance.
- Ensure that building promoters, particularly inexperienced ones, obtain good value for money and a trouble free procurement process.
- Attach great importance on qualified training and continuing education as well as registration based on competence in the planning professions for the above-mentioned reasons.
- Ensure that building materials are from renewable resources or recyclable even for other purposes.
- Develop the understanding that a Genius loci is a basic factor for the creation of architectural values and that coherence of the built environment is more important for the public interest than an ambitious individual design not corresponding to the local situation.

These proposals contain values that serve as a standard and that an architectural policy must immediately realise. Any vision or action which does not consider all these issues is either incomplete or inadmissible. But the main target of any architecture policy has to be the proper shelter of the poorest and the protection of natural resources.

2.7 New Ways of Thinking

A new approach to how we make the built environment is both possible and necessary.

To enunciate the challenge now facing us makes it clear that a change in patterns of thought – a paradigm shift – is needed. To develop this paradigm, we have to agree on values as standards for decisions. Without moral values, standards, sustainable political and planning strategies a human built environment can neither be designed nor achieved.

To make a contemporary environment of the same quality as has occasionally been achieved before, and is now economically within our grasp, involves shaping the legal and economic conditions within which such an environment can come about. The shift in public consciousness which will result in the achievement of social and cultural coherence, correct hierarchy, a sustainable built environment, widespread built beauty, and at the same time as all this, the best value for the promoter's money, involves political choices with far-reaching consequences.

Such choice implicates law, regulation, and policies on the part of each Member State, and of municipal authorities. It involves building promoters, designers and constructors, educationalists, politicians and administrators. Every person must have the necessary information, knowledge and freedom to be included in the public dialogue and to make this choice.

3
Architecture and Society

3.1 Social and Cultural Life Today

Life and the built environment exhibit many tensions. The most urgent need is for humane, people-centred sustainable surroundings.

Contemporary life is becoming predominantly urban, whether in town or city, and often culturally diverse. Yet, whether in the city or in the country, people often not only earn less than they can live on, but are also insulated from contact with those of other social classes or outlook. They feel powerless over the battle for existence and the forces influencing their life: food eaten, television watched, curricula studied, environment inhabited. The locus of business and financial decisions becomes remote: New York, Brussels, Tokyo... At the same time, there exists a conscious striving for local community, the urban village and ecosystem, a healthy environment, the decentralisation of power, the right to a say in a matter. These attempts indicate that individuals are willing to take on more responsibility in public life and the community.

The built environment mirrors life and its potential for good, or ill, is evident. On one hand, the anonymous office sheltering the large company; social housing by bureaucracy; synthetic materials; architectural styles closer to those in other parts of the world than to those in the neighbouring village; the consumption of new styles of building analogous to new styles of clothing. And on the other hand, the desire for environmentally friendly materials and systems, a search for cultural coherence in architecture, the growth of user participation in social housing design.

In the environment, today's overriding "soft" fact is a universal need for humane, people-centred surroundings considering good social and adequate cultural conditions.

3.2 Building and Architecture

Architecture means more than building alone. Architecture enables high aesthetic and functional qualities and a variety in the buildings that surround us. Architecture offers added value to the building.

We all dwell in buildings. Most of our waking hours and all our sleeping ones are spent in shelter. We cannot avoid seeing where we live. At all scales, from the intimate to the greatest, for better or worse, we interact with our buildings. The manner of their achievement is a necessary and worthwhile debate. If the world could achieve good building, that would be much; but not enough. Building does not equal architecture. Architecture is more.

Architecture is about 'idea' given shape in built form. 'Idea' goes further than the optimum assembly of construction components. Architecture is about environmental quality: warmth and coolness, light and shade; about human scale, about the appropriate use of building materials and structure. It is about social appropriateness: spaces which support people living or working together or being alone, and which foster and give meaning to peoples' tasks or activities. It is about ecological and functional appropriateness: long life, sustainable materials, low energy consumption, flexibility in use. It is about economic appropriateness: value for money in a cost conscious age. It is about aesthetic appropriateness: proportion of form and line, solid and void, silhouette; and about cultural expression or appropriateness: respect for a city or landscape context, a vision of the future or an expression of respect for the past. Architecture can symbolise or solidify some of life's most important truths. Architecture is about coherence: coherence of culture, function, technique, environment, scale, aesthetics; coherence in a fragmented built environment. The best architecture works at many of these levels at the same time. Architecture expresses the invisible: atmosphere, mood, spiritual conditions, new ideas and tendencies, lifestyle and beauty – sometimes the image of a generation or a time.

Great architecture infuses buildings with 'idea', and gives it life at scales from the great to the small. 'Idea' runs from the scale of the doorknob, to the building, to the street, sometimes even to the scale of a landscape, town or city. Not every building is a work of architecture. Good architecture can be large or small, lavish or simple, traditional or contemporary in origin. The promoter's contribution, and responsibility, in the making of good architecture is of fundamental importance. Good architecture can develop as a co-operation under the overall control of the architect as a specialist for the whole.

3.3 The Interests of Society

Society needs quality in building, and quality in architecture. This involves consideration of the user's needs and the social context.

The interest of society in the built environment is to obtain the highest possible quality at all its scales, from the intimate to the urban, and in building and architecture alike: built quality in everyday life. Society's primary concern regarding the design and construction process is how this process influences the quality of the built outcome. This concern endures long after the promoter's concerns – themselves equally valid – relating to time span or cost have passed.

Quality in building results in buildings which are functional and economic in energy consumption, free of construction defects, robust in engineering, accessible to all, of potentially long life, responsive to the promoter's concerns.

Quality in architecture creates the opportunity, through all aspects of the physical environment, to live fully in its varied forms: for encounter, predictable or otherwise, participation or withdrawal, access to clean water, clean air and silence. In the city, it means being able to get to work reasonably easily, quickly and inexpensively. It means a safe environment, by day and night, in the week or at weekends. It means quality in all levels of experience and expression possible through the built environment. This is of importance to all citizens, but especially the poor, who cannot escape their everyday environment. A pleasant living environment could prevent many city-dwellers from taking refuge in vacation.

Building and architecture go hand in hand. Good architecture cannot properly exist unless in good building. Today we need good building. The planet demands it; economic competitiveness demands it; people need it for healthy indoor lives, and so as to be able to live without worrying about their shelter. In today's throwaway society of consumerism on one side and the lack of nearly everything on the other side, we urgently need good and sustainable architecture. Without it, the world will have more industrialised, placeless, anonymous, over scaled, ugly, well built disasters of the kind which already surround us abundantly.

In the name of the public user, public policy should now shift to:

- Strengthen the voice of both building user and citizen, and better respect the user's legitimate rights and aspirations alongside those of the other actors in the process.
- Foster consumer groups, environmental groups and societies of all kinds, and strengthen appropriate government agencies, so that the voice of the public can be heard.
- Promote public debate and information on the built environment and how best to shape its future, for architecture is a part of culture, of building culture, that involves everybody every day.
- Make beauty, as the harmonious expression of the reality of its time, the ultimate aim of building culture.

4
Architecture and Environment

4.1 The Countryside

Only when the last tree has died and the last river has been poisoned and the last fish has been caught, will we realise that we cannot eat money.

North American Cree Indians

In the countryside, the former symbiotic relationship between building and environment has been lost. Buildings were once constructed of local materials, responded to microclimate, and related to farm or village. Today, despite world-wide effort, much of the countryside is losing its population. Farmers are leaving their fields trying their luck in big cities. Formerly cultivated landscapes are falling back to wilderness. At the same time, new construction for tourism and business – the out-of-town hypermarket or placeless industrial facility – imposes alien values, consumes often scarce resources and is rapidly eroding the few remaining unspoilt areas. Agricultural areas are being abandoned or merchandised as artificial landscapes (for example: golf courses, ski-runs, foreshore developments). This destruction of countryside and natural resources, once wrought, is impossible to repair. Once built upon, the countryside is impossible to replace. Leisure developments in poor countries frequently don't bring the necessary economic benefit.

The design of our rural environment is important for work, social life and recreation. Even if building in the countryside does not take on urban shape, the rural style of building has an impact on one's spirit and the communal sense of country-dwellers. Sufficient infrastructure, controlled urban development and wide-ranging services offered, can make the life of people living in the country easier and have a positive influence on them. But large rural areas are suffering from depopulation, outdated infrastructures and excessive building activity. The continuing uncontrolled urban spread in the countryside has a considerable impact not only on the rural, but also on the urban environment (suburbia). A deeper analysis of forestry, growing methods, rural eco-systems, traffic infrastructure etc. would exceed the scope of this paper; however, housing schemes for the countryside must provide viable and durable solutions. This includes the preservation and simultaneous upgrading of built areas (of e.g. abandoned industrial estates) as well as the maintenance and preservation of land developed and cultivated by man that is no longer used agriculturally. To remake the rural environment, public policy should now shift to:

- Reinforce measures for the conservation of significant and as yet unspoilt landscape.
- Maintain the functions and manifestations of land developed and cultivated by man, despite structural changes in agriculture.
- Promote rural development which is sensitive to local climate, culture, and resources.
- Ensure that building for agriculture and rural tourism meets the needs of the settled population, is sustainable in its working, and enhances the landscape.
- Promote sustainable locations for settlement: dwelling and industry, related to existing settlement so as to claim as little countryside and infrastructure as possible.
- Plant the margins of many highways and roads, particularly with indigenous species, so as to promote shelter, reduce the environmental impact of the traffic those roads carry and create a homogeneous landscape.
- Agree measures to limit the physical and social impact of mass tourism on the countryside and its resources in rural areas.
- Conserve and reconstruct failing social and economic infrastructure.
- Include landscape as a part of urban infrastructure such as the metropolis and industrial zones.

4.2 The City

Cities have always been places of exchange, of tolerance, of culture. Understanding the global tasks of the future presupposes urban forms of living. The city is the single most significant repository of our common cultural heritage and, as a symbol, the quintessence of the coexistence of different peoples and opinions. Only urban structures enable the variety of cultures. Our civic tradition seeks to make society comprehensible and present to its citizens through its organisation of the built environment, allowing them to participate in a rational, democratic and ultimately fully human way. The city combines a mix of functions: residential, working, leisure, civic and religious, in a well-understood hierarchy. It controls built form for reasons of fire, hy-giene, and aesthetics: there have been controls on facade and building design since the beginning of civic life. It promotes cultural continuity: a matching of ends and means, an appropriate response to local climate, tradition, and cultural context and a correct visionary synthesis.

The huge majority of people lives in an urban environment. The more freedom the individual claims and the stronger social differences appear, the more difficulties arise in urban communities. The city develops out of a variety of individual actions and the buildings of individuals have an impact on all of them. Urban planning is not the sum of decisions made by individual investors. As in the case of individual buildings, the grouping of buildings impacts on the civic environment on both the functional and the aesthetic level.

And today?

The reality of social life in common is denied. A few years ago, the Prime Minister of a large European State said: "there is no such thing as Society, there are just individuals." The city is the battleground for self-interested profit and the place where the homeless drift. Despite advances in the engineering of traffic, of noise and air quality – none of which can in any way be taken for granted – the reality for many is lack of participation and social exclusion.

Urban zoning ordinances and master plans – designed for nineteenth century industry – have made monofunctional deserts of large parts of our cities, whether of residences, offices, or industry. This dismembering of functional integration has increased travel demand, travel stress, facilitated crime, and reduced social interaction. Ordinances controlling development density impose low densities at the urban periphery, increasing dependence on private transport, impacting on energy consumption and global pollution levels. At the same time, they permit intense development of commercial centres, which overloads infrastructure, destroys historical scale, and fills every open space with building; and they fail to control the shoddy back lands development which surrounds the approaches to many towns and cities.

Suburbia is not only a problem in cities of affluent societies – unrestrained growths (favelas, squatters, slums) surround the almost uncontrollable megalopolises of developing countries. Uncontrolled immigration, inadequate urban planning, insufficient infrastructures lead to inhumane living conditions especially for the poor. Social tensions, a high crime rate, problems of supply and disposal as well as intolerable hygienic conditions are the consequence.

To obtain better urban quality, public policy should now shift to:

- Design for the good of our future society, not just for the benefit of individual building promoters.
- Supply sufficient infrastructure, clean water, clean air and urban quiet
- Develop better models of control so that contemporary cities can achieve the mixture of uses, density of form, and control of expression of our historic centres, which will ensure the safety and security of all.
- Promote measures for freer social interaction, including more and better social facilities.
- Prepare detailed action plans for every city and town, which focus on the small scale and promote humane, three-dimensional, culturally-literate urban planning, with more use of local area plans and better involvement of society in the planning process.
- Devise specific political strategies that take particular care of sufficient infrastructures in large cities and metropolises.
- In urban planning and politics, tackle the problems of suburbia in all their forms in both rich and developing countries.
- Develop new political models on the communal level so as to promote the personal commitment of citizens through democratic participation.

The sustainable city is one of the key challenges facing us. The centrality and the enormity of this challenge must not obscure the fact that sustainable building is not enough. Sustainable architecture is what is truly required. To work for the sustainable city, public policy should now shift to:

- Redesign peripheries to support the development of mixed uses to focus on the many socially deprived peripheries.
- Evaluate the impact of building proposals on the surrounding microclimate and prevent the cold and windswept, or hot and glaring, open disasters of recent decades.
- Seek the longest life and the lowest energy consumption in the selection of materials, components and systems for urban paving, landscaping, servicing.
- Embrace the potential of contemporary engineering for waste recycling and treatment, and to invoke the civic responsibility of both producer and consumer for waste reduction and for a halt to conspicuous consumption.
- Integrate living, working, shopping, and cultural interaction, so that in our towns, cities, and our suburbs autonomous motor ised transport is not a necessity.

4.3 The Public Space

Outside, the buildings belong to the onlooker also. Our environment policy, urban and rural, must create conditions for the well-being of all.

What does architecture offer the urban dweller?

Our towns and cities are where we live, work, shop, play, reflect, and meet others. This built environment can be of good, indifferent or poor quality. The way others – building promoters and owners, municipal authorities – order it makes a job, shop, park, bus, train, church, or friend easier, or harder, to find; and more enjoyable, or less, to experience. The layout of built form: high or low, arcaded or open, creates shelter or wind. The urban environment can be quiet or noisy by day and by night, facilitating or hindering work, play, and rest. It can be clean or dirty in its streets, building surfaces and its air quality. It can be safe or dangerous to inhabit, as a result of crime and traffic. The external environment: roads, paths, services, can have a long or a short life, and can be easy or expensive to maintain. Despite immense advances, the public realm too often lacks delight. More planting would improve microclimate, reduce dust, screen

View into the rotunda of the Altes Museum, designed by Karl Friedrich Schinkel and built in 1824 – 1828. Photographer: Florian Profitlich

noise, refresh the eye and mind. Street furniture: signs, lighting, paving, litter bins, telephone kiosks, traffic controls, seating, is installed without regard for location or cultural coherence. Ordinary towns and cities, and many peripheries, are 'engineered environments' which lack all sense of visual quality. Outside selected conservation areas – often in the historical centre – we have neglected aesthetic controls. Everybody builds without taking stock of their neighbour. Even in historic centres, the legitimate and urgent need for protection of our built past too often results in picture-postcard recreations of a life long gone, without admitting the value of contemporary civilisation. The past hierarchy is destroyed: today, instead of civic and religious symbols, the dominant buildings are expressions of commercial and economic power, self-referential and devoid of public interest.

The most urgent functional, social, and cultural challenge is to make a contemporary environment of the same humaneness, quality and significance as has sometimes been achieved in the past, and which we can in turn make over to the future as examples of urban surroundings in which one likes to stay.

In the name of better civic culture and life, public policy should now shift to:

- Control development such that socially and culturally significant buildings regain their appropriate importance.
- Respect regional cultural tradition and promote the use of local urban landscape material.
- Promote the best facade design so that it is elegant, beautiful and regionally appropriate.
- Vastly increase the volume of urban soft landscaping, with particular regard to residential areas and to improving microclimate. The boulevards, tree-lined streets, well-designed squares and inner courtyards covered with greenery of many cities still offer excellent examples of a grammar for the quality of public spaces.
- Vastly reduce individual traffic – especially parking – in the public urban space so as to increase the quality of life for residents.
- Control the proliferation and the design of street furniture and road markings to ensure regional and visual quality.
- Visually protect the great 'set pieces of the past' not only in their construction but also in their settings.
- Foster the conservation, integration and re-use of the everyday urban environments of the past, not only to promote cultural continuity, but also to conserve resources, by restricting neglect of old buildings and by fostering preservation, refurbishment and remodelling and new uses.
- Promote authentic contemporary cultural expression, not in a few 'Grands Projets' but throughout everyday life.
- Open the public space to all urban dwellers and promote the understanding of different ways of life. Controlled access to certain facilities such as malls, galleries, shopping arcades etc. only for the well-off leads to social segregation.
- Public space must not be privatised so as to preserve the urban mixture of functions. Quite the contrary – a differentiated system of public and private urban spaces should be created.

4.4 The Building

Our buildings must better meet our functional needs, last longer, be more elegant and beautiful, use less energy and better respect their surroundings and our past. The kitchen layout can be convenient to use, comfortable to stand at, well lit, and involve a minimum of effort to permit concentration on the task in hand – or the opposite. A classroom can be a good place to learn: well lit and ventilated, with enough space to concentrate and a patch of sunlight to brighten an otherwise dull lesson, without distractions from next door - or the opposite. A hospital ward can be a good place to recover from illness: to allow privacy through the use of curtains or partitions, or a chat with fellow-sufferers when appropriate; it can beckon people from bed with a reminder of the beauty of the outside world, or it can do none of these things.

At the scale of the building, too, quality of architecture involves suitability, firmness and durability, and delight. Suitability for use: rooms which are the correct size and scale for individual and group use; a place to sleep which is quiet, warm or cool as appropriate for the time of year; with fresh air and somewhere secure to rest; an office which is bright, practical, and which allows us to do good work; a suitable place for ritual or social interaction; a place which can adapt over time to changing functions and needs; and a place which is ecologically sound, with a healthy built environment. Durability in performance: buildings must be dry, economic in energy consumption and mainten-ance, last for a satisfactory life span, and function without defect. Delight: elegance of proportion, a joy in good craftsmanship, an awareness of the possibilities of colour, light and shade, form and outline, and cultural appropriateness and significance through respect for the past and for regional identity, and through belief in the cultural legitimacy of the present.

And today?

As regards suitability: Many contemporary dwellings have inadequate space standards. In the European Union, the average living space per inhabitant amounts to approx. 30 sq.m, which is three times as much as the corresponding space in the People's Republic of China – not to mention the fact that millions of people have no roof over their head. We build offices and shops without natural light where it is impossible to comprehend the time of day, much less the changing seasons. As regards durability in performance: Believing in untrammelled and potentially infinite growth, we pull buildings down after 20 years because they cannot adapt to changes of function. The related waste of resources, capital and material is unacceptable. To obtain more suitable building, public policy should now shift to:

· Provide everyone with a simple and healthy place in which to live, as an absolutely fundamental criterion.
· Ensure that building properly meets our needs for sunlight, privacy, conviviality, durability.
· Obtain qualitative control over building design, not control which merely deals with reducing constructional defects or with saving energy, welcome and necessary though such strategies also are. Promote the quality of life.
· Discuss, agree, and promote design quality, so our buildings are truly suitable for use. A value system is required which insists on long life, loose fit buildings, to stand for 100 years, capable of sensible adaptation for unpredictable use. This requires more generous plan sizes and floor to ceiling heights, adaptable construction, natural light and ventilation, and an examination of the relationship between cost and value for money.

As regards durability: today, despite technical regulation, the profligate use of resources abounds. Building promoters renew commercial interiors twice a decade. Fitted furniture is changed as citizens used to renew their clothing. Buildings consume almost half of the primary energy and have already greatly worsened global warming. Air conditioning has irreparably damaged the ozone layer. Synthetic materials have lowered indoor air quality. The potential of information technology to optimise building energy performance is under-used. Poor design and construction techniques have resulted in concrete repair work becoming a major industry. In industrialised countries, the percentage of building rubble of the entire waste amounts to approx. two thirds, and the majority of demolition material still does not return to the material cycle.

To obtain more durable and sustainable building, public policy should now shift to:

· Promote the construction and design of buildings which employ sustainable servicing, supply and waste techniques, use fewer synthetic materials, deliver healthy environments, and consume as little energy as is now possible with skilled design and proper regulation.
· Significantly reduce the need for maintenance through far-sighted design. Regional building technologies of the past can show us today how to make economical use of material and energy. The durability combines financial, economical and environmental sustainability.
· Make buildings which respond to the exigencies of regional climate in a passive way, to thus express regional difference as happened in the past.
· Design buildings in view of their planned service life. Easily recyclable elements are required if the building is to be used only temporarily or if it is to be shut down early on.
· Calculate the cost of buildings not only in terms of investment costs. We need to determine the global cost – including maintenance and demolition as well as the cost for non-recyclable material which is lost after demolition.
· Develop and continually control a cyclical economy of energy and material for the investment and planning of sustainable buildings.
· Integrate the results of sustainable building methods in practice into building research and compare their effects with the global cost of "classical" building methods so as to develop political (fiscal) incentives for sustainable building for the general public.

As regards delight: the paradox of contemporary regulation is that as society puts more effort into technical building improvements – safety in fire, structural stability – buildings become more ugly, individually and collectively. We use industrialised construction which looks the same from Tokyo to Naples and from Sydney to Berlin and beyond. This debases at many levels: the regional, the cultural, the aesthetic, the human. Furthermore, nothing built under an impulse to incessant change can be of enduring value. The buyer or the user who is as yet unknown at the planning stage has to bear the cost of inferior planning and execution after short warranty periods. Those who have already examined the very different quality of buildings for private use and buildings that are "commodities" know what we are talking about.

Today, despite considerable advances in conservation policy and techniques, many ordinary old buildings, authentic expressions of local and regional culture, are neglected and demolished through a regulatory or financial system which ignores the old and demands the new, or which condones speculative short term expedients to capitalise on novel but ephemeral tastes. At the same time, many urban planning authorities demand pastiche design which debases our genuine heritage and which mocks that heritage through crude contemporary detail. There is no justification for not at the same time promoting better building and better architecture: more elegant, more culturally significant, and where appropriate, more contemporary. This is not merely a question of aesthetics; it is not a question of excluding technology; it is not a question of squandering economic resources which – though more abundant than ever – are finite. Rather, it is a question of providing citizens with a place fit to live in. This touches upon issues of how building culture can provide an identity and how a society views itself.

To obtain more delight in building in times of globalisation, public policy should now shift to:

- Foster a holistic version of high-quality, innovative and sustainable architecture.
- Conserve for present and future reuse the everyday architecture of the past as well as the set pieces.
- Benefit from knowledge of the past concerning material cycles, our environment and the careful use of energy and capital.
- Favour the highest standards of contemporary expression over and above mediocre reproduction.
- Make architectural efforts to provide people with a sense of identity and self-confidence on the basis of their built environment.
- Only building culture can help to transform our feelings and impressions, the sense of spatial identity, the spirit of our time and the relation to our past into a unique architecture – this is the architects' true task.

4.5 The Detail

We need policy changes to obtain suitability, durability, elegance and beauty, at the smallest scale – the detail. This may sound superfluous for the poorest of the poor but it represents an essential for human living in the built environment. At its smallest scale, architecture infuses building interiors and fittings with ideas about personal use and comfort, durability in functioning, and cultural appropriateness. For example: A window shutter should repel intruders, conserve heat in winter and exclude it in summer, be of sustainable materials which stand up to sunlight and wear and tear, durable and easy to maintain, well proportioned, and consistent in its cultural expression with the building where it is used. This is to be considered for all buildings which are more than only a necessary shelter over ones head. Quality of architecture at the scale of the fixture or fitting involves suitability for use, durability in performance, and visual delight. Suitability for use involves ergonomic considerations, especially for those who are not able bodied or strong; and the correct selection of materials, related to the functions they will support. Durability of performance involves proper length of life, taking all costs into account including the environmental. Delight for daily human life derives from elegance, style, and the contribution to the building's architecture made by even the smallest details. Well-designed details communicate the love of things and respect for the skill of the person who created them. They have something to do with sustainability in a larger sense – in the best sense of refining, which one would attribute to being rather than mere human existence.

And today?

In view of buildings, which are more than a simple and modest accommodation, the following things are to be considered:

Too often, short-life components and fittings are used without thought for waste of resources or for the environmentally damaging effects of their production. The lowest initial price is preferred to the lowest life cycle cost – the real cost. Despite advances in designing and building for persons with physical disability, too many buildings have sanitary and access fittings which the elderly and disabled cannot use. The over promotion of standardised components and the elimination of local handicraft has reduced cultural and regional diversity in windows, kitchen fittings, doors, ironmongery, shutters, joinery work, paving, and building materials generally. The loss of design quality is often compensated by superficial design, whose fashionable appearance rapidly becomes trivial and which leads to the ever more frequent replacement of things with which we do not like to identify after a short time. It simultaneously signifies a loss of identity. The less wealthy, especially in the developing countries, have to content themselves with the lowest quality of export products of the industrialised countries.

To improve the detail design of our built environment, public policy should now shift to:

- Properly take into account the ergonomic needs of the poor, the young, the old, and the handicapped, not merely at the level of sanitary facilities but throughout the building.
- Promote the use of local components, craftsmanship and materials so as to foster and promote pride in cultural and regional differences while seeking the highest standards of technical performance.
- Evaluate all building materials and components as to sustainability in production, transportation, use and reuse; and foster the manufacture and use of sustainable materials.
- Promote a design and construction process which encourages proper detail design especially in less industrialised countries.
- Change our decision making process so that the governing criterion of lowest direct financial cost is replaced by more sophisticated measures, which take intrinsic quality and environmental appropriateness fully into consideration.
- For the poorest of the poor help or self-help should be offered to achieve a good quality of housing. Do not forget that the poor can teach us! Make – do housing often gives a better result than contrived architecture. In economically weak and developing countries there also exists a demand for quality and sustainability.
- Superficial design must not hide insufficient quality. The frequent exchange of things leads to a throw-away mentality, the disregard of utility values and the loss of respect for values – even one's own

5
Architecture and Economic Life

5.1 Economic Life Today

The contemporary (western-) economic system has achieved much, but at high cost. The present challenge is to improve the construction process and obtain a better built environment. In recent times the command economy has been discredited and State intervention increasingly questioned. The open economy is the global economic paradigm, involving a drive for profit maximisation and cost reduction, seen on the supply side in trends to increasing scale of operation; to increased industrialisation and specialisation; to broader freedoms in market access, competition, and consumer choice; and, on the demand side, in higher levels of consumer protection. The free market paradigm dominates State economic thinking, with the effect that more people now have access to greater quantities of goods and commercially available services than ever before.

And yet, at the same time as it promotes the interests of large enterprises through product standardisation and technical change, the global free market sits uneasily with ordinary peoples' aspirations. Current manifestation of globalisation have lead to a hitherto unknown power and influence of the "global players". At times, they take over the existing positions of power of elected governments (e.g. in the communication and media sector) and determine more and more the political decisions in countries selling natural products and raw materials.

Unrestricted competition concentrates industry, reduces economic diversity, and increases the numbers of migrant or unemployed people. Fewer people than ever can afford personal service from craftsmen and professionals. Many achievements of the growth-oriented market economy have been attained by externalising and postponing social and environmental costs. The free market shapes the construction sector as all others. The impetus to growth or to avoid stagnation, profit maximisation, price-based selection and externalisation of environmental cost are all visible including many infrastructure projects in developing countries (e.g. financed by the World Bank). Economic challenges facing the construction sector include increasing global competition, the need to improve delivery standards and reduce costs, to quicken delivery time, and many others.

Economic studies, produced in response to these issues, often concentrate on supply-side economics and on sectoral interests. They do also examine demand issues – but generally from the promoter's viewpoint only. They neglect any proper consideration of the wider public interest in the built environment, whether of the user or of the citizen. Hence, propositions abound, which are of limited scope; fundamental questions are often unasked, and developments and their consequences are not seen in their complexity, let alone from an holistic point of view.

5.2 The Interests of the Building Promoter

The building promoter's legitimate interests must be fostered and reconciled with those of society, culture and heritage.

The promoter – the legal or moral person who commissions design and construction work – is central to the free market construction sector. Without the promoter, there is no employment, investment, accommodation, or built environment. The promoter requires accommodation or investment return, or both. The promoter may also be the user: an owner-occupier, but usually is not. The interests of the promoter centre on the design and construction process. They differ from and permanently risk contradicting those of the user.

The promoter's interests are: procurement of the desired accommodation at the appropriate level of quality – not always the highest level attainable; within an agreed time scale and as rapidly as possible; for an agreed budget and as cheaply as possible; and, for commercial owners – developers – the highest rate of interest return. The promoter's interest involves a trouble-free design and construction process where the outcome is in line with requirements and predictions. The working out of the promoter's interests implicates the promoter's freedom of choice, freedom of competition among the suppliers of services, and the assurance of quality, with protection from the unexpected: major defect, time or cost overrun, or contractor bankruptcy.

In the expensive, risky, and slow matter of procuring buildings, the building promoter's interests must be protected and promoted. These interests are not only those of the private sector promoter, but also, in the public sector, of the State promoter and ultimately the taxpayer. This involves the design team optimising the design solution, and developing the design so as to ensure thorough, fair and proper competition among contractors. It involves performance guarantees, consumer protection, and proper control of project cost and time scale. It involves competent advice on cost control and life cycle costing – often provided by related professions to architecture. At the same time, these interests must also be reconciled with those of the user and of society.

5.3 Society, Economy and the Building Process

The quality of the built result should guide all effort at improving the design and construction process.

Our shared social and cultural framework for living, the built environment, comes into being through a private desire for accommodation or profit, through competition and market forces. Private interest influences public good in a lasting and profound way. The contemporary economic paradigm has a profound influence on the construction process and its outcome. It is absurd to consider changes in the construction process with a view to improving its economic efficiency, without considering how these changes will affect social and cultural life, both in terms of the built product and in terms of the construction sector itself.

Has the working of the market economy improved the built environment in recent decades? In general, no. Peripheral and rural sprawl, the destruction of the urban core, declining housing space standards, increased urban noise, energy consumption and pollution all stem from an inadequately regulated market economy.

We should regulate the process by which buildings come into existence, in order at the same time to foster the public need for the best quality built environment and to negotiate between the private interests of building promoters and their responsibility vis-à-vis the public, society and the environment.

5.4 Scale

The size of the enterprise does not determine the quality of the result. The enormous number of small enterprises, independent craftsmen and the independent professions should be protected and fostered.

The size of the construction sector mesmerises politicians and the economy. In view of global markets, economies of scale including standardised components and systems as well as larger enterprises are preached. Public procurement directives and international aid projects tend to favour large organisations and the bureaucratic model. And large scale has advantages in many aspects of manufacturing, research and development and services delivery.

However, not everything advocated is desirable. An increase in the scale of design, manufacturing and contracting enterprises increases concentration, and results in a reduction of choice and hence of diversity and overlooks detailed local knowledge. It promotes specialist operation within large enterprises, which leads to loss of overview and of a total quality solution. It sometimes reduces costs, but this does not equate to an increase in quality.

In reality, there is no homogeneous global construction sector, but many. Many enterprises employ five people or less, work in local markets and have no intention of establishing transnational organisations and working abroad. Unlike in other sectors, many such enterprises are not sub-suppliers: they have a direct relationship with the promoter. The small or indeed micro-enterprise – whether in design or construction – whose owner has a personal stake in proper performance, can be equalled but rarely bettered in providing personal service to individual and many corporate clients. Small enterprises facilitate mobility and are more adaptable in meeting changing demand. Many world-famous architectural firms, undertaking the most prestigious commissions employ a handful of staff. Locally-made building components and locally-sourced materials have for centuries shaped local culture and responded to local climate.

In developing countries many (often primitive) houses are built by their inhabitants. "Barefoot-architects" are needed to advise and help them. But also in the richer countries the "Do-it-yourself" industry plays an important role – also these people need services for financial, technical, organisational and formal know-how: an hitherto neglected field for architects

We all desire improved quality and performance. It does not follow that, because the industry is fragmented and the scale of operation often small, that scale must be increased. Neither small nor large operational scale induces capability of itself. Rather, it is organisational competence, co-ordination and co-operation which matter. Furthermore, the issue of scale is different as between design and construction enterprises: small design enterprises are often very well suited to undertake large design commissions.

Despite restricted intervention, the construction sector can only profit from differentiated regulations. At the present time, large enterprises are those best fitted to look after themselves and must be encouraged to do so in their own drive for improved quality. Yet, public policy should aim to focus on the development of smaller enterprises to foster regional differences, maintain employment, reduce centralisation, reduce transport and environmental costs and locally distribute profit. It will bring suppliers closer and make them more responsive to their client. This will advance not only society's, but also the promoter's interest.

Public policy on the construction sector scale should be innovative and shift to take account of all the relevant interests:

- Concentrate State care on small and medium-sized enterprises to promote training and development of their enterprises and workforces so as to obtain the highest standards from even the smallest firms.
- Encourage small manufacturers to produce the highest quality components.
- Ensure that procurement procedures allow the smallest firms to compete on an equal footing for all commissions.
- Consult small and medium-sized enterprises on the development of norms and quality mechanisms so that harmonisation

no longer favours large scale.
- Ensure that fiscal policy encourages the conscientious small contractor and not the black economy which is damaging to consumer protection.
- Educate "barefoot" architects how to advise the poorest to get a roof over their head and educate architects to be able to serve the "do-it-yourself" customers.
- Maximise local expenditure on development and programmes.

5.5 Industrialisation

Construction sector industrialisation has improved profitability, but is widely recognised as having caused immense damage. Policy should focus on built and unbuilt environmental quality and aim at a generally accepted building cultural standard.

Industrialisation has pervaded the construction sector. This can be seen at all scales and stages of the design and construction process. Facade construction and installation technology have already become high-tech areas in the construction business.

Industrialisation and pre-construction have many advantages. They have speeded up the construction process: projects which 30 years ago took 18 months are now completed in nine, a significant benefit to a promoter borrowing finance. Pre-fabrication has facilitated high quality components: today's window loses less heat, is less draughty and requires less maintenance than yesterday's. In this context, it is easy to understand calls for more industrialisation.

However, industrialisation is not always beneficial. It progressively reduces labour content as trades are removed from site. It promotes uniformity and neglects craft skills. It damages local tradition, which is nowhere more important than in the built environment. Traditional components and skills are vital in an industry where over half the work is to maintain and repair old buildings, and must be fostered so as to protect valued buildings and environments. When employed at a larger scale, standardised heavyweight construction techniques – such as prefabricated panel construction – have often proved to be insufficiently tested as regards simple component performance: accidental collapse; reinforced concrete corrosion.

At the same time, and harder to undo, those same techniques have often had a devastating impact on the neighbourhoods where they are employed. In recent decades, massive industrialisation in the building sector has resulted in an ideological, technical and architectural disintegration. Thousands of social housing blocks built in the second half of the twentieth century bear witness to this.

How did this happen?

We should not judge by today's standards. Enormous works of social accommodation – streets, housing, schools, hospitals – were built from an urgent need in many countries. The use of industrialised processes to maximise output is understandable in this context. Recalling the situation before industrialisation and in extremely poor countries, the advances in accommodation have been enormous. However, this has not always been the case, and profit-driven development in an unregulated climate also caused immense damage..

Finally, it must be said that powerful profit-driven economic interests in the construction sector promote industrialisation and increases in scale for their own ends, under the guise of improved efficiency and lower costs. This is understandable from the economic point of view, but not acceptable from a social point of view.

Several lessons concerning public architectural policy must be learned from this account of recent industrialisation. Firstly, what was necessary in the past is no longer necessary or desirable. Circumstances have substantially changed. Today, there are many needs, but the most urgent is to rethink our industrialised environment. Secondly, what happened before can happen again unless the lessons of many twentieth century built environments and the damage to both nature and the quality of human lives, plain to see, are learned. Thirdly, much of what was done can now also be undone. We must stitch together our torn built fabric. Finally, we must apply the benefits of industrialised techniques in a way which does not damage society's interests.

Public policy in construction sector industrialisation should now shift to:

- Promote industrialisation only where this improves the built environment and benefits society as a whole, not the constructor, manufacturer or promoter alone.
- Avoid industrialisation where it leads to unnecessary standardisation, neighbourhoods of inhuman scale or destruction of local culture.
- Foster good craftsmanship, through proper apprentice schemes, craft training, public awards and registers of competent tradesmen.
- Foster reuse and rehabilitation and the use of appropriate traditional materials and methods through legal and other incentives.
- Carefully reflect the use of prefabricated, industrialised building components in historically preserved ensembles and buildings.
- Undertake a search for construction methods which are technically rational and architecturally adequate to cultural, social and individual adaptation and which are also economically

View of a part of the new Federal Chancellery, designed by Axel Schultes Architekten, built in 1997 – 2001. Photographer: Florian Profitlich

sustainable.

- Create a network of compatible industrialised building components so as to reduce those interfaces on building sites that are prone to damage.
- Strengthen good practice in building design, and foster more research, not only at the technical level of improved and healthier components and cities, but also at the cultural and social level of architecture which responds to the needs of society.
- Socially audit all proposals for innovation and merger, whether or not they lead to monopoly.
- Develop an understanding, especially by architects and other designers, for the correct place of industrialisation and of the damage inappropriate industrialisation has wrought in the past in construction sector education.

5.6 Specialisation

The separation of the design from the construction process often neglects the interests of the user and of society. The generalist is needed to achieve the requisite overview.

As elsewhere in economic life, specialisation in the design-construction process grows from increased sectional complexity, and from the desire for focused efficiency and profitability. Construction materials have proliferated, increasing perhaps fifty-fold in number in a century. A medium size project involves perhaps six design firms and twenty construction companies; a large project three times as many. The emergent project manager's goal is to control quality, time and cost in the promoter's interest.

The benefits are clear: a greater competence in defined areas; a closer focus on complex issues.

At the same time, however, specialisation dilutes responsibility for the quality of the overall result. Each person is concerned with only a small portion. Design specialisation promotes fragmentary and not holistic solutions: in building design for example, the use of specialist-engineered air conditioning instead of passive natural cooling. In the urban environment, fragmentation of responsibility is evident in the public realm, where traffic measures, soft landscaping, urban sculpture, lighting, paving and facades are all 'designed' in an uncoordinated manner. On the construction site, a myriad of contractors, unfamiliar with the works, reduces safety. The specialist project manager is in constant danger of neglecting holistic concerns, detail quality and the interests of the user or the public.

A coherent built environment of high quality can only be realised to the extent that the design-construction process is managed with that end in view. This is one of the original tasks of urban planners and architects who are trained as specialists for the whole.

Public policy on specialisation in the construction sector should now take account of all relevant interests:

- Focus building and urban design regulation more clearly on the overall quality of the result and not on the quality of individual parts.
- Ensure that the design-construction process, at all levels and scales, is led by competent generalists who combine technical, artistic, social and managerial competencies and vision, capable of understanding the long-term effect on society and the built environment of decisions taken at the design stage.
- Help all involved in the process to understand that its social purpose is to make good buildings and good architecture, and that the process is not an end in itself.
- Help all involved in the process better understand the contribution each specialist makes, but not by overloading already crowded curricula.
- Incorporate an understanding of the social and cultural significance of the built environment in all related technical and specialist education.
- Self-help and know-how should be offered to and encouraged in developing countries.

5.7 Freedom of Market Access

Deregulation, the objective of public policy and of some companies, frees access to the market, but is often incompatible with the real interests of the user, of the society and of cultural traditions.

Recent trends in the framework of the international negotiations to broaden freedom of establishment and market access are visible in the construction sector as elsewhere. Within this, two particular pressures are visible. On the supply side, large enterprises seek to loosen restrictions on market access in architecture, engineering and some specialist trades. Secondly, political agreements seek to broaden market access for public procurement of small and medium enterprises and to promote transparency of selection procedures.

For many years, customs barriers and professional regulations have restricted market access in architecture, in some engineering and specialist trades and in some design and build work. The reason for such restrictions varies, and can be caused by different levels of formation. It can be a desire for a better quality of design, for safe performance or for consumer protection. In some instances, restriction on market access is in the public interest, sometimes in the promoter interest and sometimes in both. The need for such restriction has in no way lessened in recent years.

Directives on Public Procurement in the framework of international agreements, intended to foster transparency and protect promoter interest, unintentionally damage the broader public interest. Difficulties of market access for small enterprises, for even medium size design and construction work, arise because of how transparency seeks to adopt objective criteria and reduce the possibility of new market entrants or of small enterprises participating in many public sector projects. This will reduce innovation and cultural diversity, undermine small enterprises especially in developing countries and centralise the design process. It is also unnecessary, thanks to insurance-backed guarantees.

Contracting and other enterprises, invariably of large size, promote design-and-build strategies on the basis that this particular combination of services best fits the promoter's need. The danger of self-interest as a basis for such pressures is evident, but in any event, the promoter's needs are rarely truly met by such strategies.

The quality of design and life-cycle performance in such projects is almost inevitably lower than occurs where design and construction are kept separate. Even the State, which usually regards itself as an experienced promoter, is frequently incapable of adequately assessing design and life-cycle quality as part of a package deal where lowest initial total cost inevitably, for reasons of transparency, becomes a major consideration.

Transparency cannot deal with the 'soft facts'. Contractor-led design focuses on maximum efficiency during construction, which promotes mechanisation, coarsens detail design, neglects life cycle issues of best design and best performance and ignores user needs. All questions normally pertaining to careful design are ignored. None of this is in the promoter's, the buyer's, the user's or society's interest.

Even where the promoters' interests can be said to be met, those interests are not coterminous with society's. For example, a promoter desiring a short term investment return will not provide the long life building and aesthetic quality society now needs. In such situations, it is inevitable that the choice of architecture is primarily on a cost basis, and the danger of the architect becoming, in the process, a mere agent of the contractor and overlooking the user's interest is ever present.

Freedom of market access, in relation to activities which impact on people not part of the private agreement, must continue to be regulated in the common good. Intellectual services must not be subjected to economic competition. Inexperienced promoters, in particular, need to be protected from quality-diminishing planning and construction processes.

Public policy – e.g. in the GATT/GATS negotiations – on construction sector market access should now shift to:

· Recognise that the traditional procurement method of separate design and construction, with independent professional advice to the promoter at the design stage may not always be the best to ensure high design quality and low life-cycle costing. Shape both public and private procurement policy accordingly, using value management techniques, involving the whole project team, including the client and the user, to add value to the project and ensure that the project is suitable for the end user.
· Review Directives on Public Procurement so that the adopted selection criteria allows a firm of any size, if competent, to be commissioned for a project. Those criteria should represent personal competence and responsibility, and not confuse size of enterprise with quality of performance. This will foster great competition between small and large firms, cultural diversity and also small and medium enterprises.
· Regulate market access in sensitive areas with regulation criteria based on tested ability and competence, on an understanding of the public nature of the built environment, and on ability and concern to deliver a correct level of service where this is not readily accessible.

5.8 Freedom of Supplier Competition

Free competition is essential, but may reduce quality if only based on price. Competition must be based also on the supplier's competence and on the quality of service. Competition must be comprehensive and not based only on price.

Construction must be competitive, for global economic reasons and also to benefit society in general. Free competition in the sector produces an abundance of goods, technical innovation, lower prices, higher standards, and larger choice. Competition benefits not only the building promoter, but also taxpayers, tenants and purchasers. The duty of authorities to foster greater competition is clear.

Today, however, this pursuit is often undertaken without regard to its negative implications. New procedures to produce professional services may radically change the process by which we create our built environment, but they may also be innovative and gain advantages unthought of in traditional procurement methods . When restrictions on competition are removed, often the result is that neither prices, nor scope of services, nor con-

View into the atrium of the new DZ-Bank at Pariser Platz designed by Frank O. Gehry, built in 1996 – 1999. Photographer: Florian Profitlich

tracts can be codified, but there may be a net gain as the project is refined and for example, more areas are shared by some of the users. The UNESCO/UIA Charter for Architectural Education provides the principles and competencies required for architectural study programmes and the UIA Recommended International Standards of Professionalism in Architectural Practice and its accompanying Guidelines reflect new ways of practising architecture. These should be consulted when in doubt or for more information on educational requirements, a method or procedure.

Only price-based designer competition leads to appointment on a lowest cost basis. If not quality based also, this may limit the time available for optimisation of design and may result in poor long-term solutions. This ignores the needs of the many promoters, private, commercial, and institutional, who build once or twice in a lifetime, who do not know the process involved and who cannot adequately pre-specify their requirements, particularly as regards the 'soft facts'. It also ignores the needs of promoters, both public and private, for a comprehensive service without cost increases during its execution. It leads to inadequate project management, resulting in construction defects, project cost and time overruns, and poorly performing buildings in use. It leads to fraternising between architect and contractor and hence to a lack of independent technical advice and supervision for the promoter, with a weakening of consumer protection.

The standpoint that national markets are too closed and that they must be opened to more competition is not the correct starting point. Rather, we must judge to what extent opening of the markets will create better architecture and act accordingly. The open market built neither Chartres Cathedral nor the Sydney Opera House, nor the Pyramids or Brasilia.

In the long term, therefore, it is not the promoter, designer, contractor or purchaser who pays for a given underbid, but society in general. Competition policy must be reconciled with social, cultural and environmental policy. The paramount issue is that while one person commissions a building, a thousand work in it and ten thousand view it. The problem is wider than financial saving. To promote competition only on price does not secure the best built environment as it precludes proper consideration of quality. This is why architects' codes of professional conduct in many countries restrict competition on price in favour of the intellectual competition based on quality.

How, therefore, should competition be promoted among building designers?

The ultimate public objective of competition must be highest quality and not merely lowest cost as a matter of private interest. We must promote competition in both the building promoter's and society's interest. Promoters must receive maximum value for money, through life cycle costing, global costs for society and environment, cost estimating at the design stage and good design which exploits the context to its best architectural potential. Innovative design and construction techniques, allowance for alternative bids, maybe the use of different materials or construction methods to achieve the required project, all of which still give quality and value for money, will be ways of achieving this. The three themes of Humanity, Quality and Ability must always be part of any design process and its result. It is difficult, however, to completely quantify or pre-specify architectural quality. The problem cannot be solved on the basis of quantity, but only by way of experienced jury experts on the basis of quality. Quality is not only the highest, but the best method of obtaining durability and sustainability according to local needs and possibilities.

Accordingly, the best framework for competition, so as to foster the delivery at the same time of good architecture and of proper project management, is through a well-informed promoter selecting a good architect for a fair fee with an agreed and comprehensive scope of services to be provided. This combination, upheld in law by some countries, assures the promoter of a full service for a clearly-defined fee, an assurance not available in a predatory business relationship. The promoter has free choice. If all prices are equal, the person who exhibits the best combination of good design and high service will obtain the commission. This arrangement ensures best results for both promoter and the environment.

In the public sector selection of architects, the competitive interview, making use of references, evidence of design imagination, design and management ability, quality of service and personal commitment with transparency of appointment and accountability for selection decisions will produce the best long-term results.

The architectural design competition offers advantages to all participants: promoter, public and architect alike. Specific management procedures should be established in each case. The local architects' associations should co-operate with the clients and architects to find the best method of conducting each competition.

To take account of the potentially conflicting interests, public policy on construction sector designer competition should now shift to:

- Promote competition only on the combined basis of quality of work, standard of service and price.
- Foster competition on this basis, the criterion of lowest price

must be combined with standards of quality and service.
- Exclude competition based only on price. This involves both ethical restrictions and assessment mechanisms which judge quality as well as price.
- Ensure that inexpert clients of design or of construction receive a comprehensive, high quality service, free from hidden cost increases.
- Promote competition on design quality by the correct use of architectural competitions, by encouraging and celebrating good design. Award-winning works should be published to advertise outstanding architecture and its author and to stimulate informed public discussions about building culture.

5.9 Freedom of Customer Choice

Freedom of consumer choice must be based on the public interest, which has to take account of broader issues – for example preservation of regional culture and competency-based qualification systems.

Customer freedom of choice is central to the open economy. Lack of choice, whether in types of goods or in suppliers of services, raises costs and lowers standards, while freedom of choice is a fundamental benefit. Today, the construction sector promoter has tremendous freedom of choice. The range of suppliers of services, building products and procurement routes has never been wider.

Those who seek greater freedom of choice say that selection of building designers, of building materials and of contractual arrangements is the promoter's private affair. Taken to the extreme of being allowed to choose anybody, no matter their skill, or any material no matter its provenance, this argument ignores the resultant effects on the built environment for those excluded from the process, whether building users or the general public.

Governments have long restricted choice of construction sector suppliers in certain fields. The grounds for so doing are not only aesthetic, but also for public safety. Freedom of choice has long resulted in trade of building products: stone, slate, lead, good timber. However, to allow total freedom of choice of building products ignores the effect on the local built environment of abandoning local materials and techniques. This has long been understood by some municipal control authorities.

Total freedom of choice in procurement routes neglects consideration of the impact different procurement methods have on the quality of the built result. The lessening of design and life cycle quality inherent in package deal contracts, where lowest cost considerations inevitably introduce coarser design, greater industrialisation and an over-emphasis on the 'hard realities' has already been discussed.

Public policy on freedom of construction sector choice should take account of all relevant interests. It should:

- Manage the market in construction products to promote local materials of cultural significance. This will foster diversity and reduce environmental cost.
- Develop urban design codes, technical standards and promotional literature to foster local materials where culturally relevant.
- In conjunction with the persons concerned, develop registers of competent designers, craftsmen and general contractors so as to help the promoter, especially the inexperienced one, choose appropriate suppliers of service.
- Ensure that competent persons carry out those building tasks which impact on the public interest, even if the market in building design services is unregulated.
- Ensure that the use of materials corresponds to the principles of sustainability, to the careful use of resources, recyclability and material cycles.

5.10 Consumer Protection

To achieve the best levels of consumer protection, public policy must foster both constructional, architectural and service quality, backed by sound guarantees.

The consumer, cornerstone of the market economy, has an ambiguous identity in the construction sector. Is it the 'consumer of the service' – the promoter, or the 'consumer of the building' – the citizen? In the present context, the consumer is the promoter, particularly the inexpert promoter or purchaser of the product. The open market seeks to protect this person by developing quality assurance programmes with improved technical standards and by greater supplier liability and warranties. Consumer protection is an important and necessary aim.

However, it must be remembered that the consumer is not the whole of society and that society must take into account a variety of interests. It has already been shown that promoter protection and user protection require different measures. More regard for the consumer is an inadequate protection of society's interest in the built environment, necessary though such protection is.

Quality assurance programmes assure quality to a pre-agreed standard and hence provide consumer protection in the risky business of construction procurement. However, such programmes relate to the management of the process, not necessarily to the quality of the end product; moreover, they do not promote quality as an end in itself, but merely set targets to achieve a pre-defined quality level, be this high or low. The meaning of 'quality' has been unacceptably distorted. Furthermore, the quality assurance process as it has developed to date derives from a manufacturing background and tends to invoke quantifiable criteria. When used in the architectural design process, such criteria do not promote architectural quality which is not easily quantifiable. Total quality management is a more useful concept in this regard.

How, therefore, might quality assurance best be used to promote both consumer protection and the best design?

It has already been said that the best consumer protection is a well-trained professional advocate acting to protect the consumer's interests. This aside, improving design and construction quality must be an a priori objective when discussing consumer protection. The achievement of quality in architecture must be based on the architects' professional competence and ethical attitude, on the transparency of the process, on a correct definition of the scope of services to be provided and on the proper resourcing of the architect for the tasks involved.

Public policy on construction sector quality assurance should broaden and develop to:

- promote appropriate quality assurance for architectural design, which – in contrast to the results – recognises that some of the process deals with unquantifiable issues and which considers not only the quality level desired by the promoter, but also that required for society.
- promote appropriate quality assurance for engineering and construction, which can involve considering more issues than QA programmes have comprised heretofore, including risk management standards.
- promote the preparation and delivery of building operation and maintenance manuals on all projects.

Technical standards protect the consumer with regard to normative matters: for example, glazing thickness', insulation levels, and structural performance. The open market seeks to harmonise such standards to reduce costs and to facilitate and increase observance. This is in the interest of the promoter, both public and private, and is to be welcomed and supported.

However, standards are a necessary and powerful tool for good building, but rarely deal with quality in architecture. Once adequate construction standards are attained, measures which cannot be quantified are the most significant. For example, plastic window frames perform excellently in many respects: long life, lack of maintenance, good thermal insulation, bright colour; with correct design, they can even be recycled. But what of the impact of the same plastic window in an old building, in a hitherto coherent work of architecture, or in many traditional environments, whether rural or urban? Inappropriate regulation can create conditions where good architecture cannot be procured; but, on the other hand, proper regulation can create a milieu where good architecture can flourish.

Public policy on construction sector standards should shift to:

- Develop technical standards which are appropriate to a personal, local, once-off, unique social and cultural activity and not always based on models appropriate to a repetitive manufacturing process.
- Develop standards to actively promote better and more durable, recyclable components and sustainable building components and systems.
- Promote the use of 'Best available technology' and 'Most appropriate technology' among promoters in general and among public sector promoters in particular. This will result in longer component life, lower running costs, lower life cycle costs and overall better value for money.
- Broaden the scope of regulation to include not only normative considerations, whether at the level of building or district, but also qualitative considerations: the 'soft facts' of social life.

Parallel to the concern for consumer protection, proponents of the open market advocate harmonisation of liability. In seeking this harmonisation, it is unnecessary to impose uniformity on varying legal systems.

Neither inadequate or excessive liability levels will promote consumer protection. An equilibrium between the broader long-term interests must be sought. The best consumer protection is obtained where liability levels are clearly defined.

Warranties from those involved in the design-construction process that their work will meet predefined standards and offer powerful consumer protection. To date, the provision of mandatory guarantees or insurance is not widespread, but necessary. To back these guarantees with substance requires that those offering warranties which can relate to either design or construction or both, be competent to generally achieve the necessary performance standards. For this reason, consumer protection implies restriction of market access: registration of designers or constructors working in areas of consumer importance such as house building.

The potential of the architect-client contract to protect the promoter so as to specify acceptable performance standards, to undertake mutually satisfactory liability levels and to provide wide-ranging guarantees – often backed by insurance – cannot be overstated.

Public policy on construction sector consumer protection should develop:

- Encouragement by the construction sector to provide worthwhile guarantees to consumers. Whether this either involves insurance backed guarantees or mandatory insurance for professionals, it will raise design and construction standards, reduce defects and promote consumer confidence. Public fostering of guarantees – both for design and for construction – from small enterprises will promote small and medium enterprises and discourage the black economy.
- Promotion of consumer protection through state regulation of architectural fees and services. This is no obstacle to freedom of movement at all, for, given the opening of international markets, free trade in services will be established.
- Design and construction contracts for consumers which protect the consumer by delivering proper quality of design and construction, and in controlling project time and cost, through transparent agreements both for project management and for the resolution of disputes.
- Improvement in the equilibrium between the imposition of liability on designers and suppliers, the promotion of consumer protection and the reaction of defensive design and overspecification. To this end, the architectural profession has played an active role

View from the new Federal Foreign Office designed by Müller & Reimann (1997 – 2000) to the Friedrichswerdersche Kirche. Photographer: Florian Profitlich

6
Architecture and Political Life

6.1 Conflicting Realities and Intentions

Public authority has the duty to reconcile different social interests – for instance those of promoter, user and society.

The wishes of the user and of society for our built environment could be listed to: maximise suitability in performance; maximise building durability; minimise environmental cost; facilitate human potential through social interaction or personal privacy; obtain new and retain existing aesthetic, environmental and cultural appropriateness, good proportions and delight. These wishes and needs have already been termed architecture's 'soft facts'. In general, these are quality-oriented, open to subjective assessment and not always quantitative.

The promoter's wishes could be listed to: correlate quality of result with that of planning; reduce uncertainty; minimise construction defects; maximise value for money; minimise life-cycle cost; and minimise design-construction time. These considerations influence the making of architecture through budgets, timescale, regulation, contractual relationships and guarantees and might be termed the 'hard reality' of the building process. These are in general quantitative and open to objective assessment.

Good building and architecture emerge through optimising the 'hard reality' and 'soft facts' relationship, quality and quantity. Collective political action is needed if building is to have economic, social, functional and cultural validity and if the history of building culture is to be continued as the expression of social and cultural identity. It is the duty of politicians and administrators to reconcile the different aspirations of those involved.

6.2 The Need for Overview

The making of a coherent and sustainable environment requires an overview of all interests involved. This is the particular skill and duty of the architect.

The skills and activities of specialists are no doubt necessary in their specific sector; yet specialists are generally not capable of recognising comprehensive factors and integrating them so as to create more than the sum of all the parts – architecture. Coherence requires overview not only of the building design and construction process, but also of the built result. All involved benefit.

The architect as a creative co-ordinator, a specialist for the whole, needs the requisite overview, needs to be able to manage the process and co-ordinate the specialists' disparate work. The architect also needs to be competent to understand the technical content of the proposals and the promoters' needs. This requires an understanding of the needs of the user and of society, to incorporate these in a project proposal which also embodies economic, technical and aesthetic qualities. In other words, to achieve the optimum holistic solution from the given conditions. Such designs or balanced wholes offer more than the sum of their components.

The architects' generalist formation, competence and outlook is a unique strength in the construction sector, especially in an age of specialists. For this reason, the International Union of Architects (UIA) believes that the pursuit of a coherent built environment and the co-ordination of the design and construction process is primarily a task, skill and duty; the legal and cultural environment must encourage the pursuit of this goal.

6.3 A Basis for a Coherent Environment

Coherence will not come about through regulation. Public policy must support holistic, sustainable solutions that are viable for the future. In creating our everyday environment, we seek – in imitation of nature – to combine correct functioning with a fine shape in a single organism: detail, building or environment. A happy result is achieved through coherence of idea and goals and of responsibilities and skills, applied both to process and product. We see such coherence at high moments in architectural and cultural history, where promoter, designer, user and builder understood and encouraged the principal purpose: the best possible built environment.

The conditions for shaping our environment are never optimum. The design of buildings and cities must seek to resolve conflicting wishes. Today, the lack of coherence of intention among promoter, user and society results too often in unsatisfactory building and poor architecture. The desire for technical and aesthetic quality collides with that for low design or construction cost. The desire for economic development conflicts with the wish to preserve an unspoiled site. The wish to physically conserve a historic centre, foster its everyday urban life and to simultaneously open it to mass tourism is a major contemporary 'irreconcilable'. Today, the place of the user and also that of society itself is frequently vacant at the design stage. The architects' responsibility to consider the user's and society's needs while being paid by the promoter involves significant conflict. A set of completely reconcilable interests is rarely found.

Through regulation, authority seeks to control and reconcile conflicting intentions as between the promoter on the one hand and the user and society on the other. However, this process as it now operates is often far from automatically achieving happy results. Building regulation is normative and cannot result in architectural quality. Attempts to procure architectural quality through controlling plot ratio, zoning, site coverage and set back rarely succeed.

As a coherent environment does not imply uniformity of appearance and cannot deny either complex social, cultural, economic and ecological reality or the unexpected, the following principles should underlie the development of public policy on a coherent environment:

- Strive to reconcile the conflicting aspirations of user, society and promoter and seek a 'win-win' situation.
- Look to the present and future and not to models of the past alone.
- Given present day realities, strengthen and support the voice of the user and of society in their need for architectural quality.
- Recognise that coherence and its pursuit are dynamic quantities. Pursuit involves change, and a changing society continually invokes changing relationships.
- Architects as specialists for the whole should use their skills to direct and control all planning processes and thus act as generalists and in the same comprehensive way be responsible to society and the environment.

6.4 Fostering Coherence: Authority and Society

Authority must foster coherence and innovation by promoting debate on architecture and the construction process, by changing regulatory systems and by acting as an exemplary promoter of the highest quality work on its own behalf.

An environmental conscience is the responsibility of every citizen, but national governments act in the name of all. National legislation determines the central parameters for architects, in which they need to take account of general factors in design and construction. Several states have recently adopted policies on architecture. Many educational programmes seek to impart an understanding of the comprehensive role of architects.

But there is more to be done. Authority must now seek to regulate the construction process in the interest of the common good

- the best possible quality of building and of architecture – and to foster the client's interests – efficiency of the processes, quality of the result, value for money – at the same time. Conflicting intentions can best be articulated and then reconciled through debate and dialogue. Authority must promote debate and dialogue about architecture and the construction process. It must also develop an understanding of extraordinary potential in a coherent built environment and must seek to foster this.

To promote coherence of intention in relation to the built environment, public policy should move and broaden to:

- Develop a mutual understanding of intentions, rights and responsibilities in the education of citizen, designer, promoter, client and user.
- Develop guidelines which facilitate discussion and transparent decision making in the building process; foster user responsiveness and small-scale intervention.
- Accept and foster the promoter's rights and speed the design-construction regulatory process, where socially acceptable and well-designed proposals are made.
- Stop building procurement where promoters look for excessive short-term profit or create an built environment which does not fulfil social, ecological and cultural requirements.
- Administrators and designers charged with making decisions about the built environment should have a thorough understanding of architecture and environmental problems and of peoples' needs for a good built environment.
- Promote public evaluation of architecture as a value for the built environment in the media at every level.
- Introduce simple principles of building culture as a basis of economics, ecology, social needs and culture in schools and other educational institutions.

The importance of the public sector as a construction sector promoter in its own right is often mentioned. Authority has a duty to promote a coherent built environment through its management not only of the building design and construction process, but also in the management of the transport and energy infrastructures as well as the social regulations, which it undertakes. The vistas of many urban peripheries – blocks of single-use housing, isolated between motor highways under a sky scarred by electricity pylons, with the adjacent 'space left over after planning' abandoned to weeds and dereliction – are a product of incoherent state intervention.

Proposals for designer selection on the basis of qualitative considerations (quality of previous work, commitment to the proposed job, imagination, cultural sensitivity, understanding of the milieu and social responsibility, general management and architectural design skills) have already been made. The means for government to achieve transparency (pre-qualification processes not based on lowest cost, design competitions with publication of results, and above-all public scrutiny of and debate over the built result) have also been canvassed.

To foster coherence, the policy of the public sector, as a construction promoter, should shift to:

· Ensure that those engaged to undertake construction design in general and building design in particular understand the social and cultural importance of the built environment.
· Examine its many-sided activities and ensure that the overall result is greater than the sum of the individual interventions.
· In all building and infrastructure initiatives, audit environmental impact, insist on a coherent result and do not allow the lowest initial individual project cost to determine lifetime environmental quality.

6.5 The Architectural Profession

Architects must remember their core skills and conceive of their profession as an art: the art of adding value through invention, innovation, promoting environmental quality and contributing to building culture.

The pursuit of a coherent built environment has important implications for the architectural profession in general, and for its leaders in practice and education in particular. The profession must ensure that those entrusted with the design of architecture and the management of the building project have a generalist and humanist formation. They must be able to interpret social and cultural needs and to give these needs technical and constructional expression. They must be able to positively engage with society's and the users' aspirations and to dialogue with all those concerned with the built environment. At the same time, they must have an adequate training to protect and foster the promoter's interests of project quality, cost and time management.

This state has already been reached in large measure, at least as far as design as the architects' most important area of training is concerned. Moreover, architectural schools need to teach financing, project and facility management and administration alongside the more traditional skills. Advisory committees on education and training in architecture have made important studies and recommendations in many areas. Many professional institutes have sizeable programmes of ongoing development. Education must offer a scale of values to the students to enable them to select wisely between all the possibilities of today.

Today more than ever, the architectural profession should recall its core skills and duties:

· Remind itself to achieve coherence among different interests, this demands more than planning knowledge, building skill and high technical standards. It demands commitment to the beauty of things and respect for the individual and for the planet. Remember that in the right hands, a window may grow from being a hole in the wall to being a source of that wondrous element – light, which gives birth to the indefinable notion of space.
· Remind itself that the architects' most valuable contribution is likely to be the most characteristic: an ability to add value through intervention, to sketch and to explore options for a future, which is not even known; and to combine this with the respect of the past and the site in working for and with individuals so as to provide a democratic response to aspirations for use.
· Remind itself that through a unique set of competencies, the architect is the person best placed in the construction industry to act as an environmental conscience for buildings, promote the quality of the built environment, provide professional advice uncontaminated by commercial interests, champion the promoter's and client's rights and facilitate dialogue between promoter, user and public.
· Act with according responsibility vis-à-vis economic demands and possibilities, the resources of our environment, the demands of society and of cultural identity.

These skills have to be based on moral and ethical values like responsibility, consciousness, respect for limited resources, general formation, compassion, bravery, uprightness, fairness, honesty and civil courage as a measure for decisions.

View of the main facade of the Nationalgalerie, designed by August Stüler and built 1866 – 1876 on the Museumsinsel. Photographer: Florian Profitlich

7
Proposals for the Future

7.1 The Future of Architecture in the Age of Globalisation

The economic goal of maximum production efficiency must be replaced by goals of improved creativity better human relationships, better quality of life, a sustainable, harmonious human-scaled built environment and a design and construction process serving the interests of all concerned.

Life is today too diverse and multi-dimensional for unitary solutions. Modern society is pluralistic and too complex for state secrecy: we need public transparency of action so as to encourage people to cooperate. The level of autonomous national states has long been left due to international interdependencies. A comprehensive consciousness of common traits and shared responsibilities on and for our planet is evolving. Yet life on this planet is too complex for continentalism or globalism: a federal system of regions will be the model for the future. Business is too powerful to be self-centred: we need new codes of business ethics and of regulation for the common good. Our problems are too large to exclude people from their solution: we need democratic inclusion and participation in political, economic and social life. We need to remind ourselves that life is not a virtual, but a very tangible reality. The economic goal of profit is too trivial to be placed before the pursuit of human happiness. Above all, the freedom, dignity, creativity and intelligence of people have too often been called into question for us to continue to do so – democratic processes should be adopted everywhere.

A new paradigm involves a fundamental shift in values. This holistic paradigm must be based on a set of demand-led aspirations, i.e. aspirations which centre on peoples' needs rather than on maximising production of goods and services. These aspirations are notably: an increase in individual freedom as well as in human and material diversity; a higher valuing of craftsmanship and of personal service, and, out of respect for each other and the globe, a higher value on human relationships in the sense of mutual responsibility.

Many areas throughout the world have been forever destroyed by buildings, being devastated or destroyed. But this environment must be transformed to become an acceptable environment for all citizens. It must be a built environment fit for human beings born out of a coherence of intention, where the timeless themes of human co-operation, good education, quality of life, joie de vivre, moral standards religious tolerance and the responsibility for society, traditions and environment are restated.

The ethical standards of a future global architectural policy have to measure up to the following demands:

- A built environment where building embraces and supports both privacy and conviviality, with interiors which are carefully detailed, good to live and work in; which respects individual human scale; where the prominent buildings are those of social and not merely economic significance; and where both user and citizen have more say in what is constructed and demolished.
- A built environment with durable, solid buildings and high quality details in terms of finish and craftsmanship. All materials should be used in an environmentally friendly way and return to the material cycle of resources.
- A built environment with proper sanitation, clean air, shelter and quiet; where settlement and building make motorised transport less necessary; where building respects and admits the natural environment; building which the planet can sustain, with waste pollution and energy consumption minimised.
- A built environment which is well proportioned and whose positive atmosphere contributes to human conditions; where the contribution of the past is valued; where contemporary culture has its proper say; where regional characteristics continue and where the diversity of metropolitan culture finds its appropriate expression.
- A built environment where international investment enhances the lives of all who it touches.
- Political action is required to supply the millions of inhabitants of today's megalopolises with the urban qualities mentioned above. This also necessitates efforts of the international community.
- Architects have to remember that the thoughtless imitation of so-called international models leads to anonymity, the exchangeability of architectural language and eventually to a loss of building culture.
- Architects in developing countries should co-operate with the engineering profession for the homeless and help create the necessary infrastructure and human shelter.

Architects should innovate and lead in constructing self-made buildings; they can adopt a significant position in society and advocate in co-operation with other disciplines for the poor and those without shelter.

A changed intellectual climate should also improve the procurement process to satisfy the often diverging goals of promoter and public. Tomorrow's construction process must be a truly integrated one where

- The participants, including users, promoters, designers, constructors, government and society generally better understand each others' views and priorities, and act co-operatively together in a environment of trust.
- Construction quality is maximised and defects minimised.
- Construction site accidents are virtually eliminated.
- Value for money and efficiency are maximised and cost minimised.

· Disputes are resolved fairly, quickly and simply, with less litigation than at present.
· The freedom, creativity and intelligence of all are used to the greatest benefit.
· The eradication of poverty, sufficient shelter and hygiene and the promotion of education in the developing countries are among the main goals of architecture and building culture.
· Moral and ethical standards replace the speculative pursuit of profit.

To implement the propositions in this paper will shift the creation and maintenance of our built environment towards meeting lasting and fundamental concerns of globalisation. The following paragraphs synthesise the policy propositions of previous chapters and suggest a new architectural policy. Authority has a main role, but architects must also consider their own profession if they are to prescribe for others.

7.2 International Policy on Architecture

An intended coherent policy of international organisations on architecture should start with the improvement of the built environment.

International organisations like the United Nations (UN) and its departments in conjunction with the International Union of Architects (UIA) and its proposals, accords and charters, through its Member Sections should aim to:

· Stimulate and formulate an international architectural policy that takes into account the needs of citizens, users, promoters, designers, the building industry and the crafts and which conveys high standards for society and the environment.
· Review the various national laws on public procurement of architectural services to give expression to the primacy of the public interest of the best possible quality of building design and promote small and medium-sized enterprises.
· Help to carry through architectural competitions that promote competition on design ability and quality and that ban competition on a lowest price basis.
· Reorient construction sector standards to foster methods to ensure the delivery of the best quality architecture.
· Introduce national and regional measures to promote strategies of regional cultural significance.
· Suggest possibilities for reducing energy demand in the construction sector by a third, by the year 2010, for example through the use of fiscal and other measures to promote this aim in both new building and in reconstruction.
· Indicate possibilities of saving resources through material cycles so as to use the same amount of energy and material in construction as can be reintroduced into the material cycle through modification or demolition – albeit in a different form.
· Make proposals for the sustainable design of a humane, acceptable built environment that is viable for the future. For example by promoting low energy, passive solar and sustainable design techniques; by increased dissemination of information on best practice; by reorienting professional formation; by reducing emissions; by developing strategies for waste reduction and recycling; by modifying infrastructure, by undertaking studies into sustainable urban community.
· Insist on maintaining regional characteristics in the GATS negotiations and urge for the introduction of human, moral and ethical standards that can serve as the basis for professional standards.
· Audit all building and infrastructure proposals seeking UN subvention or World Bank funding for architectural quality and environmental impact. Criticise applications for funding of low quality projects.
· Devise a policy with financial measures where appropriate for the remaking of those areas including city centres, peripheries and access nodes devastated by poor architecture and lacking in infrastructure.
· Devise a policy for the remaking natural landscapes devastated by inappropriate development.
· Devise a policy, with regulation where appropriate, for the promotion of informed cultural tourism and with control of mass tourism.
· Develop and suggest a policy of consumer protection which promotes building culture.
· Support political administrators charged with making decisions and ensure that they have a thorough understanding of architecture and of peoples' needs; building culture should be a medium to ensure cultural identity and the quality of the built environment.
· Become a best practice promoter of the highest quality architecture and use life cycle and environmental costing procedures in all public commissions.

7.3 National Policy on Architecture

Coherent national policy on architecture of those countries represented in the UIA starts with the improvement of the built environment and with the process of its realisation.

To the extent that they have not done so already, UIA Member Sections should encourage the authorities at national level in their countries to:

· Prepare and adopt a national policy on architecture in conjunction with all interested parties: citizens, preservationists, users, consumer groups, public and private sector promoters, contractors, engineering professionals, artists and especially architects. The policy should include appropriate measures on education at all levels, promotion, information, public sector procurement and professional formation.

- Ensure that non-building infrastructures are designed to take account of overall environmental quality.
- Promote sustainable development, for example by developing a national land use policy, by banning further extension of urban peripheries, out-of-town hypermarkets and office parks; by implementing pilot sustainable urban development programmes with a view to wide implementation in the medium term; by researching the relationship between land use zoning, developmental sprawl, transportation modes, energy consumption, pollution and global warming; and by seeking to integrate environmental costs into economic accounting.
- Ensure that fiscal and construction regulatory policies do not discriminate against preservation and reconstruction as opposed to new building.
- Establish statutory powers to safeguard heritage and promote contemporary architecture, with a broad base of interested representation.
- Devise a policy for construction sector liability, with a framework for construction sector guarantees or mandatory insurance for consumers.
- Ensure that building design and conservation is carried out by technically and culturally competent persons.
- Draw up registers of construction craftsmen and small construction enterprises with admission restricted by competence.
- Focus all training efforts on small and medium-sized enterprises in both the design and the construction sectors to promote high standards of safety, design and construction and to ensure changing practice is internalised.
- Improve the design and construction of dwellings, for example, not only by raising but also by improving space standards, by improving environmental performance and by minimising unwanted sound transmission.
- Review legislation to ensure that holistic architectural values are given primacy of value in local development control, for example, by removing legislation permitting development after quantitative assessment by way of plot ratio, site coverage or zoning; and by drafting national guidelines for qualitative substitutes.
- Publicly state that a built environment of the highest quality is worth having and worth paying for – and act accordingly.
- Ensure that national administrators charged with making decisions about the built environment have a thorough understanding of architecture and of peoples' needs for the best built environment.
- Become a best practice promoter of the highest quality architecture and use life cycle and environmental costing procedures in all public commissions.

7.4 Regional and Local Policy on Architecture

Coherent regional or local policy on architecture starts with the improvement of the built environment and with the process of its realisation.

To the extent that it has not done so already, the local architectural profession should encourage authorities at regional and local level to:

Prepare regional and local area plans which include physical, social and cultural infrastructure, employment opportunities and democratic involvement.

- Draw up three dimensional frameworks for development control, for example, by providing architectural guidelines for elements of public impact: setbacks, heights, open spaces, block size, facades. To ensure correct decisions, foster public debate on proposals for development. To foster coherence of intention, encourage the meeting of promoters, conservationists and citizens.
- Prepare detailed three dimensional action plans for the reconstruction of urban areas devastated by mono-functional development and by inappropriate architecture. This will involve extensive reconstruction, audit proposals as regards density and use mix so as to reduce dependence on private motorised transport.
- Prepare detailed action plans for the reconstruction of rural areas damaged by inappropriate development, including tourism, transport, energy and communications buildings and infrastructure as well as mono-activity.
- Re-examine development control methods in the light of the experience of the past decades. Scrap quantitative restrictions which have led to over-concentration in some districts and over-dispersion in others. Prohibit development in the few remaining unbuilt areas.
- In the light of international research on the sustainable city, foster the development of sustainable neighbourhoods. Extend measures to maximise recycling and reuse of glass, plastics, water, domestic refuse, soils and waste. Introduce local taxes on glass, metals and plastics packaging. Promote closed systems of waste treatment.
- Devise appropriate fiscal measures to limit urban development, especially of the 'comprehensive' type, to an appropriate size, for example by the use of land and property taxes to penalise against overscaled or inappropriate development, derelict buildings and vacant sites.
- Promote city centre renewal, for example by surveying vacant upper floors in commercial urban areas, examining the reasons for such and by taking steps to maximise its residential re-use, including fiscal incentives, promotion of good design and provision of physical and social infrastructure.

· Ensure social housing is of the highest constructional and architectural quality, providing adequate space standards, incorporating sustainable components and strategies, and by insertion in a sustainable environment.
· Ensure the use of regionally appropriate materials in urban landscaping and public spaces.
· Review design of traffic engineering components to ensure visual quality.
· Increase tree planting and soft landscaping by a substantial factor in urban and suburban areas, public spaces and along highways.
· Promote social interaction and cohesion through increased provision of public parks, crèches, meeting halls, libraries and other cultural and educational facilities.
· Become a best practice promoter of the highest quality architecture and use life cycle and environmental costing procedures in all public commissions.

7.5 The Architectural Profession

The architectural profession will continue its own policy shifts in order to foster the interests of promoter, user and society alike in the built environment.

To the extent that it has not already done so, the International Union of Architects (UIA) and its Member Sections pledge themselves to:

· Work with the engineering profession and the other disciplines of the building profession to promote a better built environment.
· Promote mandatory continuing professional development, to include, for example, emphasis on developing management and communication skills, and on fostering knowledge of building conservation and sustainable design.
· Take steps to improve the broad level of design and project management standards and processes for example, by making incompetent performance in these areas a matter of professional concern.
· Encourage dialogue with the user and with society, for example, by reinforcing consultation with user and conservationist groups, promoting public exhibition of design work and by promoting involvement of conservationists, user and consumer groups and non-professionals in publications, tribunals, design juries and education.
· Research and put into practice procedures to promote the quality of the built environment for private promoters of limited means.
· Foster consumer protection, for example, through the promotion of mandatory professional guarantees or insurance.
· Ensure that the interests of society in good architecture are properly explained to all involved in the design and the construction processes.
· Ensure all students of architecture receive a solid and comprehensive professional grounding in matters of concern to the promoter and user, for example project management, including time, cost and quality control; in communication skills, and in sustainable design strategies as well as in renovating old buildings and building maintenance and financing, as set out in the UNESCO - UIA Charter for Architectural Education.
· Strengthen architectural education, for example, by improving links between architectural schools and art colleges, between the schools and the profession, by studying the user's needs, by nurturing the understanding of regional characteristics and cultural traditions and by assisting cross-border exchanges, as proposed in the UNESCO-UIA Validation System for Architectural Education.
· Reinforce co-operation among professional architectural organisations, for example, by promoting cross-border professional collaboration and by promoting and diffusing standard conditions of appointment and other professional documentation, as has already been done in the UIA Accord on Recommended International Standards of Professionalism in Architectural Practice.

8 Epilogue

8.1 Policy for an Architecture of Tomorrow

Architecture and Policy are allies for better quality of life in the built environment.

Some basic conditions providing a built environment of high quality have to be the perspectives and goals of a future sustainable policy for architecture:

- A carefully planned built environment respecting society, culture, resources and nature should be a general human right.
- Legislation should secure the quality of urban design, public space and architecture with special regard to the regional and local environment.
- Public Procurement Systems should provide intellectual competition as the basic criteria for selecting designers for urban and building projects, and not price or cost.
- Educational systems should foster the awareness of the public and the authorities for regional and local identity as well as the quality of the built and natural environment.

Some steps in this direction have already been taken by single countries like Finland and by international co-operation for a common Policy of Architecture in the European Union.

8.2 Duty of Architects for an Architecture of Tomorrow

Architects are able to offer the most comprehensive professional service for the built environment.

The Architectural Profession is complex and comprehensive. Architects serve with a holistic responsibility towards society and the environment. They are educated to be competent and to integrate into their work the necessities of economic reason, ecological measures, regional requirements, social and public context as well as cultural values and the contributions of other disciplines working in the building sector.

Architects offer their services with responsibility to the public, to their clients, to politicians, users, inhabitants, to all citizens defending their right for a qualified built environment and to other disciplines. They feel obliged to offer their knowledge and help also, to those who have no shelter. Architects are willing to submit their proposals to public intellectual competitions in favour of finding the best specific solution for projects in urban design and architecture.

The Architectural Profession feels politically responsible and offers its co-operation in developing a sustainable Policy for Architecture to make the architecture of today the cultural heritage of tomorrow.

8.3 Architecture: Yesterday, Today and Tomorrow

Architectural Building Culture, the Built Environment and Monuments give witness to the imagination and creativity of the human spirit in the past, in our times and in the future.

The world is at a time of shifting social and cultural values, of economic restructuring and of vast environmental challenge. Now, as always, architecture continues to possess real power: that of joining people, of inviting encounter and of excluding confrontation, of fostering global thought while encouraging local action, and of creating and nurturing a sense of place and of regional and cultural identity.

It lies within our power to take a decision to set about creating a contemporary built environment equal to the best of our past. Such an environment will not be achieved without cost, but the greatest cost will be one of change of attitude: many of the proposals in these pages can be implemented at negligible financial expense by creative and innovative thinking.

Such an environment will not be achieved overnight, but steps to implement many of the proposals in these pages can be initiated immediately by changing our mind and actions. By promoting the use of the UNESCO/UIA Charter for Architectural Education, the UIA Accord on Recommended International Standards of Professionalism in Architectural Practice and its related Guidelines and encouraging the use of the proposed UNESCO-UIA Validation System for Architectural Education the architectural profession has developed instruments to implement the values of building culture in tomorrow's human life.

The outcome of the policy changes proposed in this paper will be to protect our environment through energy efficiency and sustainable performance. It will promote employment work in the careful maintenance and repair of the existing building stock. It will promote economic competitiveness through encouraging building economy, flexibility, long life and better value for money for the promoter. It will foster sustainable tourism through contemporary and widespread architectural beauty, cultural coherence and controlled development, both urban and rural.

This White Book will confirm the role of architecture as the ongoing joining factor of civilisation. It will promote mutual understanding through respect for each others traditions and skills. It will promote engaged dialogue by information.

View from the new Kranzler-Eck, designed by Murphy/Jahn and built in 1998 – 2001 to the Theater des Westens (1895 – 1897) at Kantstrasse. Photographer: Florian Profitlich

Photo Credits/Bildnachweis

Page/Seite	Copyright
16	Bitter+Bredt
24	Stephanus Schmitz
33	Archive UIA Berlin 2002
59	Reinhard Görner
66	Archive UIA Berlin 2002
79	Bitter+Bredt
89	Peter Lorenz, Weijen Wang
92	Archive UIA Berlin 2002
96	Stefan Müller
111	Peter Raffelt
114	Archive UIA Berlin 2002
123	Archive UIA Berlin 2002
128	Archive UIA Berlin 2002
133	Stephanus Schmitz
149	Archive UIA Berlin 2002
155	G. Berengo Gardin
160	Archive UIA Berlin 2002
167	Waltraud Krase
182	Archive UIA Berlin 2002
189	Archive UIA Berlin 2002
195	Susanne Hofmann
205	Waltraud Krase
213	John Linden
218	Rainer Mader
222	Nigel Young/Foster and Partners
250, 262, 269, 272, 277, 281, 287	Florian Profitlich

Cover (M)

Conversion of a residential building with europe's largest Photovoltaik Facade, Gewers Kühn + Kühn;
Residential building constructed in traditional building method, archive UIA Berlin 2002

All further images were provided by the authors.
Photo credits are given where the photographer was made known to the publisher.

Editing

Andreas Denk
Art Historian
*1959
Lives in Bonn
Since 1990 Correspondent of "Kunstforum International"
Since 2000 Chief Editor of the Magazine "Der Architekt"

Anne Ruf
Dipl.-Ing. Architecture
*1974
Lives in Berlin
Member of the Organising Office of the
XXI World Congress of Architecture UIA Berlin 2002

Carl Steckeweh
Economist
*1947
Lives in Bonn and Berlin
Since 1984 Federal Managing Director of
Bund Deutscher Architekten BDA
Since 2000 General Secretary of the
XXI World Congress of Architecture UIA Berlin 2002

Sophie Pache
Dipl.-Ing. Architecture
*1974
Lives in Berlin
Member of the Organising Office of the
XXI World Congress of Architecture UIA Berlin 2002

Esther Schwöbel
Dipl.-Ing. Architecture
*1975
Lives in Berlin
Member of the Organising Office of the
XXI World Congress of Architecture UIA Berlin 2002